M (909) 303-2309

Max Stress — 3 indep criteria 1) $X_{C,T}$ 2) $Y_{C,T}$ 3) S

Longitudinal

$$\sigma_{x_{max}} = \frac{X_{T \text{ or } C}}{\cos^2\theta}$$

Transverse

$$\sigma_{x_{max}} = \frac{Y_{T \text{ or } C}}{\sin^2\theta}$$

Shear

$$\sigma_{x_{max}} = \frac{S}{\sin\theta\cos\theta}$$

OFF AXIS TENSION

We Know How it Fails!

θ	Trig Factor $\cos^2\theta / \sin^2\theta / \cos\theta\sin\theta$	X_{fail}	Y_{Fail}	S_{Fail}
0	1 / 0 / 0			
30	.75 / .25 / .433			
45	.5 / .5 / .5			
60	.25 / .75 / .433			
90	0 / 1 / 0			

3 main probs.

1) @ small θ - Physically improbable that strength ↑ w/theta for small theta.

2) cusps

3) No accountability for interaction.

Given stress state, transform Mohr's, compute $\begin{bmatrix} \sigma_1 \\ \sigma_2 \\ \tau_{12} \end{bmatrix}$
$\sigma_x, \sigma_y, \tau_{xy}$

1) Compare σ_1 w/ X_T & X_C
2) ✓ σ_2 w/ Y_T & Y_C
3) ✓ τ_{12} w/ S

Tsai-Hill Criteria p 108

$\angle 1$ Good
$= 1$ OK
> 1 Fails

$$\frac{\sigma_1^2}{X^2} - \frac{\sigma_1 \sigma_2}{X^2} + \frac{\sigma_2^2}{Y^2} + \frac{\tau_{12}^2}{S^2} = 1$$

$$X = S_L$$
$$Y = S_T$$
$$S = S_{LT}$$

OFF AXIS
tension

$\sigma_1 = \sigma_x \cos^2\theta$ In Tension
$\sigma_2 = \sigma_x \sin^2\theta$
$\tau_{12} = -\sigma_x \sin\theta\cos\theta$

Off Axis
In shear

$\sigma_1 = 2\tau_{xy}\sin\theta\cos\theta$
$\sigma_2 = -2\tau_{xy}\sin\theta\cos\theta$
$\tau_{12} = \tau_{xy}(\cos^2\theta - \sin^2\theta)$

Thermal Stresses

Mat'l coord   Global

$$\begin{bmatrix} \varepsilon_1 \\ \varepsilon_2 \\ \gamma_{12} \end{bmatrix} = \begin{bmatrix} \alpha_1 \Delta T \\ \alpha_2 \Delta T \\ 0 \end{bmatrix} \qquad \begin{bmatrix} \varepsilon_x \\ \varepsilon_y \\ \frac{\gamma_{xy}}{2} \end{bmatrix} = [T]^{-1} \begin{bmatrix} \varepsilon_1 \\ \varepsilon_2 \\ \frac{\gamma_{12}}{2} \end{bmatrix}$$

$$\begin{bmatrix} \varepsilon_x \\ \varepsilon_y \\ \frac{\gamma_{xy}}{2} \end{bmatrix} = \begin{bmatrix} c^2 & s^2 & -2cs \\ s^2 & c^2 & 2sc \\ cs & -cs & c^2 - s^2 \end{bmatrix} \begin{bmatrix} \alpha_1 \Delta T \\ \alpha_2 \Delta T \\ 0 \end{bmatrix}$$

$$\varepsilon_x = \alpha_1 \Delta T c^2 + \alpha_2 \Delta T s^2$$
$$\varepsilon_y = \alpha_1 \Delta T s^2 + \alpha_2 \Delta T c^2$$
$$\gamma_{xy} = 2\{\alpha_1 \Delta T sc - \alpha_2 \Delta T sc\}$$

McGraw-Hill Series in Mechanical Engineering

Consulting Editors

Jack P. Holman, *Southern Methodist University*
John R. Lloyd, *Michigan State University*

Anderson: *Modern Compressible Flow: With Historical Perspective*
Arora: *Introduction to Optimum Design*
Bray and Stanley: *Nondestructive Evaluation: A Tool for Design, Manufacturing, and Service*
Culp: *Principles of Energy Conversion*
Dally: *Packaging of Electronic Systems: A Mechanical Engineering Approach*
Dieter: *Engineering Design: A Materials and Processing Approach*
Eckert and Drake: *Analysis of Heat and Mass Transfer*
Edwards and McKee: *Fundamentals of Mechanical Component Design*
Gebhart: *Heat Conduction and Mass Diffusion*
Gibson: *Principles of Composite Material Mechanics*
Hamrock: *Fundamentals of Fluid Film Lubrication*
Heywood: *Internal Combustion Engine Fundamentals*
Hinze: *Turbulence*
Howell and Buckius: *Fundamentals of Engineering Thermodynamics*
Hutton: *Applied Mechanical Vibrations*
Juvinall: *Engineering Considerations of Stress, Strain, and Strength*
Kane and Levinson: *Dynamics: Theory and Applications*
Kays and Crawford: *Convective Heat and Mass Transfer*
Kelly: *Fundamentals of Mechanical Vibrations*
Kimbrell: *Kinematics Analysis and Synthesis*
Kreider and Rabl: *Heating and Cooling of Buildings*
Martin: *Kinematics and Dynamics of Machines*
Modest: *Radiactive Heat Transfer*
Norton: *Design of Machinery*
Phelan: *Fundamentals of Mechanical Design*
Raven: *Automatic Control Engineering*
Reddy: *An Introduction to the Finite Element Method*
Rosenberg and Karnopp: *Introduction to Physics*
Schlichting: *Boundary-Layer Theory*
Shames: *Mechanics of Fluids*
Sherman: *Viscous Flow*
Shigley: *Kinematic Analysis of Mechanisms*
Shigley and Mischke: *Mechanical Engineering Design*
Shigley and Uicker: *Theory of Machines and Mechanisms*
Stiffler: *Design with Microprocessors for Mechanical Engineers*
Stoecker and Jones: *Refrigeration and Air Conditioning*
Ullman: *The Mechanical Design Process*
Vanderplaats: *Numerical Optimization: Techniques for Engineering Design, with Applications*
White: *Viscous Fluid Flow*
Zeid: *CAD/CAM Theory and Practice*

McGraw-Hill Series in Aeronautical and Aerospace Engineering

Consulting Editor

John D. Anderson, Jr., *University of Maryland*

Anderson: *Fundamentals of Aerodynamics*
Anderson: *Hypersonic and High Temperature Gas Dynamics*
Anderson: *Introduction to Flight*
Anderson: *Modern Compressible Flow: With Historical Perspective*
D'Azzo and Houpis: *Linear Control System Analysis and Design*
Donaldson: *Analysis of Aircraft Structures: An Introduction*
Gibson: *Principles of Composite Material Mechanics*
Kane, Likins, and Levinson: *Spacecraft Dynamics*
Katz and Plotkin: *Low-Speed Aerodynamics: From Wing Theory to Panel Methods*
Nelson: *Flight Stability and Automatic Control*
Peery and Azar: *Aircraft Structures*
Rivello: *Theory and Analysis of Flight Structures*
Schlichting: *Boundary Layer Theory*
White: *Viscous Fluid Flow*
Wiesel: *Spaceflight Dynamics*

Max Strain Criterion

Ex

$2°$ $30°$ $75°$

$$\sigma_x < \frac{S_L(+)}{\cos^2\theta - \nu_{12}\sin^2\theta} =$$

$$\sigma_x < \frac{S_T +}{\sin^2\theta - \nu_{21}\cos^2\theta}$$

$$\sigma_x < \frac{S_{LT}}{\sin\theta\cos\theta}$$

$$\sigma = \sigma_0 e^{i\omega t} = (E' + iE'')\mathcal{E}$$

$$\sigma = E_v \mathcal{E} + \eta_v \left(\frac{d\mathcal{E}}{dt}\right)$$

$$\mathcal{E} = \frac{\sigma_0 e^{i\omega t}}{E' + iE''}$$

$$\mathcal{E} = \frac{i\omega \sigma_0 e^{i\omega t}}{E' + iE''}$$

$$\sigma_0 e^{i\omega t} = \frac{E_v \sigma_0 e^{i\omega t}}{E' + iE''} + \eta_v \frac{i\omega \sigma_0 e^{i\omega t}}{E' + iE''}$$

$$E' + iE'' = E_v + i\omega \eta_v$$

$$\frac{1}{G_{12}} = \frac{V_f}{G_{f_2}} + \frac{V_m}{G_m} \qquad \tau_c = \tau_t = \tau_m$$

$$\Delta = \gamma w$$

$$\gamma = \frac{\tau}{G_m}$$

$$\Delta = \Delta_m + \Delta_f$$

P 70
67

PRINCIPLES OF COMPOSITE MATERIAL MECHANICS

Ronald F. Gibson

Department of Mechanical Engineering
Wayne State University
Detroit, Michigan

McGraw-Hill, Inc.

New York St. Louis San Francisco Auckland Bogotá
Caracas Lisbon London Madrid Mexico City Milan
Montreal New Delhi San Juan Singapore Sydney Tokyo Toronto

This book was set in Times Roman.
The editors were John J. Corrigan and Eleanor Castellano;
the production supervisor was Louise Karam.
The cover was designed by Joseph Gillians.
R. R. Donnelley & Sons Company was printer and binder.

PRINCIPLES OF COMPOSITE MATERIAL MECHANICS

This book is printed on recycled, acid-free paper containing a minimum of 50% total recycled fiber with 10% postconsumer de-inked fiber.

2 3 4 5 6 7 8 9 0 DOC DOC 9 0 9 8 7 6 5 4

ISBN 0-07-023451-5

Library of Congress Cataloging-in-Publication Data

Gibson, Ronald F.
  Principles of composite material mechanics/Ronald F. Gibson.
    p.   cm.—(McGraw-Hill series in mechanical engineering)
  (McGraw-Hill series in aeronautical and aerospace engineering)
  Includes bibliographical references and index.
  ISBN 0-07-023451-5
  1. Composite materials—Mechanical properties.   I. Title.
II. Series.   III. Series: McGraw-Hill series in aeronautical and
aerospace engineering.
  TA418.9.C6G53   1994
  620.1.'1892—dc20   93-22119

ABOUT THE AUTHOR

Ronald F. Gibson is a Professor of Mechanical Engineering and Director of the Advanced Composites Research Laboratory at Wayne State University. Dr. Gibson received his B.S. degree in Mechanical Engineering from the University of Florida, his M.S. in Mechanical Engineering from the University of Tennessee, and his Ph.D. in Mechanics from the University of Minnesota. He has held full-time faculty positions at Iowa State University, the University of Idaho, and Wayne State University, and visiting faculty positions at the University of Florida and Michigan State University. He has been a Development Engineer for Union Carbide Corporation and a Summer Faculty Fellow at the NASA Langley Research Center.

Dr. Gibson is an active member of numerous professional societies, including the American Society of Mechanical Engineers, the American Society for Composites, the American Society for Testing and Materials, the Society for Experimental Mechanics, and the Society for the Advancement of Material and Process Engineering. He has been the recipient of the Hetenyi Award for Best Research Paper of the Year from the Society for Experimental Mechanics and the College of Engineering Outstanding Faculty Award from the University of Idaho. The results of his research have been published in numerous scholarly articles and presented at a variety of national and international meetings.

To my wife Mary Anne,

My daughter Tracy,

And the memory of my parents,
Jim and Lora Gibson

CONTENTS

Preface   xv

1 Introduction   **1**
1.1 Basic Concepts   1
1.2 Constituent Materials for Composites   6
     1.2.1 Fiber Materials   7
     1.2.2 Matrix and Filler Materials   11
1.3 Structural Applications of Composites   13
1.4 Fabrication Processes   21
1.5 Elements of Mechanical Behavior of Composites   28
Problems   31
References   33

2 Lamina Stress–Strain Relationships   **34**
2.1 Introduction   34
2.2 Effective Moduli in Stress–Strain Relationships   35
2.3 Symmetry in Stress–Strain Relationships   39
2.4 Orthotropic and Isotropic Engineering Constants   43
2.5 The Specially Orthotropic Lamina   46
2.6 The Generally Orthotropic Lamina   48
Problems   58
References   61

3 Effective Moduli of a Continuous Fiber-Reinforced Lamina   **62**
3.1 Introduction   62
3.2 Elementary Mechanics of Materials Models   67
     3.2.1 Longitudinal Modulus   69
     3.2.2 Transverse Modulus   73
     3.2.3 Shear Modulus and Poisson's Ratio   75

xi

3.3	Improved Mechanics of Materials Models	77
3.4	Elasticity Models	83
3.5	Semiempirical Models	90
	Problems	93
	References	97

4 Strength of a Continuous Fiber-Reinforced Lamina   **99**

4.1	Introduction	99
4.2	Multiaxial Strength Criteria	101
	4.2.1 Maximum Stress Criterion	103
	4.2.2 Maximum Strain Criterion	106
	4.2.3 Quadratic Interaction Criteria Tsai-Hill	108
4.3	Micromechanics Models for Lamina Strength	114
	4.3.1 Longitudinal Strength	115
	4.3.2 Transverse Strength	122
	4.3.3 In-plane Shear Strength	125
	Problems	126
	References	128

5 Analysis of Lamina Hygrothermal Behavior   **131**

5.1	Introduction	131
5.2	Hygrothermal Degradation of Properties	132
5.3	Lamina Stress–Strain Relationships Including Hygrothermal Effects	144
5.4	Micromechanics Models for Hygrothermal Properties	149
	Problems	152
	References	154

6 Analysis of a Discontinuous Fiber-Reinforced Lamina   **156**

6.1	Introduction	156
6.2	Aligned Discontinuous Fibers	157
6.3	Off-Axis Aligned Discontinuous Fibers OMIT	169
6.4	Randomly Oriented Discontinuous Fibers	173
	Problems	186
	References	188

7 Analysis of Laminates   **190**

7.1	Introduction	190
7.2	Theory of Laminated Beams in Pure Flexure	192
7.3	Theory of Laminated Plates with Coupling	201
7.4	Stiffness Characteristics of Selected Laminate Configurations	207
	7.4.1 Symmetric Laminates	208
	7.4.2 Antisymmetric Laminates	210
	7.4.3 Quasi-Isotropic Laminates	212
7.5	Derivation and Use of Laminate Compliances	215
	7.5.1 Inversion of Laminate Force-Deformation Equations	215
	7.5.2 Determination of Lamina Stresses and Strains	217

Test 2

	7.5.3	Determination of Laminate Engineering Constants	220
	7.5.4	Comparison of Measured and Predicted Compliances	223
7.6	Hygrothermal Effects in Laminates		226
	7.6.1	Hygrothermal Degradation of Laminates	227
	7.6.2	Hygrothermal Stresses in Laminates	227
	7.6.3	Laminate Hygrothermal Expansion Coefficients	231
7.7	Interlaminar Stresses		232
7.8	Laminate Strength Analysis		237
	7.8.1	First Ply Failure Due to In-Plane Stresses	238
	7.8.2	Delamination Due to Interlaminar Stresses	244
7.9	Deflection and Buckling of Laminates		249
	7.9.1	Analysis of Small Transverse Deflections	250
	7.9.2	Buckling Analysis	255
7.10	Selection of Laminate Designs		258
	Problems		262
	References		267

8 Analysis of Viscoelastic and Dynamic Behavior   270

8.1	Introduction		270
8.2	Linear Viscoelastic Behavior of Composites		274
	8.2.1	Boltzmann Superposition Integrals for Creep and Relaxation	275
	8.2.2	Differential Equations and Spring-Dashpot Models	279
	8.2.3	Quasi-Elastic Analysis	288
	8.2.4	Sinusoidal Oscillations and Complex Modulus Notation	291
	8.2.5	Elastic-Viscoelastic Correspondence Principle	295
	8.2.6	Temperature and Aging Effects	300
8.3	Dynamic Behavior of Composites		306
	8.3.1	Longitudinal Wave Propagation and Vibrations in Specially Orthotropic Composite Bars	307
	8.3.2	Flexural Vibration of Composite Beams	311
	8.3.3	Transverse Vibration of Laminated Plates	315
	8.3.4	Analysis of Damping in Composites	321
	Problems		330
	References		335

9 Analysis of Fracture   338

9.1	Introduction		338
9.2	Fracture Mechanics Analyses of Through-Thickness Cracks		339
	9.2.1	Stress Intensity Factor Approach	341
	9.2.2	Strain Energy Release Rate Approach	345
9.3	Stress Fracture Criteria for Through-Thickness Notches		349
9.4	Interlaminar Fracture		356
	Problems		367
	References		369

10 Mechanical Testing of Composites and Their Constituents

374

10.1 Introduction   374
10.2 Measurement of Constituent Material Properties   375
     10.2.1 Fiber Tests   375
     10.2.2 Neat Resin Matrix Tests   378
10.3 Measurement of Basic Composite Properties   382
     10.3.1 Tensile Tests   382
     10.3.2 Compressive Tests   387
     10.3.3 In-Plane Shear Tests   390
     10.3.4 Interlaminar Shear Tests   393
     10.3.5 Flexure Tests   394
     10.3.6 Interlaminar Fracture Tests   395
     10.3.7 Fiber/Matrix Interface Tests   397
10.4 Measurement of Viscoelastic and Dynamic Properties   399
     10.4.1 Creep Tests   399
     10.4.2 Vibration Tests   402
Problems   408
References   410

Index   417

PREFACE

Composite materials is truly an interdisciplinary subject, and the number of students taking courses in this area is steadily increasing. Books on the subject tend to emphasize either the mechanics or the materials science aspects of composites. *Principles of Composite Material Mechanics* is mechanics-oriented. Composite materials technology is new enough so that many working engineers have had no training in this area, so a textbook in composite material mechanics should be useful not only for the education of new engineers, but also for the continuing education of practicing engineers and for reference. The high level of interest in composite materials, the interdisciplinary nature of the subject, the need to reeducate practicing engineers, and the need for a new composite mechanics textbook at the introductory level all led to my decision to write this book.

Chapters 1 through 7 form the basis of a one-semester senior/graduate-level course in Mechanical Engineering, which I have taught for the last 15 years. Chapters 8 through 10, along with selected papers from technical journals and student research projects/presentations, form the basis of a second one-semester course, which is taken only by graduate students, and which I have taught for the last 4 years. The book could also be the basis for a two-quarter sequence by omitting some topics. Prerequisities for the course are Mechanics of Materials, Introduction to Materials Engineering, and Ordinary Differential Equations, and previous exposure to linear algebra is highly desirable. For some of the graduate-level material earlier courses in Advanced Mechanics of Materials, Elasticity, and Partial Differential Equations are recommended, but not required.

Some of the basic elements of composite mechanics covered in this book have not changed since the first books on the subject were published in the 1960s and 1970s, and, where possible, I have tried to use the accepted terminology and nomenclature. For example, the coverage of stress–strain relationships and transformation of properties for aniso- tropic materials in Chap. 2 and the Classical Lamination Theory in Chap. 7

is consistent with that of previous textbooks such as the *Primer on Composite Materials* by Ashton, Halpin, and Petit and *Mechanics of Composite Materials* by Jones. However, rather than beginning the study of laminates by jumping directly into the Classical Lamination Theory, I have concluded that a better pedagogical approach is to introduce first basic laminate concepts by using the simpler theory of laminated beams in pure flexure. Also, I believe that the concept of an effective modulus of an equivalent homogeneous material, which had previously been covered only in advanced books such as *Mechanics of Composite Materials* by Christensen, is essential for the proper development of heterogeneous composite micromechanics. Thus, effective modulus concepts are emphasized from their introduction in Chap. 2 to their use in the analysis of viscoelastic and dynamic behavior in Chap. 8.

Although many basic concepts have been presented in earlier textbooks, numerous new developments in composite mechanics over the last two decades have made it increasingly necessary to supplement these books with my own notes. Thus, I have added coverage of such important topics as hygrothermal effects in Chap. 5, discontinuous fiber composites in Chap. 6, viscoelastic behavior and dynamic behavior in Chap. 8, fracture in Chap. 9, and mechanical testing in Chap. 10. The coverage of experimental mechanics of composites has been expanded to include summaries of the most important ASTM standard test methods, many of which did not exist when the early mechanics of composites books were published. A variety of example problems and homework problems, a number of them related to practical composite structures, is also included.

The contents of this book represent the cumulative effects of more than 25 years of interactions with colleagues and students, and I would be remiss if I did not mention at least some of them. My fascination with composites began in 1965 with my first engineering position in what is now part of Oak Ridge National Laboratory in Tennessee, where I was involved in the design and development of high-speed rotating equipment. At that time I realized that the advantages of using composites in rotating equipment are numerous, as is the case in many other applications. My experiences working with Dean Waters and other colleagues in the mechanical development group in Oak Ridge have had a strong influence on my later career decision to emphasize composites research and education. My doctoral research on vibration damping characteristics of composites with Robert Plunkett at the University of Minnesota further cemented my desire to continue working in the composites area and ultimately led to my career in university teaching and research.

After beginning my academic career at Iowa State University in 1975, I began a long and productive association with C. T. Sun, and later had the pleasure of spending a one-year leave working with C. T. and his colleagues Robert Sierakowski and Shive Chaturvedi at the University of Florida. I owe much of my understanding of composite mechanics to interactions with them. The notes leading to this book were developed by teaching composite mechanics courses at Iowa State University, the University of Idaho, the University of Florida, Michigan State University, and Wayne State University, and I am indebted to the students who took my classes and helped me to "debug" these notes over the years. Most recently, my students at Wayne State University have been particularly effective at finding the inevitable errors in my notes. Interaction with my graduate students over the years has contributed immeasurably to my understanding of composite mechanics, and the work of several of those students has been referred to in this book. I am particularly indebted to Stalin Suarez, Lyle Deobald, Raju Mantena, and Jimmy Hwang, all former graduate students at the University of Idaho.

Serious work on this book actually began during a sabbatical leave at Michigan State University in 1987, and I am indebted to Larry Drzal and his colleages for our many stimulating discussions during that year. Particularly important was the interaction with Cornelius Horgan, with whom I team-taught a course on advanced mechanics of composites. Most recently, my collaboration with John Sullivan and his colleagues of the Ford Scientific Research Laboratory has proved to be very rewarding, and I am indebted to John for his careful review of the manuscript and helpful comments. I am grateful to Carl Johnson, also of the Ford Scientific Research Laboratory, for his encouragement and support and for providing several of the figures in Chap. 1. The strong support of Wayne State University, which made it possible to establish the Advanced Composites Research Laboratory there in 1989, is gratefully acknowledged. The support and encouragement of my department chairman, Ken Kline, has been particularly important. Generous support for my composites research from numerous funding agencies over the years has also helped to make this book possible. Grants from the Air Force Office of Scientific Research, the National Science Foundation, the Army Research Office, the Boeing Company, and the Ford Motor Company have been particularly important.

McGraw-Hill and I would like to thank the following reviewers for their many helpful comments and suggestions: Charles W. Bert, University of Oklahoma; Olivier A. Bauchau, Rensselaer Polytechnic Institute; Shive Chaturvedi, Ohio State University; Vincent Choo, New Mexico State University; John M. Kennedy, Clemson University; Vikram K. Kinra, Texas A & M University; C. T. Sun, University of Florida; and Steven W. Yurgartis, Clarkson University.

Finally, my wife and best friend, Mary Anne, has been my strongest supporter as I labored on this project, and there is no way that I could have done it without her love, encouragement, patience, and understanding.

Ronald F. Gibson

PRINCIPLES OF COMPOSITE
MATERIAL MECHANICS

CHAPTER
1

INTRODUCTION

1.1 BASIC CONCEPTS

Structural materials can be divided into four basic categories: metals, polymers, ceramics, and composites. Composites, which consist of two or more separate materials combined in a macroscopic structural unit, are made from various combinations of the other three materials. Although many man-made materials have two or more constituents, they are generally not referred to as composites if the structural unit is formed at the microscopic level rather than at the macroscopic level. Thus, metallic alloys and polymer blends are usually not classified as composites (this may soon change, however, with the development of "molecular composites" consisting of oriented rodlike polymer molecules in a polymer binder material).

The relative importance of the four basic materials in a historical context has been presented by Ashby [1.1] as shown schematically in Fig. 1.1, in which the steadily increasing importance of polymers, composites, and ceramics and the decreasing role of metals is shown clearly. Composites are generally used because they have desirable properties which could not be achieved by either of the constituent materials acting alone. The most common example is the fibrous composite consisting of reinforcing fibers embedded in a binder, or matrix material. Particle or flake reinforcements are also used, but they are not so effective as fibers.

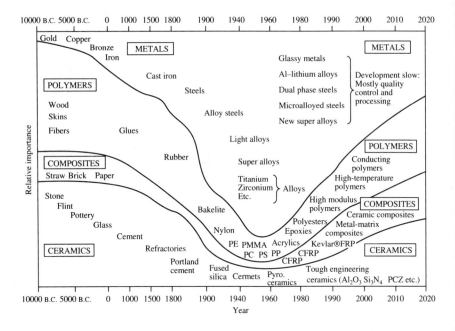

FIGURE 1.1

The relative importance of metals, polymers, composites, and ceramics as a function of time. The diagram is schematic and describes neither tonnage nor value. The time scale is nonlinear. (*From Ashby* [1.1].)

Although it is difficult to say with certainty when or where humans first learned about fibrous composites, nature provides us with numerous examples. Wood consists mainly of fibrous cellulose in a matrix of lignin, whereas most mammalian bone is made up of layered and oriented collagen fibrils in a protein-calcium phosphate matrix [1.2]. The book of Exodus in the *Old Testament* recorded what surely must be one of the first examples of man-made fibrous composites, the straw-reinforced clay bricks used by the Israelites. The early natives of South and Central America apparently used plant fibers in their pottery. These early uses of fibrous reinforcement, however, were probably based on the desire to keep the clay from cracking during drying rather than on structural reinforcement. Much later, humans developed structural composites such as steel-reinforced concrete, polymers reinforced with fibers such as glass and graphite, and many other materials.

Fibrous reinforcement is so effective because many materials are much stronger and stiffer in fiber form than they are in bulk form. It is believed that this phenomenon was first demonstrated scientifically in 1920 by Griffith [1.3], who measured the tensile strengths of glass rods

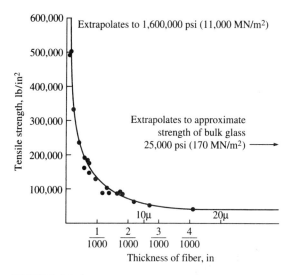

FIGURE 1.2
Griffith's measurements of tensile strength as a function of fiber thickness for glass fibers.
(*Data from Griffith* [1.3], *as analyzed by Gordon* [1.4].)

and glass fibers of different diameters. Griffith found that as the rods and fibers got thinner, they got stronger (see Fig. 1.2 from Ref. [1.3], as shown in Ref. [1.4]), apparently because the smaller the diameter, the smaller the likelihood that failure-inducing surface cracks would be generated during fabrication and handling. By extrapolating these results, Griffith found that for very small diameters the fiber strength approached the theoretical cohesive strength between adjacent layers of atoms, whereas for large diameters the fiber strength dropped to near the strength of bulk glass.

Results similar to those published by Griffith have been reported for a wide variety of other materials. The reasons for the differences between fiber and bulk behavior, however, are not necessarily the same for the other materials. For example, polymeric fibers are stronger and stiffer than bulk polymers because of the highly aligned and extended polymer chains in the fibers and the randomly oriented polymer chains in the bulk polymer. A similar effect occurs in crystalline materials such as graphite. In addition, a single crystal tends to have a lower dislocation density than a polycrystalline solid, so single crystal "whisker" materials are much stronger than the same material in polycrystalline bulk form. Indeed, "whiskers" are currently the strongest reinforcing materials available.

There can be no doubt that fibers allow us to obtain the maximum tensile strength and stiffness of a material, but there are obvious

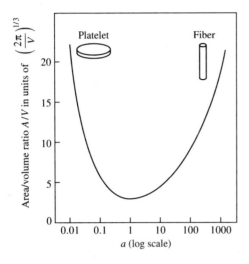

FIGURE 1.3
Surface area-to-volume ratio A/V of a cylindrical particle of given volume plotted vs. particle aspect ratio $a = l/d$. (*From McCrum et al.* [1.5]. *Copyright* © 1988 *by Oxford University Press.*)

disadvantages of using a material in fiber form. Fibers alone cannot support longitudinal compressive loads and their transverse mechanical properties are generally not so good as the corresponding longitudinal properties. Thus, fibers are generally useless as structural materials unless they are held together in a structural unit with a binder or matrix material and unless some transverse reinforcement is provided. Fortunately, the geometrical configuration of fibers also turns out to be very efficient from the point of view of interaction with the binder or matrix. As shown in Fig. 1.3 from Ref. [1.5], the ratio of surface area to volume for a cylindrical particle is greatest when the particle is in either platelet or fiber form. Thus, the fiber/matrix interfacial area available for stress transfer per unit volume of fiber increases with increasing fiber length-to-diameter ratio. The matrix also serves to protect the fibers from external damage and environmental attack. Transverse reinforcement is generally provided by orienting fibers at various angles according to the stress field in the component of interest. Filler particles are also commonly used in composites for a variety of reasons, such as weight reduction, cost reduction, flame and smoke suppression, and prevention of ultraviolet degradation due to exposure to sunlight.

The need for fiber placement in different directions according to the particular application has led to various types of composites, as shown in Fig. 1.4. In the continuous fiber composite laminate [Fig. 1.4(*a*)] individual continuous fiber/matrix laminae are oriented in the required directions and bonded together to form a laminate. Although the continuous fiber laminate is used extensively, the potential for delamination, or separation of the laminae, is still a major problem because the

[handwritten: Delamination is a problem]

(a) Continuous fiber composite

[handwritten: No delamination strength & stiffness sacrificed (fibers not straight)]

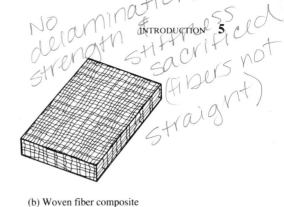

(b) Woven fiber composite

[handwritten: Low cost poor prop.]

(c) Chopped fiber composite

[handwritten: mixed.]

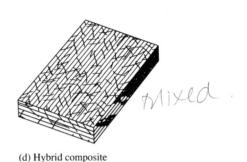

(d) Hybrid composite

FIGURE 1.4
Types of fiber-reinforced composites.

interlaminar strength is matrix-dominated. Woven fiber composites [Fig. 1.4(b)] do not have distinct laminae and are not susceptible to delamination, but strength and stiffness are sacrificed due to the fact that the fibers are not so straight as in the continuous fiber laminate. Chopped fiber composites may have short fibers randomly dispersed in the matrix, as shown in Fig. 1.4(c). Chopped fiber composites are used extensively in high-volume applications due to low manufacturing cost, but their mechanical properties are considerably poorer than those of continuous fiber composites. Finally, hybrid composites may consist of mixed chopped and continuous fibers, as shown in Fig. 1.4(d), or mixed fiber types such as glass/graphite. Another common composite configuration, the sandwich structure (Fig. 1.5), consists of high strength composite facing sheets (which could be any of the composites shown in Fig. 1.4) bonded to a lightweight foam or honeycomb core. Sandwich structures have extremely high flexural stiffness-to-weight ratios and are widely used in aerospace structures. The design flexibility offered by these and other composite configurations is obviously quite attractive to designers, and the potential now exists to design not only the structure, but also the structural material itself.

[handwritten: → I-beam like.]

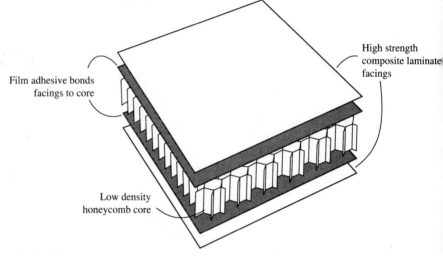

High strength
composite laminate
facings

Film adhesive bonds
facings to core

Low density
honeycomb core

FIGURE 1.5
Composite sandwich structure.

1.2 CONSTITUENT MATERIALS FOR COMPOSITES

Fiberglass-reinforced plastics were among the first structural composites. Composites incorporating glass or other relatively low modulus fibers (less than about 12×10^6 psi) are used in many high-volume applications such as automotive vehicles because of their low cost, and are sometimes referred to as "basic" composites. The so-called "advanced" composites made from graphite, silicon carbide, aramid polymer, boron, or other higher modulus fibers are used mainly in more exotic applications such as aerospace structures where their higher cost can be justified based on improved performance. The growing use of composites is shown in Fig. 1.6, along with a breakdown of current shipments of "basic" and "advanced" composites in different industries.

The advantages of advanced fibers over glass fibers and conventional bulk metallic materials are seen by the comparison of selected properties in Table 1.1. The main advantages are higher modulus and lower density. In many applications such as aerospace and automotive structures, structural weight is very important. Depending on whether the structural design is strength-critical or stiffness-critical, the material used should therefore have a high strength-to-weight ratio (or specific strength) or a high stiffness-to-weight ratio (or specific stiffness). As shown in Figs. 1.7 and 1.8 from Ref. [1.6], advanced fibers and their

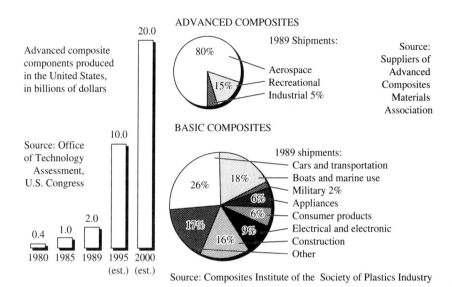

FIGURE 1.6
The growing use of composite materials. (*From* The New York Times, *March* 25, 1990, *Sec. F, p.* 14. *Copyright* © *by* The New York Times *Company. Reprinted by permission.*)

composites have a tremendous advantage over conventional materials in this area; and this is the principal reason that composites will be used with increasing frequency in future aerospace and automotive structures.

1.2.1 Fiber Materials

Glass fibers consists primarily of silica (silicon dioxide) and metallic-oxide-modifying elements and are generally produced by mechanical drawing of molten glass through a small orifice. E-glass (named for its electrical properties) accounts for most of the glass fiber production and is the most widely used reinforcement for composites. The second most popular glass fiber, S-glass, has roughly 30 percent greater tensile strength and 20 percent greater modulus of elasticity than E-glass (Table 1.1) but is not as widely used because of its higher cost. S-glass actually has greater strength than most advanced fibers, but its relatively low modulus limits its application. Glass/epoxy and glass/polyester composites are used extensively in applications ranging from fishing rods to storage tanks and aircraft parts.

　　Graphite or carbon fibers are the most widely used advanced fibers, and graphite/epoxy or carbon/epoxy composites are now used routinely

TABLE 1.1
Selected properties of fibers and bulk metals

Material	Tensile strength 10^3 psi (MPa)	Tensile modulus 10^6 psi (GPa)	Density lb/in^3 (g/cm^3)
Bulk 6061T6 Aluminum	45.0 (310)	10.0 (69)	0.098 (2.71)
Bulk SAE 4340 Steel	150.0 (1034)	29.0 (200)	0.283 (7.83)
E-glass fibers	500.0 (3448)	10.5 (72)	0.092 (2.54)
S-glass fibers	650.0 (4482)	12.5 (86)	0.090 (2.49)
Carbon fibers (PAN precursor)			
AS-4 (Hercules)	580.0 (4000)	33.0 (228)	0.065 (1.80)
IM-7 (Hercules)	785.0 (5413)	40.0 (276)	0.064 (1.77)
T-300 (Amoco)	530.0 (3654)	33.5 (231)	0.064 (1.77)
T-650/42 (Amoco)	730.0 (5033)	42.0 (290)	0.064 (1.77)
Carbon fibers (pitch precursor)			
P-55 (Amoco)	250.0 (1724)	55.0 (379)	0.072 (1.99)
P-75 (Amoco)	300.0 (2068)	75.0 (517)	0.072 (1.99)
P-100 (Amoco)	325.0 (2241)	100.0 (690)	0.078 (2.16)
Aramid fibers			
Kevlar® 29 (Dupont)	550.0 (3792)	9.0 (62)	0.052 (1.44)
Kevlar® 49 (Dupont)	550.0 (3792)	19.0 (131)	0.053 (1.47)
Boron fibers			
0.004″ diameter (Textron)	510.0 (3516)	58.0 (400)	0.093 (2.57)
0.0056″ diameter (Textron)	510.0 (3516)	58.0 (400)	0.090 (2.49)
Silicon carbide fibers			
0.0056″ diameter (Textron)	500.0 (3448)	62.0 (427)	0.110 (3.04)

in aerospace structures. Unfortunately, the names "carbon" and "graphite" are often used interchangeably to describe fibers based on the element carbon. These fibers are usually produced by subjecting organic precursor fibers such as polyacrylonitrile (PAN) or rayon to a sequence of heat treatments, so that the precursor is converted to carbon by pyrolysis. The major difference is that graphite fibers are subjected to higher temperature pyrolysis than carbon fibers. The result is that carbon fibers typically are less than 95% carbon, whereas graphite fibers are at least 99% carbon [1.7]. Although carbon fibers were once prohibitively expensive, the cost has dropped significantly as production capacity and demand has increased over the past 20 years (Fig. 1.9 from Ref. [1.8]). Development of new carbon and graphite fibers continues at a rapid

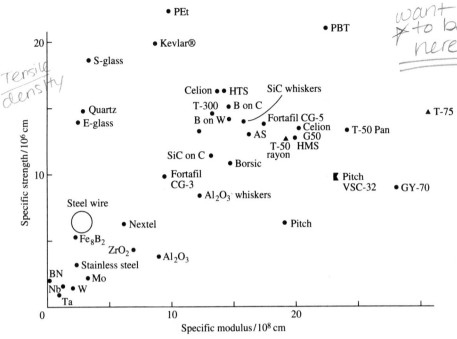

FIGURE 1.7

Specific strength vs. specific modulus for a variety of fibers. The specific value is the value of the property divided by the density. (*From Kelly* [1.6].)

pace. For example, fibers based on a pitch precursor (P-100) are now available with a modulus of 100×10^6 psi (690 GPa), more than three times that of steel.

Aramid polymer fibers, produced primarily by E. I. duPont de Nemours & Company under the tradename "Kevlar®," were originally developed for use in radial tires. Kevlar® 29 is still used for this purpose, but a higher modulus version, Kevlar® 49, is used more extensively in structural composites. The density of Kevlar® is about half that of glass and its specific strength is among the highest of currently available fibers. Kevlar® also has excellent toughness, ductility, and impact resistance, unlike brittle glass or graphite fibers.

Boron fibers are actually composites consisting of a boron coating on a substrate of tungsten or carbon, and the diameter of boron fibers is among the largest of all the advanced fibers, typically 0.002–0.008 in (0.05–0.2 mm). Boron fibers have much higher strength and stiffness than graphite (Table 1.1), but they also have higher density. Boron/epoxy and boron/aluminum composites are widely used in aerospace structures, but high cost still prevents more widespread use.

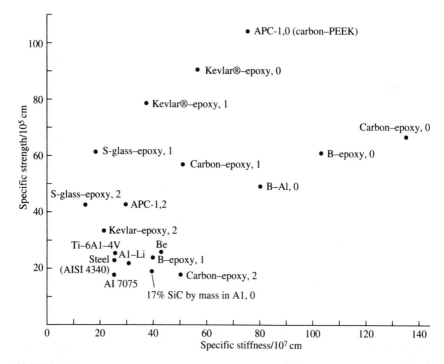

FIGURE 1.8

Specific strength vs. specific modulus of some isotropic materials and of fiber composites. The designation 1 on the composite means the following arrangement of fibers: 50% at 0°, 40% at ±45°, and 10% at 90° to the stress; 2 denotes balanced laminates with equal proportions at 45°, 90°, and 135°; 0 means aligned fibers in the specified matrix. The volume fractions of fibers in the various composites are not the same in the different systems. They vary between 40 and 60%. (*From Kelly* [1.6].)

Silicon carbide (SiC) fibers are used primarily in high-temperature metal and ceramic matrix composites because of their excellent oxidation resistance and high-temperature strength retention. At room temperature the strength and stiffness of SiC fibers are about the same as those of boron. SiC whisker-reinforced metals are also receiving considerable attention as alternatives to unreinforced metals and continuous fiber-reinforced metals. SiC whiskers are very small, typically 8–20 μ in (20–51 nm) in diameter and about 0.0012 in (0.03 mm) long, so that standard metal-forming processes such as extrusion, rolling, and forging can be easily used [1.7].

The list of advanced fibers is steadily increasing, and it is not feasible to discuss all of them here. Other fibers such as highly oriented polyethylene (Spectra® by Allied Corporation) and aluminum oxide (FP® by duPont Company) have found a number of applications where

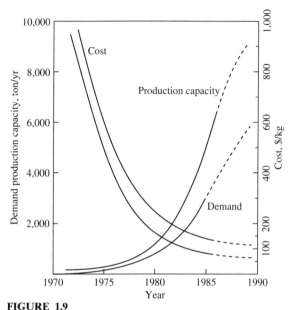

FIGURE 1.9
PAN-base carbon fibers are the major reinforcement used in today's advanced polymer matric composites. Fiber cost has and will continue to drop as production capacity and demand increases. (*From Reinhart*[1.8].)

their particular properties have proved to be beneficial. Hybrids consisting of mixed fiber materials can be used when a single fiber material does not have all the desired properties. More complete descriptions of fiber materials and their properties can be found in several composites handbooks [1.7, 1.9–1.12]. Further discussion of fiber properties, including anisotropic behavior, will be given later in Chap. 3.

1.2.2 Matrix and Filler Materials

Polymers, metals, and ceramics are all used as matrix materials in composites, depending on the particular requirements. The matrix holds the fibers together in a structural unit and protects them from external damage, transfers and distributes the applied loads to the fibers, and in many cases contributes some needed property such as ductility, toughness, or electrical insulation. A strong interface bond between the fiber and matrix is obviously desirable, so the matrix must be capable of developing a mechanical or chemical bond with the fiber. The fiber and matrix materials should also be chemically compatible, so that undesirable reactions do not take place at the interface. Such reactions tend to be more of a problem in high-temperature composites. Service temperature is often the main consideration in the selection of a matrix material. Thus, the materials will be discussed below in order of increasing temperature capability.

Polymers are unquestionably the most widely used matrix materials in modern composites. Polymers are described as being either thermosets (e.g., epoxy, polyester, phenolic) or thermoplastics [e.g., polyimide (PI), polysulfone (PS), polyetheretherketone (PEEK), polyphenylene sulfide (PPS)]. Upon curing, thermosets form a highly cross-linked, three-dimensional molecular network which does not melt at high temperature. Thermoplastics, however, are based on polymer chains which do not cross-link. As a result, thermoplastics will soften and melt at high temperature, then harden again upon cooling.

Epoxies and polyesters have been the principal polymer matrix materials for several decades, but advanced thermoplastics such as PEEK and PPS are now receiving considerable attention for their excellent toughness and low moisture absorption properties, their simple processing cycles, and their higher temperature capabilities. Aerospace grade epoxies are typically cured at about 177°C (350°F) and are generally not used at temperatures above 150°C (300°F), whereas advanced thermoplastics such as PPS, PI, and PEEK have melting temperatures in the range 315–370°C (600–700°F). At this time it appears that polymer matrix materials for use up to 425°C (800°F) are feasible. For higher temperatures metal, ceramic, or carbon matrix materials are required.

By using lightweight metals such as aluminum, titanium, and magnesium and their alloys and intermetallics such as titanium aluminide and nickel aluminide, operating temperatures can be extended to about 1250°C (2280°F). Other advantages of metal matrices are higher strength, stiffness, and ductility than polymers at the expense of higher density. Ceramic matrix materials such as silicon carbide and silicon nitride can be used at temperatures up to 1650°C (3000°F). Ceramics have poor tensile strength and are notoriously brittle, however, and there is a need for much research before these materials can be routinely used. Finally, carbon fiber/carbon matrix composites can be used at temperatures approaching 2760°C (5000°F), but the cost of these materials is such that they are only used in a few critical aerospace applications. For further details on matrix materials and their properties the reader is referred to any of several handbooks [1.7, 1.9–1.12]. Matrix properties will be discussed again in Chap. 3, where properties for typical matrix materials will be given.

The third constituent material of a composite, the filler material, is mixed in with the matrix material during fabrication. Fillers are not generally used to improve mechanical properties but, rather, are used to enhance some other aspect of composite behavior. For example, hollow glass microspheres are used to reduce weight, clay or mica particles are used to reduce cost, carbon black particles are used for protection against ultraviolet radiation, and alumina trihydrate is used for flame and smoke suppression [1.11]. Fillers truly add another dimension to the design flexibility we have in composites.

1.3 STRUCTURAL APPLICATIONS OF COMPOSITES

Composite structural elements are now used in a variety of components for automotive, aerospace, marine, and architectural structures in addition to consumer products such as skis, golf clubs, and tennis rackets. Since much of the current composites technology evolved from aerospace applications, it is appropriate to begin this brief overview there.

Military aircraft designers were among the first to realize the tremendous potential of composites with high specific strength and high specific stiffness since performance and maneuverability of those vehicles depend so heavily on weight. Composite construction also leads to smooth surfaces (no rivets or sharp transitions as in metallic construction) which reduce drag. Since boron and graphite fibers were first developed in the early 1960s, applications of advanced composites in military aircraft have accelerated quickly. Composite structural elements such as horizontal and vertical stabilizers, flaps, wing skins, and various control surfaces have been used in fighter aircraft such as the F-14, F-15, and F-16 with typical weight savings of about 20 percent. The AV-8B (Fig. 1.10) has graphite/epoxy wing-box skins, forward fuselage, horizontal stabilizer, elevators, rudder and other control surfaces, and overwing fairing totaling about 26 percent of the aircraft's structural weight [1.12]. One of the most demanding applications thus far is the use of graphite/epoxy composite wing structures on the experimental forward-swept wing X-29 fighter (Fig. 1.11). Although the concept of a forward-swept wing for improved maneuverability is not new, conventional aluminum structures could not withstand the aerodynamic forces acting on such a wing, so the implementation of the concept had to wait for the development of advanced composites.

Composites applications in commercial aircraft have been steadily increasing as material costs come down (Fig. 1.9), as design and manufacturing technology evolves, and as the experience with composites in aircraft continues to build. A flight service evaluation of composite components in selected aircraft was initiated in 1973 by NASA, and the results have encouraged increased reliance on composite aircraft structures [1.13]. The Boeing 757 and 767 were among the first commercial airliners to make extensive use of composites. Table 1.2 summarizes the composite components of the Boeing 757, which has about 3000 lb

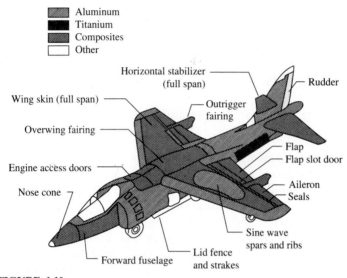

FIGURE 1.10
Composite structures on the AV-8B fighter aircraft. (*From Schwartz* [1.7]. *Copyright* © 1984 *by McGraw-Hill, Inc., Reprinted by permission.*)

FIGURE 1.11
Experimental X-29 aircraft. The forward swept wing structures are made from graphite/epoxy. (*Courtesy of Grumman Aircraft Corporation.*)

TABLE 1.2
Composite components on the Boeing 757

Description	Material	Size $T \times W \times L$	
		cm	in
Body			
Aft wing/body fairing	Aramid, carbon	$33 \times 50 \times 193$	$13 \times 20 \times 76$
Forward wing/body fairing	Aramid, carbon	$38 \times 60 \times 360$	$15 \times 24 \times 142$
Forward wing/body fairing	Aramid, carbon	$177 \times 185 \times 188$	$70 \times 73 \times 74$
Forward wing/body fairing	Aramid, carbon	$170 \times 198 \times 188$	$67 \times 78 \times 74$
Main landing gear doors	Carbon	$81 \times 170 \times 284$	$32 \times 67 \times 112$
Nose landing gear doors	Aramid, carbon	64×203	25×80
Wing			
Flap track fairing	Aramid, carbon	$35 \times 50 \times 295$	$14 \times 20 \times 116$
Aft outboard trailing edge flaps	Carbon	$8 \times 20 \times 650$	$3 \times 27 \times 256$
Aft inboard trailing edge flaps	Carbon	$13 \times 69 \times 368$	$5 \times 27 \times 145$
Aileron	Carbon	$15 \times 76 \times 460$	$6 \times 30 \times 180$
Inboard spoiler	Carbon	$8 \times 69 \times 190$	$3 \times 27 \times 75$
Outboard spoiler	Carbon	$8 \times 58 \times 163$	$3 \times 23 \times 64$
Lower wing inboard fixed leading edge panels	Aramid, carbon	$46 \times 71 \times 160$	$18 \times 28 \times 63$
Wing inboard fixed trailing edge panels	Aramid, carbon	152×229	60×90
Wing outboard fixed trailing edge panels	Aramid, carbon	36×430	14×170
Main landing gear trunnion fairing	Aramid, carbon	$51 \times 51 \times 180$	$20 \times 20 \times 72$
Engine strut fairing—forward	Aramid	$41 \times 61 \times 163$	$16 \times 24 \times 64$
Engine strut fairing—upper	Aramid	$15 \times 66 \times 160$	$6 \times 26 \times 63$
Engine strut fairing—U/W forward	Aramid	$15 \times 51 \times 66$	$6 \times 20 \times 26$
Engine strut fairing—aft	Aramid	$3 \times 58 \times 163$	$1 \times 23 \times 64$
Engine cowls	Carbon	$135 \times 135 \times 117$	$53 \times 53 \times 46$
Empennage			
Rudder	Carbon	$38 \times 234 \times 902$	$15 \times 92 \times 355$
Vertical tip fairing	Aramid	56×157	22×62
Vertical fixed trailing edge panels	Aramid, carbon	56×254	22×100
Elevators	Carbon	$23 \times 135 \times 711$	$9 \times 53 \times 280$
Horizontal tip fairings	Aramid	$23 \times 135 \times 117$	9×46

Source: From Reinhart et al. [1.12].

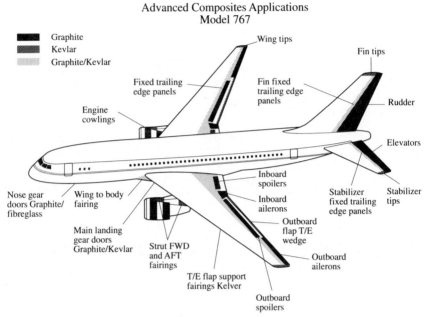

FIGURE 1.12
Composite applications on the Boeing 767 airliner. (*Courtesy of Boeing.*)

(1350 kg) composites. About 30 percent of the external surface area of the Boeing 767 (Fig. 1.12) consists of composites. While composites are currently being used conservatively in only the secondary structures in large airliners, some of the smaller business-type aircraft have composites in the primary structures as well. For example, the Beech Starship (Fig. 1.13) represents a radical new design with graphite, Kevlar, and honeycomb composite fuselage and wing structures, and is the first pressurized composite aircraft to be certified by the Federal Aviation Administration. The *Starship* shares much of its technology with the composite *Voyager,* the first aircraft to fly nonstop around the world without refueling. This record-setting flight would not have been possible with conventional structural materials. Finally, the level of sophistication that has been attained in current composite construction is strikingly illustrated by the composite helicopter rotor blade in Fig. 1.14. The construction of such a component obviously requires a multistep fabrication procedure involving many materials, and some of the fabrication processes required to do this will be discussed in the next section.

Due to the tremendous cost per unit weight to place an object in space, the value of weight saved is even greater for spacecraft than it is for aircraft. Thus, composites are extremely attractive for spacecraft

FIGURE 1.13
The Beech Starship composite business-type aircraft. (*Beech Aircraft Corporation photo.*)

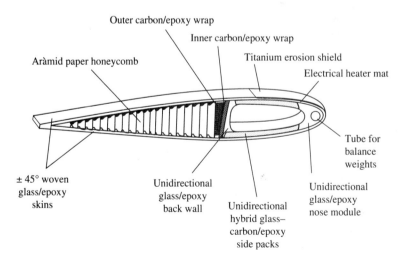

FIGURE 1.14
Composite construction of a helicopter rotor blade. (*From McCrum et al.* [1.5]. *Copyright* © 1988 *by Oxford University Press. Reprinted by permission.*)

FIGURE 1.15
Conceptual drawing of the proposed National Aero-Space Plane, which would make extensive use of high-temperature composites. (*Courtesy of National Aero-Space Plane Joint Program Office.*)

applications. The NASA Space Shuttle has a number of composite parts, including graphite/epoxy cargo bay doors and experimental graphite/epoxy solid rocket booster motor cases. For large space structures such as the proposed space station the key properties of the structural materials are high stiffness-to-weight ratio, low thermal expansion coefficient, and good vibration damping characteristics. In all three of these areas composites offer significant advantages over conventional metallic materials. The proposed National Aerospace Plane (Fig. 1.15) would not be feasible without heavy use of advanced high-temperature composites (some of which have not yet been developed) in its structure and engines.

Structural weight is also very important in automotive vehicles, and the use of composite automotive components continues to grow. Glass fiber-reinforced polymers continue to dominate the automotive composites scene, and advanced composites have still not made significant inroads. Weight savings on specific components such as leaf springs (Fig. 1.16) can exceed 70 percent compared with steel (composite leaf springs have also proved to be more fatigue resistant than steel springs). Experimental composite engine blocks have been fabricated from graphite-reinforced thermoplastics, but the ultimate goal is a ceramic

FIGURE 1.16
Liteflex® composite leaf springs for automotive vehicles. (*Courtesy of Delco Products Division of General Motors Corporation.*)

composite engine which would not require water-cooling. Chopped glass fiber-reinforced polymers have been used extensively in body panels, where stiffness and appearance are the principal design criteria. Composite primary structures such as the Ford Taurus "Tub" (Fig. 1.17) are only experimental at this point, but they offer weight reduction, fewer parts, and smaller assembly and manufacturing costs. As with airliners, the applications of composites in automotive vehicles up to this point have been mainly in secondary structural elements and appearance parts, and the full potential of composite construction remains to be explored.

Other applications of structural composites are numerous, so only a few selected examples will be given here. I-beams, channel sections, and other structural elements (Fig. 1.18) used in buildings may be made of fiber-reinforced plastic using the pultrusion process, which will be discussed in the next section. Corrosion resistance and electrical and thermal insulation are added advantages of composites compared with steel in such applications. Composite machinery components such as drive shafts (Fig. 1.19) have the advantages of faster response and better vibration-damping characteristics than metallic parts. In these examples, as well as in many of the previous examples, cost is a major consideration preventing more widespread use of composites. The fabrication process is the key to cost control, and the next section will describe the fabrication processes used to make the components described here.

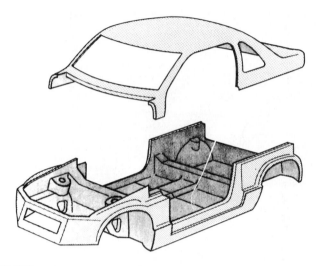

FIGURE 1.17
Ford Taurus "Tub" concept all composite body structure. (*Courtesy of Ford Motor Company, Research Staff.*)

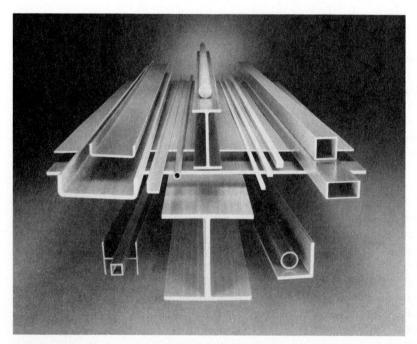

FIGURE 1.18
Fiberglass composite structural elements formed by pultrusion. (*Photo Courtesy of MMFG.*)

FIGURE 1.19
Filament wound composite power transmission shaft. (*Courtesy of Ford Motor Company, Research Staff.*)

1.4 FABRICATION PROCESSES

Although this book is concerned primarily with mechanics of composite materials, it is essential for the reader to know how these materials are made. This is because with composites, we design and build not only the structure, but also the structural material itself. The selection of a fabrication process obviously depends on the constituent materials in the composite, with the matrix material being the key (i.e., the processes for polymer matrix, metal matrix, and ceramic matrix composites are generally quite different). In this brief summary of fabrication processes only those processes that are used for polymer matrix composite fabrication will be discussed, and the reader is referred to other books for more details on metal matrix and ceramic matrix composite fabrication [1.7, 1.9–1.12, 1.14].

A summary of fabrication processes used for polymer composites with various types of fiber reinforcement is given in Table 1.3. The open mold process with hand lay-up of woven fiber mat or chopped strand mat [Fig. 1.20(a)] or spray-up of chopped fibers [Fig. 1.20(b)] is used for development work, prototype fabrication, and production of large components and relatively small quantities. A mold having the desired shape is first coated with a mold release which prevents bonding of the resin matrix material to the mold. If a smooth surface on the part is desired (i.e., boat hulls or aircraft exterior parts), a gel coat is then applied to the mold, followed by a thermosetting polymer resin and the fibers. A roller may then be used for consolidation, followed by curing of the polymer resin at the required temperature.

A major breakthrough in composite manufacturing technology occurred with the development of "prepreg tape," which is a tape consisting of fibers precoated with the polymer resin. This innovation means that the fabricator no longer has to worry about mixing the resin components in the right proportions and combining the resin with the fibers in the correct fashion. Most prepreg tape is made by the hot-melt process (Fig. 1.21). If a thermosetting resin is used, the resin coating is partially cured, and the tape must be kept refrigerated to prevent full curing until final use. If a thermoplastic resin is used, the tape can be stored at room temperature until it is melted during final use. The fabrication of a laminated structure with prepreg tape involves simply "laying-up" the tape at the required orientation on a mold, stacking layers of tape in the required stacking sequence, and then curing the assembly under elevated temperature and pressure.

Autoclave molding (Fig. 1.22) is the standard aerospace industry process for fabrication with prepreg tapes. The autoclave is simply a heated pressure vessel into which the mold (with lay-up) is placed and subjected to the required temperature and pressure for curing. The mold and lay-up are often covered with a release fabric, a bleeder cloth, and a vacuum bag. A vacuum line is then attached to the mold for evacuation of volatile gases during the cure process. Without the vacuum bagging these gases would be trapped and could cause void contents of greater than 5 percent in the cured laminate. With the vacuum bag void contents on the order of 0.1 percent are attainable. Autoclaves come in a wide range of sizes from bench-top laboratory versions to the room-size units which are used to cure large aircraft structures. The autoclave-style press cure [1.15] is often used to cure small samples for research. In this case a

TABLE 1.3
Fabrication processes for polymer matrix composites

Process	Continuous	Chopped	Woven	Hybrid
		Type of fiber reinforcement		
Open mold				
Hand lay-up		×	×	
Spray-up		×		
Autoclave	×		×	
Compression molding	×	×	×	×
Filament winding	×			
Pultrusion	×		×	
Reinforced reaction injection molding (RRIM)		×		
Thermoplastic molding	×	×	×	×
Resin transfer molding (RTM)	×	×	×	×
Structural reaction injection molding (SRIM)	×	×	×	×

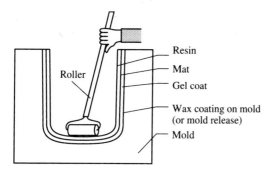

Resin
Mat
Gel coat
Wax coating on mold (or mold release)
Mold
Roller

FIGURE 1.20(a)
Open mold, hand lay-up composite fabrication.

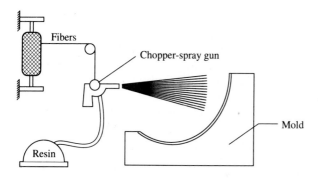

Fibers
Chopper-spray gun
Resin
Mold

FIGURE 1.20(b)
Open mold, spray-up composite fabrication.

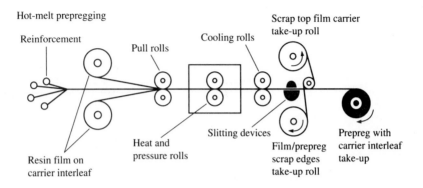

FIGURE 1.21
Hot-melt prepregging process. [*Courtesy of Suppliers of Advanced Composite Materials Association (SACMA).*]

vacuum-bagged mold assembly (Fig. 1.23) is inserted between the heated platens of a hydraulic press, and the press then generates the temperature and pressure required for curing. A vacuum press is a variation on this concept involving the use of a vacuum chamber surrounding the platen-mold assembly, and a sealed door on this chamber eliminates the need for a vacuum bag.

Sheet-molding compound (SMC) is an important innovation in composite manufacturing which is used extensively in the automobile

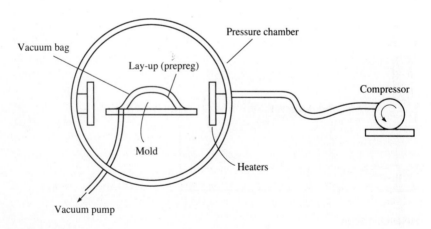

FIGURE 1.22
Autoclave molding.

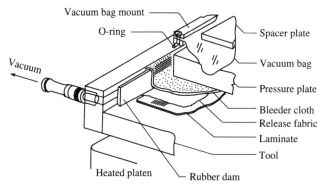

FIGURE 1.23
Lay-up sequence for autoclave-style press molding.

industry. SMC is similar to prepreg tape in that the fibers and the resin are "prepackaged" in a form that is more easily used by fabricators. SMC consists of a relatively thick, chopped fiber-reinforced resin sheet, whereas prepreg usually has continuous fibers in a thin tape. A machine for producing SMC is shown schematically in Fig. 1.24. An alternative to SMC is bulk molding compound (BMC), which consists of the chopped fiber/resin mixture in bulk form. SMC or BMC may be molded by using the matched metal die process (Fig. 1.25).

Filament winding (Figs. 1.26 and 1.27), which involves winding resin-coated fibers onto a rotating mandrel, may be used to produce any composite structure which has the form of a body of revolution. Fiber orientation is controlled by the traverse speed of the fiber winding head and the rotational speed of the mandrel. Another advantage of this process is that by controlling the winding tension on the fibers, they can be packed together very tightly to produce high fiber volume fractions. Upon completion of the winding process, the composite structure may be cured by placing the mandrel in an oven or by passing hot fluid through the mandrel itself.

Filament winding is widely used to produce such structures as rocket motor cases (Fig. 1.27), pressure vessels, power transmission shafts (Fig. 1.19), piping, and tubing. Prepreg tape is often produced by filament winding and removing the tape from the mandrel before curing. Imaginative variations on the filament winding process have produced a variety of structures such as leaf springs for automotive vehicles. A composite leaf spring may be fabricated by winding on an ellipsoidal mandrel, then cutting the cured shell into the required pieces. Experimental programs are underway to produce large, complex structures such as aircraft fuselages and automobile body structures by filament winding.

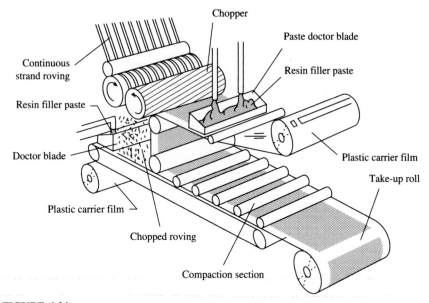

FIGURE 1.24
Machine for producing sheet-molding compound (SMC). (*From Reinhart et al.* [1.12]. *Reprinted by permission of ASM International.*)

Filament winding machines for such structures will require liberal use of computer control and robotics.

Many of the processes described above have the disadvantage of being fairly time consuming. Processes with faster production cycles are needed for high-volume applications such as automotive parts. For example, reinforced reaction injection molding (RRIM) is a very fast process which is widely used to produce such components as automobile body panels. The RRIM process (Fig. 1.28) involves injection of a chopped fiber/resin mixture into a mold under high pressure and then curing at the required temperature. "Pultrusion" (Fig. 1.29) is the

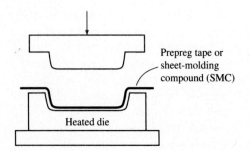

Prepreg tape or sheet-molding compound (SMC)

Heated die

FIGURE 1.25
Compression molding with matched metal dies.

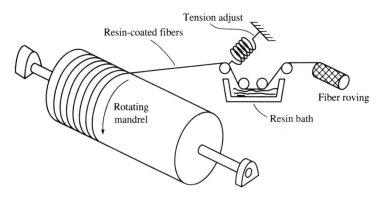

FIGURE 1.26
Filament winding process.

process of pulling a continuous fiber/resin mixture through a heated die to form structural elements such as I-beams and channel sections (Fig. 1.18). This process is relatively fast but is restricted to structures whose shapes do not change along the length. In the thermoplastic molding process (Fig. 1.30) the blank (an uncured laminate consisting of thermoplastic prepreg tape layers) is passed through an infrared oven where it is heated to near the melting point of the thermoplastic resin. The heated blank is then quickly placed in a matched metal die mold for final forming. Resin transfer molding (RTM) and structural reaction injection molding (SRIM) are attracting considerable attention because of their relatively fast production cycles and the near-net-shape of resulting parts. In both the RTM process (Fig. 1.31) and the SRIM process a "preform" consisting of fibers and possibly a foam core is first produced in the general shape of the finished part. The preform is then placed in a closed metal mold and the liquid resin is injected under pressure. The major difference between the two processes is that with RTM the resin and hardener are premixed before injection into the mold, whereas with SRIM the resin and hardener are mixed by impingement as they are injected into the mold. Three dimensionally shaped parts with foam cores can be produced with both RTM and SRIM, but SRIM tends to be faster than RTM.

In conclusion, many innovative processes exist for manufacturing polymer composites. Much of the success that composite materials have had in the past several decades is due to innovative fabrication technology, and the future success of composites will surely depend on further advances in this area. Computer-aided-manufacturing technology and robotics are expected to play important roles in the continuing drive to reduce cost and to improve the quality of composite structures.

FIGURE 1.27
Filament winding a rocket motor case. (*Copyright ASTM. Reprinted with permission.*)

1.5 ELEMENTS OF MECHANICAL BEHAVIOR OF COMPOSITES

This book is concerned with the analysis of both the micromechanical and the macromechanical behavior of fiber-reinforced composite materials. As shown schematically in Fig. 1.32, micromechanics is concerned with the mechanical behavior of constituent materials (in this case fiber and matrix materials), the interaction of these constituents, and the resulting

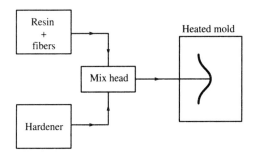

FIGURE 1.28
Reinforced reaction injection molding (RRIM) process.

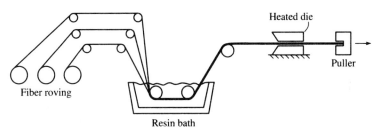

FIGURE 1.29
Pultrusion process.

behavior of the basic composite (in this case a single lamina in a laminate). Macromechanics is concerned with the gross mechanical behavior of composite materials and structures (in this case lamina, laminate, and structure) without regard for the constituent materials or their interactions. As we will see in Chap. 2, this macromechanical behavior may be characterized by averaged stresses and strains and averaged, or "effective", mechanical properties in an equivalent homogeneous material. As shown in Chaps. 3 to 6, the central problem in

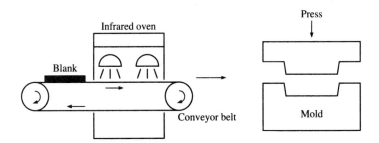

FIGURE 1.30
Thermoplastic molding process.

HIGH SPEED RESIN TRANSFER MOLDING PROCESS

COMPOSITE CROSSMEMBER

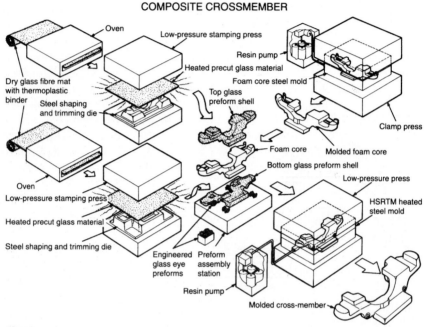

FIGURE 1.31
Resin transfer molding (RTM) process. (*Courtesy of Ford Motor Company, Research Staff.*)

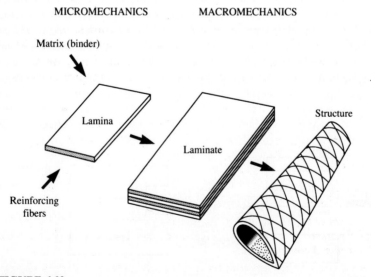

FIGURE 1.32
Micromechanics and macromechanics of composites.

micromechanics is the study of relationships between the effective composite properties and the effective constituent properties. Subsequent chapters deal with macromechanical behavior of laminates and structures.

When dealing with composite materials, we find very quickly that we can no longer draw upon the "intuition" about material behavior that we developed from years of experience with conventional metallic structural materials, and that we must learn to "think composites." Most metallic structural materials are homogeneous (properties do not vary from point to point in the material) and isotropic (properties do not depend on orientation), whereas most composites are heterogeneous and anisotropic. That is, the properties in a composite vary as we move from matrix to fiber and as we change the direction along which they are measured. For example, in a "unidirectional" composite having reinforcement in only one direction the strength and stiffness are much greater along the reinforcement direction than they are in the transverse direction.

The relationships between forces and deformations (or between stresses and strains) are much more complicated for anisotropic composites than they are for conventional isotropic materials, and this can lead to unexpected behavior. For example, in an isotropic material a normal stress induces only normal strains (extensions and/or contractions), and a shear stress induces only shear strains (distortions). In an anisotropic composite, however, a normal stress may induce both normal strains and shear strains, and a shear stress may induce both shear strains and normal strains. A temperature change in an isotropic material causes expansion or contraction that is uniform in all directions, whereas a temperature change in an anisotropic material may cause nonuniform expansion or contraction plus distortion. These so-called "coupling" effects have important implications for not only the analytical mechanics of composites, but for the experimental characterization of composite behavior as well.

It is hoped that these general observations regarding composite materials will provide motivation for further study in subsequent chapters, where the analytical and experimental characterization of mechanical behavior of composites is discussed in more detail.

PROBLEMS

1.1. For a cylindrical particle, derive the relationship between the ratio of surface area-to-volume, A/V, and the particle aspect ratio, l/d, and verify the shape of the curve shown in Fig. 1.3.

1.2. Explain qualitatively why sandwich structures (Fig. 1.5) have such high flexural stiffness-to-weight ratios. Describe the key parameters affecting the flexural stiffness-to-weight ratio of a sandwich panel.

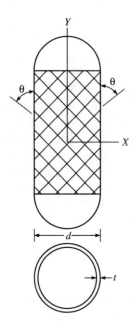

FIGURE 1.33
Filament wound composite pressure vessel.

1.3. Describe a possible sequence of fabrication processes which might be used to manufacture the helicopter rotor blade in Fig. 1.14. Note that several different materials and fiber lay-ups are used.

1.4. Which of the reinforcing fibers listed in Table 1.1 would be best for use in an orbiting space satellite antenna structure which is subjected to relatively low stresses but has very precise dimensional stability requirements? The answer should be based only on the properties given in Table 1.1.

1.5. A thin-walled filament wound composite pressure vessel has fibers wound at a helical angle θ, as shown in Fig. 1.33. Ignore the resin matrix material and assume that the fibers carry all of the load. Also assume that all fibers are uniformly stressed in tension. This gross oversimplification is the basis of the so-called "netting analysis" which is actually more appropriate for stress analysis of all-fiber textile fabrics. Using this simplified analysis, show that the angle θ must be 54.74° in order to support both the hoop (tangential) and axial stresses that are generated in a thin-walled pressure vessel. See any mechanics of materials book for the stress analysis of a thin-walled pressure vessel.

1.6. A filament wound E-glass/epoxy pressure vessel has a diameter of 50 in (127 cm), a wall thickness of 0.25 in (6.35 mm), and a helical wrap angle $\theta = 54.74°$. Using a netting analysis and a safety factor of 2, estimate the allowable internal pressure in the vessel. Compare with the allowable internal pressure in a 6061-T6 aluminum alloy pressure vessel having the same dimensions. For the aluminum vessel, assume that the tensile yield stress is 40,000 psi (276 MPa) and use the Maximum Shear Stress yield criterion. Although the netting analysis is greatly oversimplified, these approximate results should demonstrate the significant advantages of fiber composite construction over conventional metallic construction.

REFERENCES

1.1. Ashby, M. F., "Technology of the 1990s: Advanced Materials and Predictive Design," *Philosophical Transactions of the Royal Society of London,* **A322,** 393–407 (1987).

1.2. Wainwright, S. A., Biggs, W. D., Currey, J. D., and Gosline, J. M., *Mechanical Design in Organisms,* Princeton University Press, Princeton, NJ (1976).

1.3. Griffith, A. A., "The Phenomena of Rupture and Flow in Solids," *Philosophical Transactions of the Royal Society,* **221A,** 163–198 (1920).

1.4. Gordon, J. E., *The New Science of Strong Materials,* 2d ed., Princeton University Press, Princeton, NJ (1976).

1.5. McCrum, N. G., Buckley, C. P., and Bucknall, C. B., *Principles of Polymer Engineering,* Oxford University Press, New York (1988).

1.6. Kelly, A., "Composites for the 1990s," *Philosophical Transactions of the Royal Society of London,* **A322,** 409–423 (1987).

1.7. Schwartz, M. M., *Composite Materials Handbook,* McGraw-Hill, Inc., New York (1984).

1.8. Reinhart, T. J., "Polymer Matrix Composites," *Advanced Materials and Processes,* **137**(1), 33 (January 1990).

1.9. Weeton, J. W., Peters, D. M., and Thomas, K. L. (eds.), *Engineer's Guide to Composite Materials,* ASM International, Materials Park, OH (1987).

1.10. Lubin, G. (ed.), *Handbook of Composites,* Van Nostrand Reinhold Co., New York (1982).

1.11. Katz, H. S. and Milewski, J. V. (eds.), *Handbook of Fillers and Reinforcements for Plastics,* Van Nostrand Reinhold Co., New York (1978).

1.12. Reinhart, T. J. et al. (eds.), *Engineered Materials Handbook Volume 1 Composites,* ASM International, Materials Park, OH (1987).

1.13. Dexter, H. B., "Long Term Environmental Effects and Flight Service Evaluation of Composite Materials," NASA TM 89067, National Aeronautics and Space Administration (January 1987).

1.14. Strong, A. B., *Fundamentals of Composites Manufacturing,* Society of Manufacturing Engineers, Dearborn, MI (1989).

1.15. Gibson, R. F., Suarez, S. A., and Deobald, L. R., "Laboratory Production of Discontinuous-Aligned Fiber Composite Plates Using an Autoclave-Style Press Cure," *Journal of Composites Technology and Research,* **7**(2), 391–400 (1985).

CHAPTER
2

LAMINA
STRESS-
STRAIN
RELATIONSHIPS

2.1 INTRODUCTION

The basic building block of a composite structure is the lamina, which usually consists of one of the fiber/matrix configurations shown in Fig. 1.4. For the purposes of mechanics analysis, however, the "unidirectionally reinforced," or "unidirectional" lamina with an arrangement of parallel, continuous fibers is the most convenient starting point. As shown in subsequent chapters, the stress-strain relationships for the unidirectional lamina form the basis for the analysis of not only the continuous fiber composite laminate [Fig. 1.4(a)], but of woven fiber [Fig. 1.4(b)] and chopped fiber composites [Figs. 1.4(c) and (d)] as well.

A composite material is obviously heterogeneous at the constituent material level, with properties possibly changing from point to point. For example, the stress-strain relationships at a point are different for a point in the fiber material from how they are for a point in the matrix material. If we take the composite lamina as the basic building block, however, the "macromechanical" stress-strain relationships of the lamina can be expressed in terms of average stresses and strains and effective properties of an equivalent homogeneous material [2.1]. This chapter is concerned with the development and manipulation of these macromechanical stress-strain relationships without regard for the constituent materials or their interactions. The "micromechanical" relationships between the constituent material properties and the effective lamina properties will be discussed in Chap. 3.

To complicate matters further, the properties of a composite are usually anisotropic. That is, the properties associated with an axis passing through a point in the material generally depend on the orientation of the axis. By comparison, conventional metallic materials are nearly isotropic since their properties are essentially independent of orientation. Fortunately, each type of composite has characteristic material property symmetries that make it possible to simplify the general anisotropic stress-strain relationships. In particular, the symmetry possessed by the unidirectional lamina makes it a so-called orthotropic material. The symmetries associated with various types of composite laminae and the resulting lamina stress-strain relationships are discussed in this chapter, along with certain mathematical manipulations that make it easier to deal with the directional nature of composite properties.

2.2 EFFECTIVE MODULI IN STRESS-STRAIN RELATIONSHIPS

A general three-dimensional state of stress at a point in a material can be described by nine stress components σ_{ij} (where $i, j = 1, 2, 3$), as shown in Fig. 2.1. According to the conventional subscript notation, when $i = j$,

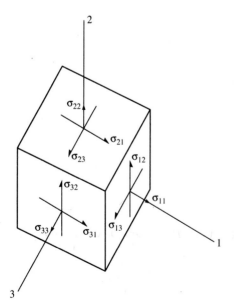

FIGURE 2.1
Three-dimensional state of stress.

the stress component σ_{ij} is a normal stress; and when $i \neq j$, the stress component is a shear stress. The first subscript refers to the direction of the outward normal to the face on which the stress component acts, and the second subscript refers to the direction in which the stress component itself acts.

Corresponding to each of the stress components, there is a strain component ϵ_{ij} describing the deformation at the point. Normal strains $(i = j)$ describe the extension or contraction per unit length along the x_i direction, and shear strains $(i \neq j)$ describe the distortional deformations associated with lines that were originally parallel to the x_i and x_j axes. It is very important to distinguish between the "tensor" strain ϵ_{ij} and the "engineering" strain γ_{ij}. In the case of normal strain the engineering strain is the same as the tensor strain, but for shear strain $\epsilon_{ij} = \gamma_{ij}/2$. Thus, the engineering shear strain γ_{ij} describes the total distortional change in the angle between lines that were originally parallel to the x_i and x_j axes, but the tensor shear strain ϵ_{ij} describes the amount of rotation of either of the lines.

In the most general stress-strain relationship *at a point* in an elastic material each stress component is related to each of the nine strain components by an equation of the form

$$\sigma_{ij} = f_{ij}(\epsilon_{11},\ \epsilon_{12},\ \epsilon_{13},\ \epsilon_{21},\ \epsilon_{22},\ \epsilon_{23},\ \epsilon_{31},\ \epsilon_{32},\ \epsilon_{33}) \tag{2.1}$$

where the functions f_{ij} may be nonlinear. For the linear elastic material, which is the primary concern in this book, the most general linear stress-strain relationships *at a point* in the material (excluding effects of environmental conditions) are given by equations of the form

$$
\begin{Bmatrix} \sigma_{11} \\ \sigma_{22} \\ \sigma_{33} \\ \sigma_{23} \\ \sigma_{31} \\ \sigma_{12} \\ \sigma_{32} \\ \sigma_{13} \\ \sigma_{21} \end{Bmatrix} =
\begin{bmatrix}
C_{1111} & C_{1122} & C_{1133} & C_{1123} & C_{1131} & C_{1112} & C_{1132} & C_{1113} & C_{1121} \\
C_{2211} & C_{2222} & C_{2233} & C_{2223} & C_{2231} & C_{2212} & C_{2232} & C_{2213} & C_{2221} \\
C_{3311} & C_{3322} & C_{3333} & C_{3323} & C_{3331} & C_{3312} & C_{3332} & C_{3313} & C_{3321} \\
\cdots & \cdots & \cdots & \cdots & \cdots & \cdots & \cdots & \cdots & \cdots \\
\cdots & \cdots & \cdots & \cdots & \cdots & \cdots & \cdots & \cdots & \cdots \\
\cdots & \cdots & \cdots & \cdots & \cdots & \cdots & \cdots & \cdots & \cdots \\
\cdots & \cdots & \cdots & \cdots & \cdots & \cdots & \cdots & \cdots & \cdots \\
\cdots & \cdots & \cdots & \cdots & \cdots & \cdots & \cdots & \cdots & \cdots \\
C_{2111} & C_{2122} & C_{2133} & C_{2123} & C_{2131} & C_{2112} & C_{2132} & C_{2113} & C_{2121}
\end{bmatrix}
\begin{Bmatrix} \epsilon_{11} \\ \epsilon_{22} \\ \epsilon_{33} \\ \epsilon_{23} \\ \epsilon_{31} \\ \epsilon_{12} \\ \epsilon_{32} \\ \epsilon_{13} \\ \epsilon_{21} \end{Bmatrix} \tag{2.2}
$$

Gen. Hooke's Law Anisotropic

where $[C]$ is a fully populated 9×9 matrix of stiffnesses or elastic constants (or moduli) having 81 components. Note that the first two subscripts on the elastic constants correspond to those of the stress, whereas the last two subscripts correspond to those of the strain. If no further restrictions are placed on the elastic constants, the material is called anisotropic and Eq. (2.2) is referred to as the generalized Hooke's law for anisotropic materials. In practice, there is no need to deal with this equation and its 81 elastic constants because various symmetry conditions simplify the equations considerably.

As shown in any mechanics of materials book [2.2], both stresses and strains are symmetric (i.e., $\sigma_{ij} = \sigma_{ji}$ and $\epsilon_{ij} = \epsilon_{ji}$), so that there are only six independent stress components and six independent strain components. This means that the elastic constants must be symmetric with respect to the first two subscripts and with respect to the last two subscripts (i.e., $C_{ijkl} = C_{jikl}$ and $C_{ijkl} = C_{ijlk}$ where $i, j, k, l = 1, 2, 3$), and that the number of nonzero elastic constants is now reduced to 36. These simplifications lead to a contracted notation that reduces the number of subscripts based on the following changes in notation [2.3–2.6]:

$$\sigma_{11} = \sigma_1 \qquad\qquad \epsilon_{11} = \epsilon_1$$

$$\sigma_{22} = \sigma_2 \qquad\qquad \epsilon_{22} = \epsilon_2$$

$$\sigma_{33} = \sigma_3 \qquad\qquad \epsilon_{33} = \epsilon_3$$

$$\sigma_{23} = \sigma_{32} = \sigma_4 \qquad 2\epsilon_{23} = 2\epsilon_{32} = \gamma_{23} = \gamma_{32} = \epsilon_4$$

$$\sigma_{13} = \sigma_{31} = \sigma_5 \qquad 2\epsilon_{13} = 2\epsilon_{31} = \gamma_{13} = \gamma_{31} = \epsilon_5$$

$$\sigma_{12} = \sigma_{21} = \sigma_6 \qquad 2\epsilon_{12} = 2\epsilon_{21} = \gamma_{12} = \gamma_{21} = \epsilon_6$$

With this contracted notation the generalized Hooke's law can now be written as

$$\sigma_i = C_{ij}\epsilon_j, \qquad i, j = 1, 2, \ldots, 6 \tag{2.3}$$

and the repeated subscript j implies summation on that subscript. Alternatively, in matrix form

$$\{\sigma\} = [C]\{\epsilon\} \tag{2.4}$$

where the elastic constant matrix or stiffness matrix $[C]$ is now 6×6 with 36 components and the stresses $\{\sigma\}$ and strains $\{\epsilon\}$ are column vectors, each having six elements. Alternatively, the generalized Hooke's law relating strains to stresses can be written as

$$\epsilon_i = S_{ij}\sigma_j; \qquad i, j = 1, 2, \ldots, 6 \tag{2.5}$$

or in matrix form as

$$\{\epsilon\} = [S]\{\sigma\} \tag{2.6}$$

where $[S]$ is the compliance matrix, which is the inverse of the stiffness matrix $([S] = [C]^{-1})$. As shown later, due to the existence of the strain energy density, the stiffness and compliance matrices are symmetric. Note that nothing has been said thus far about any symmetry that the material

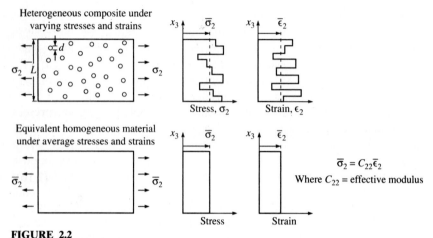

FIGURE 2.2
Concept of an effective modulus of an equivalent homogeneous material.

itself may have. All real materials have some form of symmetry, however, and no known material is completely anisotropic.

Before discussing the various simplifications of the stress-strain relationships, it is appropriate to deal with the problem of heterogeneity in the composite material. Recall that the stress-strain relationships presented up to now are only valid *at a point* in the material, and that the stresses, strains, and elastic moduli will change as we move from point to point in a composite (i.e., the elastic moduli for the matrix material are different from those of the fiber). In order to analyze the macromechanical behavior of the composite, it is more convenient to deal with *averaged* stresses and strains which are related by "effective moduli" of an equivalent homogeneous material.

As shown in Fig. 2.2, if the scale of the inhomogeneity in a material can be characterized by some length dimension, d, then the length dimension, L, over which the macromechanical averaging is to take place, must be much larger than d if the average stresses and strains are to be related by effective moduli of an equivalent homogeneous material. We now define the average stresses, $\bar{\sigma}_i$, and the average strains, $\bar{\epsilon}_i$, ($i = 1, 2, \ldots, 6$) to be averaged over a volume V, which is characterized by the dimension L, so that [2.1]

$$\bar{\sigma}_i = \int_V \sigma_i \, dv / V \qquad (2.7)$$

$$\bar{\epsilon}_i = \int_V \epsilon_i \, dv / V \qquad (2.8)$$

where $i = 1, 2, \ldots, 6$ and the σ_i and the ϵ_i are the position-dependent stresses and strains at a point, respectively. If these averaged stresses and strains are used in place of the stresses and strains at a point, the generalized Hooke's law [i.e., Eq. (2.3)] becomes

$$\bar{\sigma}_i = C_{ij} \bar{\epsilon}_j \tag{2.9}$$

and the elastic moduli C_{ij} then become the "effective moduli" of the equivalent homogeneous material in volume V. Similarly, the "effective compliances" S_{ij} may be defined by

$$\bar{\epsilon}_i = S_{ij} \bar{\sigma}_j \tag{2.10}$$

For example, in Fig. 2.2 the scale of the inhomogeneity is assumed to be the diameter of the fiber, d, and the averaging dimension, L, is assumed to be a characteristic lamina dimension such that $L \gg d$. The effective modulus C_{22} of the lamina is thus defined. *In the remainder of this book, lamina properties are assumed to be effective properties as described above.*

2.3 SYMMETRY IN STRESS-STRAIN RELATIONSHIPS

In this section the generalized anisotropic Hooke's law will be simplified and specialized using various symmetry conditions. The first symmetry condition, which has nothing to do with material symmetry, is strictly a result of the existence of a strain energy density function [2.3, 2.6]. The strain energy density function, W, is such that the stresses can be derived according to the equation

$$\sigma_i = \frac{\partial W}{\partial \epsilon_i} = C_{ij} \epsilon_j \tag{2.11}$$

where
$$W = \frac{1}{2} C_{ij} \epsilon_i \epsilon_j \tag{2.12}$$

By taking a second derivative of W, we find that

$$\frac{\partial^2 W}{\partial \epsilon_i \, \partial \epsilon_j} = C_{ij} \tag{2.13}$$

and by reversing the order of differentiation, we find that

$$\frac{\partial^2 W}{\partial \epsilon_j \, \partial \epsilon_i} = C_{ji} \tag{2.14}$$

Since the result must be the same regardless of the order of the differentiation, $C_{ij} = C_{ji}$ and the stiffness matrix is symmetric. Similarly, W can be expressed in terms of compliances and stresses, and by taking two derivatives with respect to stresses, it can be shown that $S_{ij} = S_{ji}$. Thus, the compliance matrix is also symmetric. Due to these mathematical manipulations, only 21 of the 36 anisotropic elastic moduli or compliances are independent, and we still have not said anything about any inherent symmetry of the material itself.

According to the above developments, the stiffness matrix for the linear elastic anisotropic material without any material property symmetry is of the form

$$
C_{ij} = \begin{bmatrix}
C_{11} & C_{12} & C_{13} & C_{14} & C_{15} & C_{16} \\
 & C_{22} & C_{23} & C_{24} & C_{25} & C_{26} \\
 & & C_{33} & C_{34} & C_{35} & C_{36} \\
 & \text{SYM} & & C_{44} & C_{45} & C_{46} \\
 & & & & C_{55} & C_{56} \\
 & & & & & C_{66}
\end{bmatrix}
\tag{2.15}
$$

Further simplifications of the stiffness matrix are possible only if the material properties have some form of symmetry. For example, a *monoclinic* material has one plane of material property symmetry. It can be shown that [2.3, 2.7] since the C_{ij} for such a material must be invariant under a transformation of coordinates corresponding to reflection in the plane of symmetry, the number of independent elastic constants for the monoclinic material is reduced to 13. Such a symmetry condition is not of practical interest in composite material analysis, however.

As shown in Fig. 2.3, a unidirectional composite lamina has three

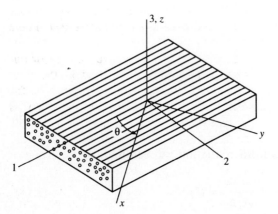

FIGURE 2.3
Orthotropic lamina with principal and nonprincipal coordinate systems.

mutually orthogonal planes of material property symmetry (i.e., the 12, 23, and 13 planes) and is called an *orthotropic* material. The term "orthotropic" alone is not sufficient to describe the form of the stiffness matrix, however. Unlike the anisotropic stiffness matrix [(Eq. (2.15)], which has the same form (but different terms) for different coordinate systems, the form of the stiffness matrix for the orthotropic material depends on the coordinate system used. The 123 coordinate axes in Fig. 2.3 are referred to as the *principal material coordinates* since they are associated with the reinforcement directions. Invariance of the C_{ij} under transformations of coordinates corresponding to reflections in two orthogonal planes [2.3, 2.7] may be used to show that the stiffness matrix for a so-called *specially orthotropic* material associated with the principal material coordinates is of the form

$$
C_{ij} =
\begin{bmatrix}
C_{11} & C_{12} & C_{13} & 0 & 0 & 0 \\
 & C_{22} & C_{23} & 0 & 0 & 0 \\
 & & C_{33} & 0 & 0 & 0 \\
 & \text{SYM} & & C_{44} & 0 & 0 \\
 & & & & C_{55} & 0 \\
 & & & & & C_{66}
\end{bmatrix}
\tag{2.16}
$$

A stiffness matrix of this form in terms of engineering constants will be obtained in the next section using observations from simple experiments. Note that there are only 12 nonzero elastic constants and 9 independent elastic constants for the specially orthotropic material.

Table 2.1 summarizes similar results for the different combinations of materials and coordinate systems used in this book. It will also be shown later that if the stress-strain relationshiips for the same orthotropic material are developed for a nonprincipal coordinate system *xyz* as shown in Fig. 2.3, the stiffness matrix is of the same form as that of the anisotropic material in Eq. (2.15). In such a nonprincipal, or off-axis coordinate system, the material is called *generally orthotropic* (see Table 2.1).

There are two other types of material symmetry that are important in the study of composites. The details will be developed in the next section, but the general forms of the stiffness matrices are given here for completeness. In most composites the fiber-packing arrangement is statistically random in nature, so that the properties are nearly the same in any direction perpendicular to the fibers (i.e., the properties along the 2 direction are the same as those along the 3 direction), and the material is *transversely isotropic*. For such a material we would expect that $C_{22} = C_{33}$, $C_{12} = C_{13}$, $C_{55} = C_{66}$, and that C_{44} would not be independent from the other stiffnesses. It can be shown [2.1] that the complete

TABLE 2.1
Elastic coefficients in the stress-strain relationships for different materials and coordinate systems

Material and coordinate system	Number of nonzero coefficients	Number of independent coefficients
Three-dimensional case		
Anisotropic	36	21
Generally Orthotropic (nonprincipal coordinates)	36	9
Specially Orthotropic (principal coordinates)	12	9
Specially Orthotropic, transversely isotropic	12	5
Isotropic	12	2
Two-dimensional case (lamina)		
Anisotropic	9	6
Generally Orthotropic (nonprincipal coordinates)	9	4
Specially Orthotropic (principal coordinates)	5	4
Balanced orthotropic, or square symmetric (principal coordinates)	5	3
Isotropic	5	2

stiffness matrix for a specially orthotropic, transversely isotropic material is of the form

$$
C_{ij} = \begin{bmatrix}
C_{11} & C_{12} & C_{12} & 0 & 0 & 0 \\
 & C_{22} & C_{23} & 0 & 0 & 0 \\
 & & C_{22} & 0 & 0 & 0 \\
 & \text{SYM} & & (C_{22}-C_{23})/2 & 0 & 0 \\
 & & & & C_{66} & 0 \\
 & & & & & C_{66}
\end{bmatrix}
\tag{2.17}
$$

where the 23 plane and all parallel planes are assumed to be planes of isotropy. In the next section a stiffness matrix of the same form will be derived, except that the so-called *engineering constants* will be used instead of the C_{ij}. Note that now there are still 12 nonzero elastic moduli but that only 5 are independent (see Table 2.1).

The simplest form of the stress-strain relationship occurs when the material is *isotropic* and every coordinate axis is an axis of symmetry. Now we would expect that $C_{11} = C_{22} = C_{33}$, $C_{12} = C_{13} = C_{23}$, that $C_{44} = C_{55} = C_{66}$, and that C_{44} again would not be independent from the other C_{ij}. The isotropic stiffness matrix is of the form [2.1] $\text{Ref}.$

$$
C_{ij} =
\begin{bmatrix}
C_{11} & C_{12} & C_{12} & 0 & 0 & 0 \\
 & C_{11} & C_{12} & 0 & 0 & 0 \\
 & & C_{11} & 0 & 0 & 0 \\
 & \text{SYM} & & (C_{11} - C_{12})/2 & 0 & 0 \\
 & & & & (C_{11} - C_{12})/2 & 0 \\
 & & & & & (C_{11} - C_{12})/2
\end{bmatrix}
$$

$$(2.18)$$

Now there are still 12 nonzero elastic constants, but only 2 are independent (see Table 2.1). Similar equations based on the engineering constants will be derived in the next section. Equations of this form can be found in any mechanics of materials book, and the design of metallic components is usually based on such formulations.

2.4 ORTHOTROPIC AND ISOTROPIC ENGINEERING CONSTANTS

In the previous section symmetry conditions were shown to reduce the number of elastic constants (the C_{ij} or S_{ij}) in the stress-strain relationships for several important classes of materials and the general forms of the relationships were presented. When a material is characterized experimentally, however, the so-called "engineering constants" such as Young's modulus (or modulus of elasticity), shear modulus, and Poisson's ratio are usually measured instead of the C_{ij} or the S_{ij}. The engineering constants are also widely used in analysis and design because they are easily defined and interpreted in terms of simple states of stress and strain. In this section, several simple tests and their resulting states of stress and strain will be used to develop the three-dimensional and two-dimensional stress-strain relationships for orthotropic and isotropic materials.

Consider a simple uniaxial tensile test consisting of an applied longitudinal normal stress, σ_1, along the reinforcement direction (i.e., the

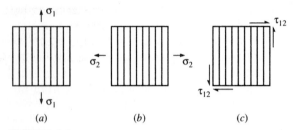

FIGURE 2.4
Simple states of stress used to define lamina engineering constants.

1 direction) of a specimen from an orthotropic material, as shown in Fig. 2.4(*a*). It is assumed that all other stresses are equal to zero. Within the linear range the experimental observation is that the resulting strains associated with the 123 axes can be expressed empirically in terms of "engineering constants" as

$$\epsilon_1 = \sigma_1/E_1$$
$$\epsilon_2 = -v_{12}\epsilon_1 = -v_{12}\sigma_1/E_1$$
$$\epsilon_3 = -v_{13}\epsilon_1 = -v_{13}\sigma_1/E_1 \tag{2.19}$$
$$\gamma_{12} = \gamma_{23} = \gamma_{13} = 0$$

where E_1 = longitudinal modulus of elasticity associated with the 1 direction

$v_{ij} = -\epsilon_j/\epsilon_i$ is the Poisson's ratio, the ratio of the strain in the *j* direction to the strain in the perpendicular *i* direction when the applied stress is in the *i* direction

Recall from mechanics of materials [2.2] that for isotropic materials no subscripts are needed on properties such as the modulus of elasticity and the Poisson's ratio because the properties are the same in all directions. This is not the case with orthotropic materials, however, and subscripts are needed on these properties because of their directional nature. For example, $E_1 \neq E_2$ and $v_{12} \neq v_{21}$. Note that, as with isotropic materials, a negative sign must be used in the definition of Poisson's ratio. A property like v_{12} is usually called a *major Poisson's ratio*, whereas a property like v_{21} is called a *minor Poisson's ratio*. As with isotropic materials, a normal stress induces only normal strains, and all shear strains are equal to zero. This lack of shear/normal interaction is observed only for the principal material coordinate system, however. For any other set of coordinates the so-called "shear-coupling" effect is present. This effect will be discussed in more detail later.

Now consider a similar experiment where a transverse normal

stress, σ_2, is applied to the same material as shown in Fig. 2.4(b), with all other stresses being equal to zero. Now the experimental observation is that the resulting strains can be expressed as

$$
\begin{aligned}
\epsilon_2 &= \sigma_2/E_2 \\
\epsilon_1 &= -v_{21}\epsilon_2 = -v_{21}\sigma_2/E_2 \\
\epsilon_3 &= -v_{23}\epsilon_2 = -v_{23}\sigma_2/E_2 \\
\gamma_{12} &= \gamma_{23} = \gamma_{13} = 0
\end{aligned}
\tag{2.20}
$$

where E_2 is the transverse modulus of elasticity associated with the 2 direction. A similar result for an applied transverse normal stress, σ_3, can be obtained by changing the appropriate subscripts in Eqs. (2.20).

Next, consider a shear test where a pure shear stress, $\sigma_{12} = \tau_{12}$, is applied to the material in the 12 plane, as shown in Fig. 2.4(c). Now the experimental observation is that resulting strains can be written as

$$
\begin{aligned}
\gamma_{12} &= \tau_{12}/G_{12} \\
\epsilon_1 &= \epsilon_2 = \epsilon_3 = \gamma_{13} = \gamma_{23} = 0
\end{aligned}
\tag{2.21}
$$

where G_{12} is the shear modulus associated with the 12 plane. Similar results can be obtained for pure shear in the 13 and 23 planes by changing the appropriate subscripts in Eq. (2.21). Again, notice that there is no shear/normal interaction (or shear coupling). As before, however, this is only true for the principal material axes.

Finally, consider a general three-dimensional state of stress consisting of all possible normal and shear stresses associated with the 123 axes as shown in Fig. 2.1. Since we are dealing with linear behavior, it is appropriate to use superposition and add all the resulting strains due to the simple uniaxial and shear tests, as given in Eqs. (2.19), (2.20), (2.21), and similar equations as described above. The resulting set of equations is given below:

$$
\begin{Bmatrix} \epsilon_1 \\ \epsilon_2 \\ \epsilon_3 \\ \gamma_{23} \\ \gamma_{31} \\ \gamma_{12} \end{Bmatrix} =
\begin{bmatrix}
1/E_1 & -v_{21}/E_2 & -v_{31}/E_3 & 0 & 0 & 0 \\
-v_{12}/E_1 & 1/E_2 & -v_{32}/E_3 & 0 & 0 & 0 \\
-v_{13}/E_1 & -v_{23}/E_2 & 1/E_3 & 0 & 0 & 0 \\
0 & 0 & 0 & 1/G_{23} & 0 & 0 \\
0 & 0 & 0 & 0 & 1/G_{31} & 0 \\
0 & 0 & 0 & 0 & 0 & 1/G_{12}
\end{bmatrix}
\begin{Bmatrix} \sigma_1 \\ \sigma_2 \\ \sigma_3 \\ \tau_{23} \\ \tau_{31} \\ \tau_{12} \end{Bmatrix}
\tag{2.22}
$$

Note that the compliance matrix is of the same form as the stiffness matrix for a specially orthotropic material [Eq. (2.16)] as it should be because $[S] = [C]^{-1}$. Note also that due to symmetry of the compliance matrix, $v_{ij}/E_i = v_{ji}/E_j$ and only nine of the engineering constants are independent.

If we now consider a simple uniaxial tensile test consisting of an applied normal stress, σ_x, along some arbitrary x axis as shown in Fig. 2.3, we find that the full complement of normal strains and shear strains are developed. The generation of shear strains due to normal stresses and normal strains due to shear stresses is often referred to as the "shear-coupling effect." As a result of shear coupling, all the zeros disappear in the compliance matrix and it becomes fully populated for the general three-dimensional state of stress associated with the arbitrary xyz axes; this is the *generally orthotropic* material. The stiffness or compliance matrices for the generally orthotropic material are of the same form as those for the general anisotropic material [Eq. (2.15)], although the material still has its orthotropic symmetries with respect to the principal material axes. Obviously then, the experimental characterization of such a material is greatly simplified by testing it as a specially orthotropic material along the principal material directions. As shown later, once we have the stiffnesses or compliances associated with the 123 axes, we can obtain those for an arbitrary off-axis coordinate system such as xyz by transformation equations involving the angles between the axes.

If the material being tested is specially orthotropic and transversely isotropic, the subscripts 2 and 3 in Eqs. (2.22) are interchangeable, and we have $G_{13} = G_{12}$, $E_2 = E_3$, $v_{21} = v_{31}$, and $v_{23} = v_{32}$. In addition, the familiar relationship among the isotropic engineering constants [2.2] is now valid for the engineering constants associated with the 23 plane, so that

$$G_{23} = \frac{E_2}{2(1 + v_{32})} \tag{2.23}$$

Now the compliance matrix is of the same form as Eq. (2.17) and only five of the engineering constants are independent.

Finally, for the isotropic material there is no need for subscripts and $G_{13} = G_{23} = G_{12} = G$, $E_1 = E_2 = E_3 = E$, $v_{12} = v_{23} = v_{13} = v$, and $G = E/2(1 + v)$. Now the compliance matrix is of the same form as Eq. (2.18) and only two of the engineering constants are independent.

2.5 THE SPECIALLY ORTHOTROPIC LAMINA

As shown later in the analysis of laminates, the lamina is often assumed to be in a simple two-dimensional state of stress (or plane stress). In this case the specially orthotropic stress-strain relationships in Eqs. (2.22) can be simplified by letting $\sigma_3 = \tau_{23} = \tau_{31} = 0$, so that

$$\left\{ \begin{array}{c} \epsilon_1 \\ \epsilon_2 \\ \gamma_{12} \end{array} \right\} = \begin{bmatrix} S_{11} & S_{12} & 0 \\ S_{21} & S_{22} & 0 \\ 0 & 0 & S_{66} \end{bmatrix} \left\{ \begin{array}{c} \sigma_1 \\ \sigma_2 \\ \tau_{12} \end{array} \right\} \tag{2.24}$$

where the compliances S_{ij} and the engineering constants are related by the equations

$$S_{11} = \frac{1}{E_1}, \qquad S_{22} = \frac{1}{E_2},$$

$$S_{12} = S_{21} = -\frac{v_{21}}{E_2} = -\frac{v_{12}}{E_1}, \qquad S_{66} = \frac{1}{G_{12}} \tag{2.25}$$

Thus, there are five nonzero compliances and only four independent compliances for the specially orthotropic lamina (Table 2.1). The lamina stresses in terms of *tensor* strains are given by Q_{11} (balanced ortho)

p 48

for 1-2 mat'l coord

$$\left\{ \begin{array}{c} \sigma_1 \\ \sigma_2 \\ \tau_{12} \end{array} \right\} = \begin{bmatrix} Q_{11} & Q_{12} & 0 \\ Q_{21} & Q_{22} & 0 \\ 0 & 0 & 2Q_{66} \end{bmatrix} \left\{ \begin{array}{c} \epsilon_1 \\ \epsilon_2 \\ \gamma_{12}/2 \end{array} \right\} \tag{2.26}$$

where the Q_{ij} are the components of the lamina stiffness matrix, which are related to the compliances and the engineering constants by

$$Q_{11} = \frac{S_{22}}{S_{11}S_{22} - S_{12}^2} = \frac{E_1}{1 - v_{12}v_{21}}$$

$$Q_{12} = -\frac{S_{12}}{S_{11}S_{22} - S_{12}^2} = \frac{v_{12}E_2}{1 - v_{12}v_{21}} = Q_{21}$$

$$Q_{22} = \frac{S_{11}}{S_{11}S_{22} - S_{12}^2} = \frac{E_2}{1 - v_{12}v_{21}} \tag{2.27}$$

$$Q_{66} = \frac{1}{S_{66}} = G_{12}$$

Note that the factor of 2 has been introduced in the Q_{66} term of Eq. (2.26) to compensate for the use of tensor shear strain. The reason for this will become apparent in the next section. As shown later, the experimental characterization of the orthotropic lamina involves the measurement of four independent engineering constants such as E_1, E_2, G_{12}, and v_{12}. Typical values of these properties for several composites are shown in Table 2.2.

The balanced orthotropic lamina shown schematically in Fig. 2.5 often occurs in practice when the fiber reinforcement is woven or cross-plied at 0° and 90°. In this case the number of independent elastic constants in Eqs. (2.24) to (2.27) is reduced to 3 because of the double symmetry of properties with respect to the 1 and 2 axes (Table 2.1).

TABLE 2.2
Typical values of lamina engineering constants for several composites

Material	E_1 Msi (GPa)	E_2 Msi (GPa)	G_{12} Msi (GPa)	v_{12}	v_f
T300/934 graphite/epoxy	19.0 (131)	1.5 (10.3)	1.0 (6.9)	0.22	0.65
AS/3501 graphite/epoxy	20.0 (138)	1.3 (9.0)	1.0 (6.9)	0.3	0.65
p-100/ERL 1962 pitch graphite/epoxy	68.0 (468.9)	0.9 (6.2)	0.81 (5.58)	0.31	0.62
Kevlar® 49/934 aramid/epoxy	11.0 (75.8)	0.8 (5.5)	0.33 (2.3)	0.34	0.65
Scotchply® 1002 E-glass/epoxy	5.6 (38.6)	1.2 (8.27)	0.6 (4.14)	0.26	0.45
Boron/5505 boron/epoxy	29.6 (204.0)	2.68 (18.5)	0.81 (5.59)	0.23	0.5
Spectra® 900/826 polyethylene/epoxy	4.45 (30.7)	0.51 (3.52)	0.21 (1.45)	0.32	0.65
E-glass/470-36 E-glass/vinylester	3.54 (24.4)	1.0 (6.87)	0.42 (2.89)	0.32	0.30

Kevlar® is a registered trademark of DuPont Company, Wilmington, Delaware; Scotchply® is a registered trademark of 3M Company, St. Paul, Minnesota; and Spectra® is a registered trademark of Allied-Signal Company, Petersburg, Virginia.

Thus, for the balanced orthotropic lamina we have $E_1 = E_2$, $Q_{11} = Q_{22}$, and $S_{11} = S_{22}$.

2.6 THE GENERALLY ORTHOTROPIC LAMINA

In the analysis of laminates having multiple laminae it is often necessary to know the stress-strain relationships for the *generally orthotropic*

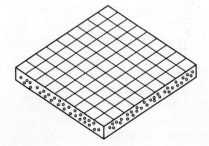

FIGURE 2.5
Balanced orthotropic lamina consisting of fibers oriented at 0° and 90°.

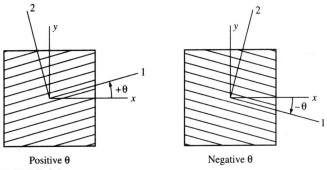

Positive θ   Negative θ

FIGURE 2.6
Sign convention for lamina orientation.

lamina in nonprincipal coordinates (or "off-axis" coordinates) such as x and y in Fig. 2.3. Fortunately, the elastic constants in these so-called "off-axis" stress-strain relationships are related to the four independent elastic constants in the principal material coordinates and the lamina orientation angle. The sign convention for the lamina orientation angle, θ, is given in Fig. 2.6. The relationships are found by combining the equations for transformation of stress and strain components from the 12 axes to the xy axes.

Relationships for transformation of stress components between coordinate axes may be obtained by writing the equations of static equilibrium for the wedge-shaped differential element in Fig. 2.7. For example, the force equilibrium along the x direction is given by

$$\sum F_x = \sigma_x \, dA - \sigma_1 \, dA \cos^2 \theta - \sigma_2 \, dA \sin^2 \theta + 2\tau_{12} \, dA \sin \theta \cos \theta = 0$$

$$(2.28)$$

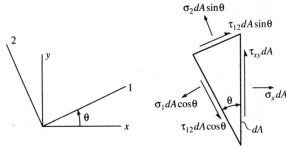

FIGURE 2.7
Differential element under static equilibrium with forces in two coordinate systems.

which, after dividing through by dA, gives an equation relating σ_x to the stresses in the 12 system:

$$\sigma_x = \sigma_1 \cos^2 \theta + \sigma_2 \sin^2 \theta - 2\tau_{12} \sin \theta \cos \theta \tag{2.29}$$

Using a similar approach, the complete set of transformation equations for the stresses in the xy-coordinate system can be developed and written in matrix form as

$$\begin{Bmatrix} \sigma_x \\ \sigma_y \\ \tau_{xy} \end{Bmatrix} = \begin{bmatrix} c^2 & s^2 & -2cs \\ s^2 & c^2 & 2cs \\ cs & -cs & c^2 - s^2 \end{bmatrix} \begin{Bmatrix} \sigma_1 \\ \sigma_2 \\ \tau_{12} \end{Bmatrix} = [T]^{-1} \begin{Bmatrix} \sigma_1 \\ \sigma_2 \\ \tau_{12} \end{Bmatrix} \tag{2.30}$$

and the stresses in the 12 system can be written as

$$\begin{Bmatrix} \sigma_1 \\ \sigma_2 \\ \tau_{12} \end{Bmatrix} = [T] \begin{Bmatrix} \sigma_x \\ \sigma_y \\ \tau_{xy} \end{Bmatrix} \tag{2.31}$$

where $c = \cos \theta$, $s = \sin \theta$, and the transformation matrix, $[T]$, is defined as

$$[T] = \begin{bmatrix} c^2 & s^2 & 2cs \\ s^2 & c^2 & -2cs \\ -cs & cs & c^2 - s^2 \end{bmatrix} \tag{2.32}$$

Methods for determining the matrix inverse $[T]^{-1}$ are described in any book dealing with matrices. It can also be shown [2.2, 2.3] that the *tensor strains* transform the same way as the stresses, and that

$$\begin{Bmatrix} \epsilon_1 \\ \epsilon_2 \\ \gamma_{12}/2 \end{Bmatrix} = [T] \begin{Bmatrix} \epsilon_x \\ \epsilon_y \\ \gamma_{xy}/2 \end{Bmatrix} \tag{2.33}$$

Substituting Eqs. (2.33) into Eqs. (2.26), and then substituting the resulting equations into Eqs. (2.30), we find that

$$\begin{Bmatrix} \sigma_x \\ \sigma_y \\ \tau_{xy} \end{Bmatrix} = [T]^{-1}[Q][T] \begin{Bmatrix} \epsilon_x \\ \epsilon_y \\ \gamma_{xy}/2 \end{Bmatrix} \tag{2.34}$$

where the stiffness matrix $[Q]$ in Eqs. (2.34) is defined in Eqs. (2.26).

Carrying out the indicated matrix multiplications and converting back to engineering strains, we find that

for x,y global coord.

$$
\begin{Bmatrix} \sigma_x \\ \sigma_y \\ \tau_{xy} \end{Bmatrix} = \begin{bmatrix} \bar{Q}_{11} & \bar{Q}_{12} & \bar{Q}_{16} \\ \bar{Q}_{12} & \bar{Q}_{22} & \bar{Q}_{26} \\ \bar{Q}_{16} & \bar{Q}_{26} & \bar{Q}_{66} \end{bmatrix} \begin{Bmatrix} \epsilon_x \\ \epsilon_y \\ \gamma_{xy} \end{Bmatrix} \tag{2.35}
$$

shear normal coupling!

where the $\bar{Q}_{ij}$ are the components of the transformed lamina stiffness matrix which are defined as follows:

$$\bar{Q}_{11} = Q_{11} \cos^4 \theta + Q_{22} \sin^4 \theta + 2(Q_{12} + 2Q_{66}) \sin^2 \theta \cos^2 \theta \qquad Q_{11}$$

$$\bar{Q}_{12} = (Q_{11} + Q_{22} - 4Q_{66}) \sin^2 \theta \cos^2 \theta + Q_{12} (\cos^4 \theta + \sin^4 \theta) \qquad Q_{12}$$

$$\bar{Q}_{22} = Q_{11} \sin^4 \theta + Q_{22} \cos^4 \theta + 2(Q_{12} + 2Q_{66}) \sin^2 \theta \cos^2 \theta \qquad Q_{22}$$

$$\bar{Q}_{16} = (Q_{11} - Q_{12} - 2Q_{66}) \cos^3 \theta \sin \theta - (Q_{22} - Q_{12} - 2Q_{66}) \cos \theta \sin^3 \theta$$

$$\bar{Q}_{26} = (Q_{11} - Q_{12} - 2Q_{66}) \cos \theta \sin^3 \theta - (Q_{22} - Q_{12} - 2Q_{66}) \cos^3 \theta \sin \theta$$

$$\bar{Q}_{66} = (Q_{11} + Q_{22} - 2Q_{12} - 2Q_{66}) \sin^2 \theta \cos^2 \theta + Q_{66}(\sin^4 \theta + \cos^4 \theta) \qquad Q_{66}$$

$$\tag{2.36}$$

Although the transformed lamina stiffness matrix now has the same form as that of an anisotropic material with nine nonzero coefficients, only four of the coefficients are independent because they can all be expressed in terms of the four independent lamina stiffnesses of the specially ortho- tropic material. That is, the material is still orthotropic, but it is not recognizable as such in the off-axis coordinates. As in the three- dimensional case, it is obviously much easier to characterize the lamina experimentally in the principal material coordinates than in the off-axis coordinates. Recall that the engineering constants, the properties that are normally measured, are related to the lamina stiffnesses by Eqs. (2.27).

Alternatively, the strains can be expressed in terms of the stresses as

$$
\begin{Bmatrix} \epsilon_x \\ \epsilon_y \\ \gamma_{xy} \end{Bmatrix} = \begin{bmatrix} \bar{S}_{11} & \bar{S}_{12} & \bar{S}_{16} \\ \bar{S}_{12} & \bar{S}_{22} & \bar{S}_{26} \\ \bar{S}_{16} & \bar{S}_{26} & \bar{S}_{66} \end{bmatrix} \begin{Bmatrix} \sigma_x \\ \sigma_y \\ \tau_{xy} \end{Bmatrix} \tag{2.37}
$$

where the $\bar{S}_{ij}$ are the components of the transformed lamina compliance matrix which are defined by equations similar to, but not exactly the same form as, Eqs. (2.36).

The lamina engineering constants can also be transformed from the principal material axes to the off-axis coordinates. For example, the modulus of elasticity associated with uniaxial loading along the x direction is defined as

$$E_x = \frac{\sigma_x}{\epsilon_x} = \frac{\sigma_x}{\bar{S}_{11}\sigma_x} = \frac{1}{\bar{S}_{11}} \tag{2.38}$$

where the strain ϵ_x in the denominator has been found by substituting the stress conditions $\sigma_x \neq 0$, $\sigma_y = \tau_{xy} = 0$ in Eqs. (2.37). By replacing $\bar{S}_{11}$ with an equation similar to the first of Eqs. (2.36) and then using Eqs. (2.25), we find that

Neglects end effects

$$E_x = \frac{1}{\dfrac{1}{E_1}c^4 + \left[-\dfrac{2v_{12}}{E_1} + \dfrac{1}{G_{12}}\right]c^2s^2 + \dfrac{1}{E_2}s^4} \tag{2.39}$$

where c and s are as defined in Eqs. (2.30). Similar transformation equations may be found for other off-axis engineering constants such as v_{xy} and G_{xy} [2.6, 2.8]. The variation of these properties with lamina orientation for a nylon fiber-reinforced elastomer composite is shown graphically in Fig. 2.8 from Ref. [2.5]. As intuitively expected, E_x varies from a maximum at $\theta = 0°$ to a minimum at $\theta = 90°$ for this particular material. It is not necessarily true that the extremum values of such material properties occur along the principal material directions, however [2.6]. What may not be intuitively expected is the sharp drop in modulus as the angle changes slightly from $0°$ and the fact that over much of the range of lamina orientations the modulus is very low. This is why transverse reinforcement is needed in most composites.

The shear-coupling effect has been described previously as the generation of shear strains by off-axis normal stresses and the generation

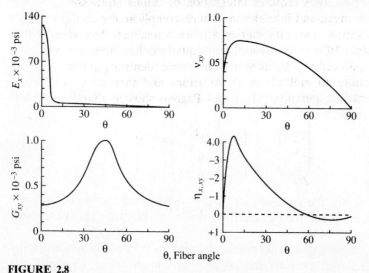

FIGURE 2.8
Variation of lamina engineering constants with lamina orientation for a nylon-reinforced elastomer composite. (*From Halpin* [2.5]. *Reprinted by permission of Technomic Publishing Co.*)

of normal strains by off-axis shear stresses. One way to quantify the degree of shear coupling is by defining dimensionless shear-coupling ratios [2.4, 2.5] or mutual influence coefficients [2.9] or shear-coupling coefficients [2.10]. For example, when the state of stress is defined as $\sigma_x \neq 0$, $\sigma_y = \tau_{xy} = 0$, the ratio

$$\eta_{x,xy} = \frac{\gamma_{xy}}{\epsilon_x} = \frac{\bar{S}_{16}}{\bar{S}_{11}} \tag{2.40}$$

is a measure of the amount of shear strain generated in the xy plane per unit normal strain along the direction of the applied normal stress, σ_x. Thus, the shear-coupling ratio is analogous to the Poisson's ratio, which is a measure of the coupling between normal strains. As shown in Fig. 2.8, $\eta_{x,xy}$ strongly depends on orientation and has its maximum value at some intermediate angle. Since there is no coupling along principal material directions, $\eta_{x,xy} = 0$ for $\theta = 0°$ and $\theta = 90°$. As the shear-coupling ratio increases, the amount of shear coupling increases. Other shear-coupling ratios can be defined for different states of stress. For example, when the stresses are $\tau_{xy} \neq 0$, $\sigma_x = \sigma_y = 0$, the ratio

$$\eta_{xy,y} = \frac{\epsilon_y}{\gamma_{xy}} = \frac{\bar{S}_{26}}{\bar{S}_{66}} \tag{2.41}$$

characterizes the normal strain response along the y direction due to a shear stress in the xy plane.

The effects of lamina orientation on stiffness are difficult to assess from inspection of stiffness transformation equations such as Eqs. (2.36) and (2.39). In addition, the eventual incorporation of lamina stiffnesses into laminate analysis requires integration of lamina stiffnesses over the laminate thickness, and integration of such complicated equations is also difficult. In view of these difficulties, a more convenient "invariant" form of the lamina stiffness transformation equations has been proposed by Tsai and Pagano [2.11]. By using trigonometric identities to convert from power functions to multiple angle functions and then using additional mathematical manipulations, Tsai and Pagano showed that Eqs. (2.36) could also be written as

$$\bar{Q}_{11} = U_1 + U_2 \cos 2\theta + U_3 \cos 4\theta$$

$$\bar{Q}_{12} = U_4 - U_3 \cos 4\theta$$

$$\bar{Q}_{22} = U_1 - U_2 \cos 2\theta + U_3 \cos 4\theta$$

$$\bar{Q}_{16} = \frac{U_2}{2} \sin 2\theta + U_3 \sin 4\theta$$

$$\bar{Q}_{26} = \frac{U_2}{2} \sin 2\theta - U_3 \sin 4\theta \tag{2.42}$$

$$\bar{Q}_{66} = \frac{1}{2}(U_1 - U_4) - U_3 \cos 4\theta$$

where the set of "invariants" is defined as

$$U_1 = \frac{1}{8}(3Q_{11} + 3Q_{22} + 2Q_{12} + 4Q_{66})$$

$$U_2 = \frac{1}{2}(Q_{11} - Q_{22})$$

$$U_3 = \frac{1}{8}(Q_{11} + Q_{22} - 2Q_{12} - 4Q_{66})$$

$$U_4 = \frac{1}{8}(Q_{11} + Q_{22} + 6Q_{12} - 4Q_{66})$$

(2.43)

As the name implies, the invariants, which are simply linear combinations of the Q_{ij}, are invariant to rotations in the plane of the lamina. Note that there are four independent invariants, just as there are four independent elastic constants. Equations (2.42) are obviously easier to manipulate and interpret than Eqs. (2.36). For example, all the stiffness expressions except those for the coupling stiffnesses consist of one constant term and terms which vary with lamina orientation. Thus, the effects of lamina orientation on stiffness are easier to interpret.

Invariant formulations of lamina compliance transformations are also useful. It can be shown [2.5, 2.10] that the off-axis compliance components in Eqs. (2.37) can be written as

$$\bar{S}_{11} = V_1 + V_2 \cos 2\theta + V_3 \cos 4\theta$$

$$\bar{S}_{12} = V_4 - V_3 \cos 4\theta$$

$$\bar{S}_{22} = V_1 - V_2 \cos 2\theta + V_3 \cos 4\theta$$

$$\bar{S}_{16} = V_2 \sin 2\theta + 2V_3 \sin 4\theta$$

$$\bar{S}_{26} = V_2 \sin 2\theta - 2V_3 \sin 4\theta$$

$$\bar{S}_{66} = 2(V_1 - V_4) - 4V_3 \cos 4\theta$$

(2.44)

where the invariants are

$$V_1 = \frac{1}{8}(3S_{11} + 3S_{22} + 2S_{12} + S_{66})$$

$$V_2 = \frac{1}{2}(S_{11} - S_{22})$$

$$V_3 = \frac{1}{8}(S_{11} + S_{22} - 2S_{12} - S_{66})$$

$$V_4 = \frac{1}{8}(S_{11} + S_{22} + 6S_{12} - S_{66})$$

(2.45)

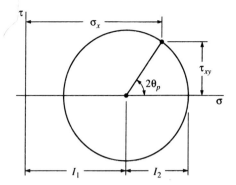

FIGURE 2.9
Mohr's circle for stress transformation.

Invariant formulations also lend themselves well to graphical interpretation. As shown in any mechanics of materials book [2.2], stress transformation equations such as Eqs. (2.30) can be combined and manipulated so as to generate the equation of Mohr's circle. As shown in Fig. 2.9, the transformation of a normal stress component σ_x can be described by the invariant formulation

$$\sigma_x = I_1 + I_2 \cos 2\theta_p \qquad (2.46)$$

where $I_1 = \dfrac{\sigma_x + \sigma_y}{2} = \text{invariant}$

$$I_2 = \sqrt{\left[\frac{\sigma_x - \sigma_y}{2}\right]^2 + \tau_{xy}^2} = \text{invariant}$$

θ_p = angle between the x axis and the principal stress axis

In this case the invariants are I_1, which defines the position of the center of the circle, and I_2, which is the radius of the circle. Note that, as with Eqs. (2.42), the invariant formulation consists of a constant term and a term which varies with orientation. Similarly, the invariant forms of the stiffness transformations can also be interpreted graphically using Mohr's circles. For example, Tsai and Hahn [2.10] have shown that the stiffness transformation equation

$$\bar{Q}_{11} = U_1 + U_2 \cos 2\theta + U_3 \cos 4\theta \qquad (2.47)$$

can be represented graphically by using two Mohr's circles, as shown in Fig. 2.10. The distance between points on each of the two circles represents the total stiffness $\bar{Q}_{11}$, whereas the distance between the centers of the two circles is given by U_1. The radius and angle associated with one circle are U_2 and 2θ, respectively, and the radius and angle associated with the other circle are U_3 and 4θ, respectively. Thus, the

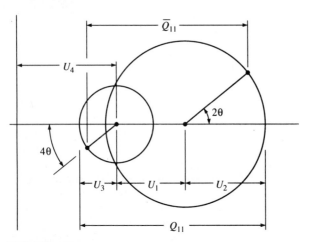

FIGURE 2.10
Mohr's circles for stiffness transformation. (*From Tsai and Hahn* [2.10]. *Reprinted by permission of Technomic Publishing Co.*)

distance between the centers of the circles is a measure of the isotropic component of stiffness, whereas the radii of the circles indicates the strength of the orthotropic component. If $U_2 = U_3 = 0$, the circles shrink to points and the material is isotropic.

Invariants will prove to be very useful later in the analysis of randomly oriented short fiber composites and laminated plates. For additional applications of invariants in composite analysis the reader is referred to books by Halpin [2.5] and Tsai and Hahn [2.10].

> **Example 2.1.** A filament wound cylindrical pressure vessel (Fig. 2.11) of mean diameter $d = 1$ m and wall thickness $t = 20$ mm is subjected to an internal pressure, p. The filament winding angle is $\theta = 53.1°$ from the longitudinal axis of the pressure vessel, and the glass/epoxy material has the following properties: $E_1 = 40$ GPa $= 40(10^3)$ MPa, $E_2 = 10$ GPa, $G_{12} = 3.5$ GPa, and $\nu_{12} = 0.25$. By the use of a strain gage, the normal strain along the fiber direction is determined to be $\epsilon_1 = 0.001$. Determine the internal pressure in the vessel.

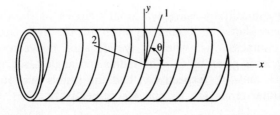

FIGURE 2.11
Filament wound pressure vessel.

Solution. From mechanics of materials, the stresses in a thin-walled cylindrical pressure vessel are given by

$$\sigma_x = \frac{pr}{2t} = \frac{0.5p}{2(0.02)} = 12.5p \qquad \tau_{xy} = 0$$

$$\sigma_y = \frac{pr}{t} = \frac{0.5p}{0.02} = 25p$$

(Note that $r = d/2 = 0.5$ m.)

These equations are based on static equilibrium and geometry only. Thus, they apply to a vessel made of any material. Since the given strain is along the fiber direction, we must transform the above stresses to the 12 axes. Recall that in the "netting analysis" in Problems 1.5 and 1.6 only the fiber longitudinal normal stress was considered. This was because the matrix was ignored, and the fibers alone cannot support transverse or shear stresses. In the current problem, however, the transverse normal stress, σ_2, and the shear stress, τ_{12}, are also considered because the fiber and matrix are now assumed to act as a composite. From Eqs. (2.31), the stresses along the 12 axes are

$$\sigma_1 = \sigma_x \cos^2 \theta + \sigma_y \sin^2 \theta + 2\tau_{xy} \sin \theta \cos \theta$$

$$= (12.5p)(0.6)^2 + (25p)(0.8)^2 + 0 = 20.5p\,(\text{MPa})$$

$$\sigma_2 = \sigma_x \sin^2 \theta + \sigma_y \cos^2 \theta - 2\tau_{xy} \sin \theta \cos \theta \quad \cdot$$

$$= (12.5p)(0.8)^2 + (25p)(0.6)^2 - 0 = 17.0p\,(\text{MPa})$$

$$\tau_{12} = -\sigma_x \sin \theta \cos \theta + \sigma_y \sin \theta \cos \theta + \tau_{xy}(\cos^2 \theta - \sin^2 \theta)$$

$$= -(12.5p)(0.8)(0.6) + (25p)(0.6)(0.8) + 0 = 6.0p\,(\text{MPa})$$

where the pressure p is in MPa. From the first of Eqs. (2.24), the normal strain ϵ_1 is

$$\epsilon_1 = \frac{\sigma_1}{E_1} - \frac{\nu_{12}\sigma_2}{E_1} = \frac{20.5p}{40(10^3)} - \frac{0.25(17.0p)}{40(10^3)} = 0.001$$

and the resulting pressure is $p = 2.46$ MPa.

Example 2.2. A tensile test specimen is cut out along the x direction of the pressure vessel described in Example 2.1. What effective modulus of elasticity would you expect to get during a test of this specimen?

Solution. The modulus of elasticity, E_x, associated with the x direction is given by Eq. (2.39) with $\theta = 53.1°$.

$$E_x = \frac{1}{\dfrac{1}{E_1}c^4 + \left[-\dfrac{2v_{12}}{E_1} + \dfrac{1}{G_{12}} \right]c^2 s^2 + \dfrac{1}{E_2}s^4}$$

$$E_x = \frac{1}{\dfrac{1}{40}(0.6)^4 + \left[-\dfrac{2(0.25)}{40} + \dfrac{1}{3.5} \right](0.6)^2(0.8)^2 + \dfrac{1}{10}(0.8)^4} = 9.33 \text{ MPa}$$

Example 2.3. A lamina consisting of continuous fibers randomly oriented in the plane of the lamina is said to be "planar isotropic," and the elastic properties in the plane are isotropic in nature. Find expressions for the lamina stiffnesses for a planar isotropic lamina.

Solution. Since the fibers are assumed to be randomly oriented in the plane, the "planar isotropic stiffnesses" can be found by averaging the transformed lamina stiffnesses as follows:

$$\tilde{Q}_{ij} = \frac{\displaystyle\int_0^\pi \bar{Q}_{ij}\, d\theta}{\displaystyle\int_0^\pi d\theta}$$

It is convenient to use the invariant forms of the transformed lamina stiffnesses because they are easily integrated. For example, if we substitute the first of Eqs. (2.42) in the above equation, we get

$$\tilde{Q}_{11} = \frac{\displaystyle\int_0^\pi \bar{Q}_{11}\, d\theta}{\displaystyle\int_0^\pi d\theta} = \frac{\displaystyle\int_0^\pi [U_1 + U_2 \cos 2\theta + U_3 \cos 4\theta]\, d\theta}{\pi} = U_1$$

Note that the averaged stiffness equals the isotropic part of the transformed lamina stiffness, and that the orthotropic parts drop out in the averaging process. Similarly, the other averaged stiffnesses can be found in terms of the invariants. The derivations of the remaining expressions are left as an exercise.

PROBLEMS

2.1. A representative section from a composite lamina is shown in Fig. 2.12, along with the transverse stress and strain distributions across the fiber and matrix materials in the section. Assuming that the dimensions of the section do not change along the longitudinal direction (perpendicular to the page), find the numerical value of the effective transverse modulus for the section.

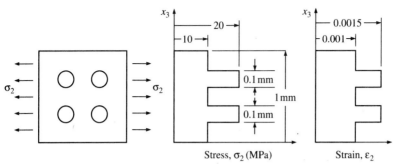

FIGURE 2.12
Transverse stress and strain distribution over a section of lamina.

2.2. Derive Eq. (2.39) for the off-axis modulus, E_x.

2.3. Find an expression for the off-axis shear modulus, G_{xy}, in terms of E_1, E_2, v_{12}, G_{12}, and θ.

2.4. Using the result from Problem 2.3:
   (a) Find the value of the angle θ (other than $0°$ or $90°$) where the curve of G_{xy} vs. θ has a possible maximum, minimum, or inflection point. $45°$
   (b) For the value of θ found in part (a), find the bounds on G_{12} which must be satisfied if G_{xy} is to have a maximum or minimum.
   (c) Qualitatively sketch the variation of G_{xy} vs. θ for the different cases and identify each curve by the corresponding bounds on G_{12} which give that curve.
   (d) Using the bounds on G_{12} from part (b), find which conditions apply for E-glass/epoxy composites. The bounds on G_{12} in part (b) should be expressed in terms of E_1, E_2, and v_{12}.

2.5. Describe a series of tensile tests that could be used to measure the four independent engineering constants for an orthotropic lamina without using a pure shear test. Give the necessary equations for the data reduction.

2.6. A balanced orthotropic, or square symmetric lamina, is made up of $0°$ and $90°$ fibers woven into a fabric and bonded together, as shown in Fig. 2.5.
   (a) Describe the stress-strain relationships for such a lamina in terms of the appropriate engineering constants.
   (b) For a typical glass/epoxy composite lamina of this type, sketch the expected variations of all the engineering constants for the lamina from $0°$ to $90°$. Numerical values are not required.

2.7. An element of a balanced orthotropic graphite/epoxy lamina is under the state of stress shown in Fig. 2.13. If the properties of the woven graphite fabric/epoxy material are $E_1 = 70$ GPa, $v_{12} = 0.25$, $G_{12} = 5$ GPa, determine all the strains along the fiber directions.

2.8. Derive Eqs. (2.27).

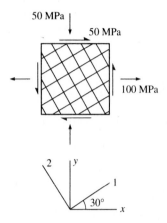

FIGURE 2.13
Stresses acting on an element of balanced orthotropic lamina.

2.9. Express the stress-strain relationships in Eqs. (2.37) in terms of off-axis
engineering constants such as the moduli of elasticity, shear modulus,
Poisson's ratios, and shear-coupling ratios.

2.10. Derive the first two equations of Eqs. (2.42).

2.11. Find all components of the stiffness and compliance matrices for a specially
orthotropic lamina made of AS/3501 graphite/epoxy.

2.12. Using the results of Problems 2.11, determine the invariants U_i and V_i for
the AS/3501 lamina, where $i = 1, 2, 3, 4$.

2.13. Using the results of Problems 2.11 or 2.12, compare the transformed lamina
stiffnesses for AS/3501 graphite/epoxy plies oriented at $+45°$ and $-45°$.

2.14. Show how the Mohr's circles in Fig. 2.10 can be used to interpret the
transformed lamina stiffness $\bar{Q}_{12}$.

2.15. Using the approach described in Example 2.3, derive the expressions for all
the averaged stiffnesses for the planar isotropic lamina in terms of
invariants. Use these results to find the corresponding averaged engineering
constants (modulus of elasticity, shear modulus, and Poisson's ratio) in
terms of invariants.

2.16. For a specially orthotropic, transversely isotropic material the "plane strain
bulk modulus," K_{23}, is an engineering constant that is defined by the stress
conditions $\sigma_2 = \sigma_3 = \sigma$ and the strain conditions $\epsilon_1 = 0$, $\epsilon_2 = \epsilon_3 = \epsilon$. Show
that these conditions lead to the stress-strain relationship $\sigma = 2K_{23}\epsilon$, and
find the relationship among K_{23}, E_1, E_2, G_{23}, and ν_{12}.

2.17. Describe the measurements that would have to be taken and the equations
that would have to be used to determine G_{23}, ν_{32}, and E_2 for a specially
orthotropic, transversely isotropic material from a single tensile test.

2.18. An off-axis tensile specimen of an orthotropic lamina with fiber orientation
$\theta = 45°$ is subjected to a known uniaxial stress, σ_x, and the resulting normal
strains ϵ_x and ϵ_y are measured with strain gages. Describe how the in-plane
shear modulus, G_{12}, can be determined from these data. Assume that no
other data are available.

REFERENCES

2.1. Christensen, R. M., *Mechanics of Composite Materials*, John Wiley & Sons, New York (1979).

2.2. Crandall, S. H., Dahl, N. C., and Lardner, T. J., *An Introduction to the Mechanics of Solids*, 2d ed. with SI units, McGraw-Hill, Inc., New York (1978).

2.3. Sokolnikoff, I. S., *Mathematical Theory of Elasticity*, McGraw-Hill, Inc., New York (1956).

2.4. Ashton, J. E., Halpin, J. C., and Petit, P. H., *Primer on Composite Materials: Analysis*, Technomic Publishing Co., Lancaster, PA (1969).

2.5. Halpin, J. C., *Primer on Composite Materials: Analysis*, Rev., Technomic Publishing Co., Lancaster, PA (1984).

2.6. Jones, R. M., *Mechanics of Composite Materials*, Hemisphere Publishing Co., New York (1975).

2.7. Vinson, J. R. and Sierakowski, R. L., *The Behavior of Structures Composed of Composite Materials*, Martinus Nijhoff Publishers, Dordrecht, The Netherlands (1986).

2.8. Agarwal, B. D. and Broutman, L. J., *Analysis and Performance of Fiber Composites*, 2d ed., John Wiley & Sons, New York (1990).

2.9. Lekhnitski, S. G., *Theory of Elasticity of an Anisotropic Body*, Mir Publishing Co., Moscow, USSR (1981).

2.10. Tsai, S. W. and Hahn, H. T., *Introduction to Composite Materials*, Technomic Publishing Co., Lancaster, PA (1980).

2.11. Tsai, S. W. and Pagano, N. J., "Invariant Properties of Composite Materials," in S. W. Tsai, J. C. Halpin, and N. J. Pagano (eds.), *Composite Materials Workshop*, pp. 233–253, Technomic Publishing Co., Lancaster, PA (1968).

—

CHAPTER
3

EFFECTIVE MODULI OF A CONTINUOUS FIBER-REINFORCED LAMINA

3.1 INTRODUCTION

In the previous chapter the concept of an effective modulus was found to be essential to the development of practical engineering stress-strain relationships for composite materials. Recall that for some representative volume element (RVE) in a heterogeneous composite the volume-averaged stresses can be related to the volume-averaged strains by the effective moduli of an equivalent homogeneous material. Chapter 2 was only concerned with the development and manipulation of macro-mechanical stress-strain relationships involving the lamina effective moduli, however, and the roles of lamina constituent materials were not examined. In this chapter we will discuss various micromechanical models for predicting the effective moduli of continuous fiber-reinforced laminae in terms of the corresponding material properties, relative volume contents, and geometric arrangements of the fiber and matrix materials.

Corresponding models for predicting strength and hygrothermal properties will be presented in Chaps. 4 and 5, respectively. Micromechanics of discontinuous fiber composites is covered in Chap. 6.

Before proceeding further, it is appropriate to discuss briefly the term "micromechanics." To a materials scientist the term may imply the study of mechanical behavior at the level of molecular or crystal structures. Since the behavior of composite material structures such as laminates is referred to as "macromechanics," it has been suggested that perhaps mechanics of composites at the constituent material level should be referred to as "minimechanics" [3.1]. In the present context, and in much of the composites literature, however, the analysis of effective composite properties in terms of constituent material properties is called "micromechanics." The terms "structure-property relationships" and "effective modulus theories" are also used in the literature. Many analytical approaches have been developed over the years, and comprehensive literature surveys have been published by Chamis and Sendeckyj [3.2], Christensen [3.3], Hashin [3.1], and Halpin [3.4].

Micromechanical analyses are based on either the mechanics of materials or the elasticity theory. In the mechanics of materials approach simplifying assumptions make it unnecessary to specify the details of the stress and strain distributions at the micromechanical level and fiber-packing geometry is generally arbitrary. Theory of elasticity models involve the solution for actual stresses and strains at the micromechanical level and fiber-packing geometry is taken into account. The elasticity approach often involves numerical solutions of the governing equations because of the complex geometries and boundary conditions. Although the simplifying assumptions used in the mechanics of materials approach violate some of the laws of elasticity theory, some of the results are sufficiently accurate that they are often used in design. A third category involves empirical solutions which are based on curve-fitting to elasticity solutions or experimental data, and some of these equations are often used along with mechanics of materials equations to formulate a complete set of simple lamina design equations.

Ideally, micromechanical models should enable us to answer quickly "What if?" questions regarding the effects of various fiber/matrix combinations without actually fabricating and testing the composites in question. On the other hand, experience has shown that there are pitfalls in such an approach and that there is no substitute for experimental characterization. Experimental data on the constituent material properties are required as input to the models, and similar data on the corresponding composite properties are required in order to assess the validity of the models. Indeed, as we will see later, some properties such as fiber transverse moduli are usually inferred from the micromechanical model and other measured properties because of the difficulty of direct measurement. Once a micromechanical model has been shown to be sufficiently accurate by comparison with experiment, however, it can become part of a powerful design methodology that enables us to design the material as well as the structure. Aside from design implications, micromechanical analysis and experimental characterization are both essential if we are to understand better "how composites work."

One of the key elements in micromechanical analysis is the characterization of the relative volume or weight contents of the various constituent materials. We will find that the micromechanics equations involve constituent volume fractions, but actual measurements are often based on weight fractions. Measurements are discussed later, but the relationships between volume fractions and weight fractions will be presented here.

For any number of constituent materials, n, the sum of the constituent volume fractions must be unity:

$$\sum_{i=1}^{n} v_i = 1 \tag{3.1}$$

where $v_i = V_i/V_c$ = volume fraction of the ith constituent

V_i = volume of the ith constituent

V_c = total volume of the composite

In many cases this equation reduces to

$$v_f + v_m + v_v = 1 \tag{3.2}$$

where v_f, v_m, and v_v are the volume fractions of the fiber, matrix, and voids, respectively. The corresponding equations for weight fractions are

$$\sum_{i=1}^{n} w_i = 1 \tag{3.3}$$

and $$w_f + w_m = 1 \tag{3.4}$$

where $w_i = W_i/W_c$, $w_f = W_f/W_c$, $w_m = W_m/W_c$, and W_i, W_f, W_m, and W_c are the weights of the ith constituent, fibers, matrix, and composite, respectively. Note that the weight of the voids has been neglected here. Substituting the product of density and volume for weight in each term of Eqs. (3.3) and (3.4) and solving for the composite density, we get the "rule of mixtures":

$$\rho_c = \sum_{i=1}^{n} \rho_i v_i \tag{3.5}$$

or $$\rho_c = \rho_f v_f + \rho_m v_m \tag{3.6}$$

where ρ_i, ρ_f, ρ_m, and ρ_c are the densities of the ith constituent, fiber, matrix, and composite, respectively. Similarly, Eqs. (3.1) and (3.2) can be rearranged as

$$\rho_c = \frac{1}{\sum\limits_{i=1}^{n} (w_i/\rho_i)} \tag{3.7}$$

and

$$\rho_c = \frac{1}{(w_f/\rho_f) + (w_m/\rho_m)} \tag{3.8}$$

Equation (3.2) can also be rearranged so that the void fraction can be calculated from measured weights and densities:

$$v_v = 1 - \frac{(W_f/\rho_f) + (W_c - W_f)/\rho_m}{W_c/\rho_c} \tag{3.9}$$

Typical autoclave-cured composites may have void fractions in the range 0.1–1 percent. Without vacuum-bagging, however, volatiles trapped in the composite during the cure cycle can cause void contents on the order of 5 percent.

In order to get some idea as to the range of constituent volume fractions that may be expected in fiber composites, it is useful to consider representative area elements for idealized fiber-packing geometries such as the square and triangular arrays shown in Fig. 3.1. If we assume that the fiber spacing, s, and the fiber diameter, d, do not change along the fiber length, then the area fractions must be equal to the volume fractions. Indeed, optical determination of area fractions is possible from micrographs. The fiber volume fraction for the square array is found by dividing the area of fiber enclosed in the shaded square by the total area of the shaded square [3.5]:

$$v_f = \frac{\pi}{4} \left(\frac{d}{s}\right)^2 \tag{3.10}$$

Clearly, the maximum theoretical fiber volume fraction occurs when

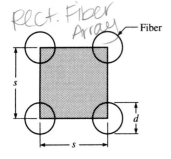

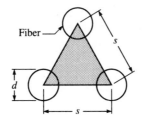

FIGURE 3.1
Representative area elements for idealized square and triangular fiber-packing geometries.

$s = d$. In this case

$$v_{f\text{max}} = \frac{\pi}{4} = 0.785 \tag{3.11}$$

A similar calculation for the triangular array shows that

$$v_f = \frac{\pi}{2\sqrt{3}} \left(\frac{d}{s}\right)^2 \tag{3.12}$$

and when $s = d$, the maximum fiber volume fraction is

$$v_{f\text{max}} = \frac{\pi}{2\sqrt{3}} = 0.907 \tag{3.13}$$

The close packing of fibers required to produce these theoretical limits is generally not achievable in practice, however. In most continuous fiber composites the fibers are packed in a random fashion as shown in Fig. 3.2, and the fiber volume fractions range from 0.5 to 0.8. In short fiber composites fiber volume fractions are usually much lower due to processing limitations (e.g., the viscosity of the fiber/resin mixture must

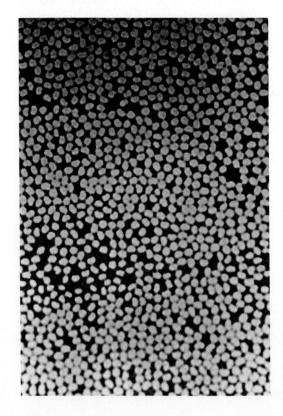

FIGURE 3.2
Photomicrograph of graphite/epoxy composite showing actual fiber-packing geometry at 400× magnification.

be controlled for proper flow during molding) and the random orientation of fibers. Since fiber-packing geometry is never entirely repeatable from one piece of material to another, we should not expect our micromechanics predictions to be exact.

Example 3.1. A graphite/epoxy composite specimen has dimensions of 2.54 cm × 2.54 cm × 0.3 cm and a weight of 2.98 g. After "resin digestion" in an acid solution the remaining graphite fibers weigh 1.863 g. From independent tests, the densities of the graphite fibers and epoxy matrix materials are found to be 1.9 g/cm³ and 1.2 g/cm³, respectively. Determine the volume fractions of fibers, epoxy matrix, and voids in the specimen.

Solution. The composite density is

$$\rho_c = \frac{2.98 \text{ g}}{(2.54)(2.54)(0.3) \text{ cm}^3} = 1.54 \text{ g/cm}^3$$

From Eq. (3.9), the void fraction is

$$v_v = 1 - \frac{(1.863/1.9) + (2.98 - 1.863)/1.2}{2.98/1.54} = 0.0122 \quad \text{or } 1.22\%$$

From Eq. (3.2),

$$v_f + v_m = 1 - v_v = 1.0122 = 0.988$$

Then, from Eq. (3.6),

$$1.54 = 1.9v_f + 1.2(0.988 - v_f)$$

Therefore, the fiber volume fraction is

$$v_f = 0.506 \quad \text{or } 50.6\%$$

and the matrix volume fraction is

$$v_m = 0.988 - 0.506 = 0.482 \quad \text{or } 48.2\%$$

Example 3.2. Assume that the graphite fibers in the specimen from Example 3.1 have been uniformly coated with an epoxy "sizing" of thickness t before bonding of the fibers and matrix together to form a unidirectional composite. If the bare fibers have a diameter $d = 0.0005$ in (0.0127 mm) and the coated fibers are assumed to be packed together in the tightest possible square array, what is the thickness of the sizing?

Solution. The fiber spacing, s, which must be equal to the coated fiber diameter, d_c, can be found from Eq. (3.10):

$$s = d_c = \sqrt{\frac{\pi d^2}{4v_f}} = \sqrt{\frac{\pi(0.0005)^2}{4(0.506)}} = 0.000623 \text{ in } (0.0158 \text{ mm})$$

The thickness is then $t = (d_c - d)/2 = 0.0000615$ in (0.00156 mm).

3.2 ELEMENTARY MECHANICS OF MATERIALS MODELS

The objective of this section is to present elementary mechanics of materials models for predicting four independent effective moduli of an

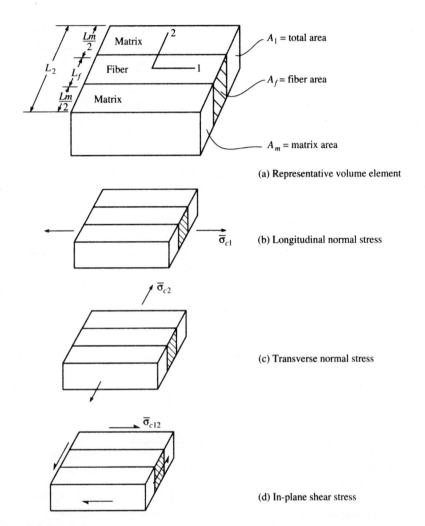

(a) Representative volume element

(b) Longitudinal normal stress

(c) Transverse normal stress

(d) In-plane shear stress

FIGURE 3.3
Representative volume element and simple stress states used in elementary mechanics of materials models.

orthotropic continuous fiber-reinforced lamina. In the elementary mechanics of materials approach to micromechanical modeling fiber-packing geometry is not specified, so that the RVE may be a generic composite block consisting of fiber material bonded to matrix material, as shown in Fig. 3.3. More sophisticated mechanics of materials models which do consider fiber-packing geometry will be discussed later.

The constituent volume fractions in the RVE are assumed to be the same as those in the actual composite. Since it is assumed that the fibers remain parallel and that the dimensions do not change along the length of the element, the area fractions must equal the volume fractions. Perfect bonding at the interface is assumed, so that no slip occurs between fiber and matrix materials. The fiber and matrix materials are assumed to be linearly elastic and homogeneous. The matrix is assumed to be isotropic, but the fiber can be either isotropic or orthotropic. Following the concept of the RVE, the lamina is assumed to be macroscopically homogeneous, linearly elastic, and orthotropic.

Micromechanics equations will be developed from either equilibrium or compatibility relationships and assumptions about either stresses or strains in the RVE which has been subjected to a simple state of stress. Since the mechanics of materials approach does not require the specification of the stresses, strains, and displacements at each point, we only deal with the corresponding volume-averaged quantities. Finally, since it is assumed that the stresses, strains, displacements, and RVE dimensions do not change along the length, we can just use area averages:

$$\bar{\sigma} = \frac{1}{V} \int \sigma \, dV = \frac{1}{A} \int \sigma \, dA \qquad (3.14)$$

$$\bar{\epsilon} = \frac{1}{V} \int \epsilon \, dV = \frac{1}{A} \int \epsilon \, dA \qquad (3.15)$$

$$\bar{\delta} = \frac{1}{V} \int \delta \, dV = \frac{1}{A} \int \delta \, dA \qquad (3.16)$$

where the overbar denotes an averaged quantity, and

σ = stress
ϵ = strain
δ = displacement
V = volume
A = area associated with the face on which loading is applied

The volume-averaging (or area-averaging) may occur over the composite lamina, the fiber, or the matrix, and the corresponding parameters will be identified by using subscripts as defined in the following derivations.

3.2.1 Longitudinal Modulus

If the RVE in Fig. 3.3(a) is subjected to a longitudinal normal stress, $\bar{\sigma}_{c1}$, as shown in Fig. 3.3(b), the response is governed by the effective longitudinal modulus, E_1. Static equilibrium requires that the total resultant force on the element must equal the sum of the forces acting on the fiber and matrix. Combining the static equilibrium condition with Eq. (3.14), we get

$$\bar{\sigma}_{c1}A_1 = \bar{\sigma}_{f1}A_f + \bar{\sigma}_{m1}A_m \tag{3.17}$$

where subscripts c, f, and m refer to composite, fiber, and matrix, respectively, and the second subscript refers to the direction. Since area fractions are equal to the corresponding volume fractions, Eq. (3.17) can be rearranged to give a "rule of mixtures" for longitudinal stress

$$\bar{\sigma}_{c1} = \bar{\sigma}_{f1}v_f + \bar{\sigma}_{m1}v_m \tag{3.18}$$

Under the assumptions that the matrix is isotropic, that the fiber is orthotropic, and that all materials follow a one-dimensional Hooke's law (i.e. Poisson strains are neglected),

$$\bar{\sigma}_{c1} = E_1\bar{\epsilon}_{c1}; \qquad \bar{\sigma}_{f1} = E_{f1}\bar{\epsilon}_{f1}; \qquad \bar{\sigma}_{m1} = E_m\bar{\epsilon}_{m1} \tag{3.19}$$

and Eq. (3.18) becomes

$$E_1\bar{\epsilon}_{c1} = E_{f1}\bar{\epsilon}_{f1}v_f + E_m\bar{\epsilon}_{m1}v_m \tag{3.20}$$

Double subscripts are used for the fiber modulus since the fiber is assumed to be orthotropic. That is, the longitudinal fiber modulus, E_{f1}, is not necessarily equal to the transverse fiber modulus, E_{f2}. For example, grahite and aramid fibers exhibit orthotropic behavior, whereas glass and boron are practically isotropic. For the isotropic case it is a simple matter to let $E_{f1} = E_{f2}$. Since the matrix is assumed to be isotropic, the matrix modulus, E_m, does not need a second subscript.

Finally, the key assumption is that the average strains in the composite, fiber, and matrix along the 1 direction are equal:

$$\bar{\epsilon}_{c1} = \bar{\epsilon}_{f1} = \bar{\epsilon}_{m1} \tag{3.21}$$

Substitution of Eq. (3.21) in Eq. (3.20) then yields the rule of mixtures for the longitudinal modulus

$$E_1 = E_{f1}v_f + E_mv_m \tag{3.22}$$

This equation predicts a linear variation of the longitudinal modulus with fiber volume fraction, as shown in Fig. 3.4. Although simple in form, Eq. (3.22) agrees well with experimental data from Ref. [3.6] [Fig. 3.4(b)] and is a useful design equation. The validity of the key assumptions leading to this equation will now be examined by using a strain energy approach.

Further insight into the micromechanics of the longitudinal loading case is possible by using a strain energy approach. Under the given state of stress the total strain energy stored in the composite, U_c, can be represented as the sum of the strain energy in the fibers, U_f, and the strain energy in the matrix, U_m.

$$U_c = U_f + U_m \tag{3.23}$$

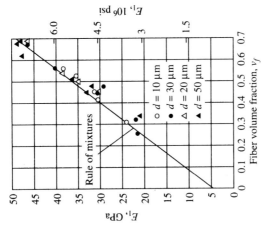

(b) Comparison of predicted and measured E_1 for E-glass/polyester. (*From Adams* [3.6]. *Reprinted with permission of ASM International.*)

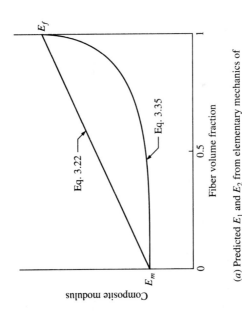

(a) Predicted E_1 and E_2 from elementary mechanics of materials models.

FIGURE 3.4
Variation of composite moduli with fiber volume fraction.

Again making the mechanics of materials assumption that the stresses and strains are uniform over the RVE and using Eqs. (3.19), the strain energy terms can be simplified as

$$U_c = \frac{1}{2} \int_{V_c} \sigma_{c1} \epsilon_{c1} \, dV = \frac{1}{2} E_1 \bar{\epsilon}_1^2 V_c \tag{3.24a}$$

$$U_f = \frac{1}{2} \int_{V_f} \sigma_{f1} \epsilon_{f1} \, dV = \frac{1}{2} E_{f1} \bar{\epsilon}_{f1}^2 V_f \tag{3.24b}$$

and

$$U_m = \frac{1}{2} \int_{V_m} \sigma_{m1} \epsilon_{m1} \, dV = \frac{1}{2} E_{m1} \bar{\epsilon}_{m1}^2 V_m \tag{3.24c}$$

In this approximation the strain energy due to the mismatch in Poisson strains at the fiber/matrix interface has been neglected [recall the assumptions leading to Eq. (3.19)]. This neglected term has been shown to be of order the square of the difference between the Poisson's ratios of the fiber and the matrix, so the approximation is justified [3.7]. It is easily shown that substitution of Eqs. (3.24) in Eq. (3.23), along with the assumption of equal strains from Eq. (3.21), again leads to the rule of mixtures given by Eq. (3.22). But the strain energy approach also allows us to ask, "What happens if the assumption of equal strains is not made?" In order to proceed, let the stresses in the fibers and the matrix be defined in terms of the composite stress as follows:

$$\bar{\sigma}_{f1} = a_1 \bar{\sigma}_{c1}; \qquad \bar{\sigma}_{m1} = b_1 \bar{\sigma}_{c1} \tag{3.25}$$

where a_1 and b_1 are constants. Substitution of Eqs. (3.25) in the rule of mixtures for stress, Eq. (3.18), leads to

$$a_1 v_f + b_1 v_m = 1 \tag{3.26}$$

Substitution of Eqs. (3.25), (3.19), and (3.24) in Eq. (3.23) leads to

$$\frac{1}{E_1} = a_1^2 \frac{v_f}{E_{f1}} + b_1^2 \frac{v_m}{E_m} \tag{3.27}$$

Note that we did not assume equal strains in fibers and matrix in order to derive these equations. To check the strain distribution, however, Eqs. (3.26) and (3.27) can be solved simultaneously for a_1 and b_1 when composite, fiber, and matrix properties are known. The ratio of the fiber strain to the matrix strain can then be found. For example, using the measured properties of an E-glass/epoxy composite [3.8]

$$E_1 = 5.05 \times 10^6 \text{ psi} \quad (34.82 \text{ GPa}); \quad v_m = 0.55$$

$$E_2 = 1.53 \times 10^6 \text{ psi} \quad (10.55 \text{ GPa}); \quad v_f = 0.45$$

$$E_{f1} = E_{f2} = 10.5 \times 10^6 \text{ psi} \quad (72.4 \text{ GPa})$$

$$E_m = 0.55 \times 10^6 \text{ psi} \quad (3.79 \text{ GPa})$$

(3.28)

we find that $a_1 = 2.0884$, $b_1 = 0.1093$, $a_1/b_1 = \bar{\sigma}_{f1}/\bar{\sigma}_{m1} = 19.1$, and $\bar{\epsilon}_{f1}/\bar{\epsilon}_{m1} = 1.00$. Thus, the assumption of equal strains which led to Eq. (3.22) is valid for this material, as it apparently is for other composites. The strain energy approach will be used again in the next section to check the validity of an assumption leading to the equation for the transverse modulus.

3.2.2 Transverse Modulus

If the RVE in Fig. 3.3(a) is subjected to a transverse normal stress, $\bar{\sigma}_{c2}$, as shown in Fig. 3.3(c), the response is governed by the effective transverse modulus, E_2. Geometric compatability requires that the total transverse composite displacement, $\bar{\delta}_{c2}$, must equal the sum of the corresponding transverse displacements in the fiber, $\bar{\delta}_{f2}$, and the matrix, $\bar{\delta}_{m2}$:

$$\bar{\delta}_{c2} = \bar{\delta}_{f2} + \bar{\delta}_{m2}$$

(3.29)

It follows from the definition of normal strain that

$$\bar{\delta}_{c2} = \bar{\epsilon}_{c2} L_2, \quad \delta_{f2} = \bar{\epsilon}_{f2} L_f, \quad \bar{\delta}_{m2} = \bar{\epsilon}_{m2} L_m$$

(3.30)

and Eq. (3.29) now becomes

$$\bar{\epsilon}_{c2} L_2 = \bar{\epsilon}_{f2} L_f + \bar{\epsilon}_{m2} L_m$$

(3.31)

Since the dimensions of the RVE do not change along the 1 direction, the length fractions must be equal to the volume fractions and Eq. (3.31) can be rearranged to get the rule of mixtures for transverse strains:

$$\bar{\epsilon}_{c2} = \bar{\epsilon}_{f2} v_f + \bar{\epsilon}_{m2} v_m$$

(3.32)

The one-dimensional Hooke's laws for this case are

$$\bar{\sigma}_{c2} = E_2 \bar{\epsilon}_{c2}, \quad \bar{\sigma}_{f2} = E_{f2} \bar{\epsilon}_{f2}, \quad \bar{\sigma}_{m2} = E_m \bar{\epsilon}_{m2}$$

(3.33)

where the Poisson strains have again been neglected. As with the longitudinal case, the inclusion of such strains would lead to a much more complex state of stress due to the mismatch in Poisson strains at the interface [3.9, 3.10]. This is another example of the difference between a mechanics of materials solution and a more rigorous theory of elasticity solution. Combining Eqs. (3.33) and Eq. (3.32), we get

$$\frac{\bar{\sigma}_{c2}}{E_2} = \frac{\bar{\sigma}_{f2}}{E_{f2}} v_f + \frac{\bar{\sigma}_{m2}}{E_m} v_m \qquad (3.34)$$

If we assume that the stresses in the composite, matrix, and fiber are all equal, Eq. (3.34) reduces to the "inverse rule of mixtures" for the transverse modulus

$$\frac{1}{E_2} = \frac{v_f}{E_{f2}} + \frac{v_m}{E_m} \qquad (3.35)$$

From the RVE in Fig. 3.3, it would seem that the assumption of equal stresses is valid because equilibrium requires that the forces must be equal for the series arrangement and both fiber and matrix blocks have equal areas normal to the 2 direction. In the actual composite, however, the fiber-packing arrangement is such that the forces and areas are not necessarily equal, and we will use a strain energy approach to show that the resulting stresses are not equal. Thus, Eq. (3.35) is generally not acceptable for design use. As shown in Fig. 3.4(a), Eq. (3.35) gives the same result as Eq. (3.22) at the extreme values of fiber volume fraction, but it predicts significant improvement in the transverse modulus only at high fiber volume fractions. This turns out to be the correct trend, but, as shown in Sec. 3.5, the experimental data falls well above the curve.

As with the longitudinal case, the strain energy approach provides additional insight into the micromechanics of the transverse loading case. We now express the fiber and matrix strains in terms of the composite strain:

$$\bar{\epsilon}_{f2} = a_2 \bar{\epsilon}_{c2}, \qquad \bar{\epsilon}_{m2} = b_2 \bar{\epsilon}_{c2} \qquad (3.36)$$

where a_2 and b_2 are constants. Substitution of Eqs. (3.36) in the compatibility expression, Eq. (3.32), yields

$$a_2 v_f + b_2 v_m = 1 \qquad (3.37)$$

By substituting Eqs. (3.36) and (3.33) in equations analogous to Eqs. (3.24) for the transverse loading case and the strain energy expression, Eq. (3.23), we find that

$$E_2 = a_2^2 E_{f2} v_f + b_2^2 E_m v_m \qquad (3.38)$$

where the strain energy due to the Poisson strain mismatch at the interface has again been neglected. It is important to note that we did not assume equal stresses in the fibers and the matrix in order to get Eq. (3.38). Using the properties for the E-glass/epoxy given in Eqs. (3.28) and solving Eqs. (3.37) and (3.38) simulaneously, we find that $a_2 = 0.432$, $b_2 = 1.465$, the strain ratio $a_2/b_2 = \bar{\epsilon}_{f2}/\bar{\epsilon}_{m2} = 0.295$, and the corresponding stress ratio is $\bar{\sigma}_{f2}/\bar{\sigma}_{m2} = 5.63$. Thus, the assumption of equal stresses in fibers and matrix which led to Eq. (3.35) is not justified for this material and is apparently not valid for most other composites as well. Alternative design equations for the transverse modulus will be discussed later.

3.2.3 Shear Modulus and Poisson's Ratio

The major Poisson's ratio, ν_{12}, and the in-plane shear modulus, G_{12}, are most often used as the two remaining independent elastic constants. The major Poisson's ratio, which is defined as

$$\nu_{12} = -\frac{\bar{\epsilon}_{c2}}{\bar{\epsilon}_{c1}} \tag{3.39}$$

when the only nonzero stress is a normal stress along the 1 direction, can be found by solving the geometric compatibility relationships associated with both the 1 and 2 directions. The result is another rule of mixtures formulation:

$$\nu_{12} = \nu_{f12}\nu_f + \nu_m \nu_m \tag{3.40}$$

where ν_{f12} = major Poisson's ratio of fiber
  ν_m = Poisson's ratio of matrix

Equation (3.40) is generally accepted as being sufficiently accurate for design purposes. As in the case for the longitudinal modulus, the geometric compatibility relationships leading to the solution are valid.

The effective in-plane shear modulus is defined as [Fig. 3.3(d)]

$$G_{12} = \frac{\bar{\sigma}_{c12}}{\bar{\gamma}_{c12}} \tag{3.41}$$

where $\bar{\sigma}_{c12}$ = average composite shear stress in the 12 plane
  $\bar{\gamma}_{c12} = 2\bar{\epsilon}_{c12}$, average engineering shear strain in the 12 plane

An equation for the in-plane shear modulus can be derived using an approach similar to that which was used for the transverse modulus. That is, geometric compatibility of the shear deformations, along with the assumption of equal shear stresses in fibers and matrix, leads to another inverse rule of mixtures:

$$\frac{1}{G_{12}} = \frac{v_f}{G_{f12}} + \frac{v_m}{G_m} \tag{3.42}$$

where G_{f12} = shear modulus of fiber in the 12 plane
G_m = shear modulus of matrix

As we might expect, this equation is not very accurate because the shear stresses are not equal as assumed. A strain energy approach similar to that used in Sec. 3.2.2 can be used here to show that the shear stresses are in fact not equal. As with the transverse modulus, we need to find better equations for estimating the in-plane shear modulus. Such equations will be discussed in the following sections.

Example 3.3. The constituent materials in the composite described in Examples 3.1 and 3.2 have the properties $E_{f1} = 32.0 \times 10^6$ psi (220 GPa), $E_{f2} = 2.0 \times 10^6$ psi (13.79 GPa), and $E_m = 0.5 \times 10^6$ psi (3.45 GPa). Estimate the longitudinal and transverse moduli of the composite. Given these fiber and matrix materials, what are the maximum possible values of E_1 and E_2?

Solution. The longitudinal modulus is given by Eq. (3.22):

$$E_1 = (32 \times 10^6)0.506 + (0.5 \times 10^6)0.482$$

$$= 16.43 \times 10^6 \text{ psi} \quad (113 \text{ GPa})$$

The transverse modulus is roughly estimated by Eq. (3.35):

$$E_2 = \frac{1}{\dfrac{0.506}{2 \times 10^6} + \dfrac{0.482}{0.5 \times 10^6}} = 0.82 \times 10^6 \text{ psi} \quad (5.65 \text{ GPa})$$

As expected, the composite is highly anisotropic, with $E_1 \gg E_2$. If the composite has the theoretical maximum fiber volume fraction of 0.907 for a close-packed triangular array [see Eq. (3.13)], the corresponding composite properties are still highly anisotropic, with $E_1 = 29 \times 10^6$ psi (200 GPa) and $E_2 = 1.56 \times 106$ psi (10.75 GPa). Note that even with this maximum fiber content, the transverse modulus is still very low. Thus, some transverse reinforcement is usually necessary in practical applications. Note also that the longitudinal modulus of the graphite/epoxy composite is now about the same as the modulus of steel, but the density of the composite is only about 20 percent of the density of steel. Composites typically have much greater stiffness-to-weight ratios than conventional metallic structural materials.

Example 3.4. For longitudinal loading of the composites in Example 3.3, compare the stresses in the fiber and matrix materials. Compare the strain energy stored in the fibers with that stored in the matrix.

Solution. From Eqs. (3.19) and (3.21), the ratio of fiber stress to matrix stress is

$$\frac{\bar{\sigma}_{f1}}{\bar{\sigma}_{m1}} = \frac{E_{f1}\bar{\epsilon}_{f1}}{E_m\bar{\epsilon}_{m1}} = \frac{E_{f1}}{E_m} = \frac{32.0}{0.5} = 64.0$$

Thus, the fiber carries most of the stress since the fiber modulus is always higher than the matrix modulus. From Eqs. (3.24b), (3.24c), and (3.21), the ratio of fiber strain energy to matrix strain energy is

$$\frac{U_f}{U_m} = \frac{E_{f1}v_f}{E_m v_m} = \frac{32.0(0.506)}{0.5(0.482)} = 67.0$$

which is almost the same as the stress ratio. If the composite had the maximum possible fiber volume fraction of 0.907, the stress ratio would remain the same since it is independent of the fiber volume fraction. The strain energy ratio would increase dramatically to 624, however, since it is proportional to the ratio of fiber volume fraction to the matrix volume fraction.

3.3 IMPROVED MECHANICS OF MATERIALS MODELS

As shown in the previous section, the elementary mechanics of materials models for E_1 and v_{12} are good enough for design use. The corresponding models for E_2 and G_{12} are of questionable value, however, because they are based on invalid assumptions and agreement with experimental results is generally poor. We will now discuss several refinements of the elementary mechanics of materials models.

Due to the simplified RVE that was used for the elementary mechanics of materials approach (Fig. 3.3), the resulting equations were not tied to any particular fiber-packing geometry. Since the results for E_1 and v_{12} were so favorable, we can conclude that those properties must be essentially independent of fiber-packing geometry. By the same reasoning, it appears that E_2 and G_{12} may be more sensitive to fiber-packing geometry. Thus, the assumption of a specific fiber-packing array is one possible refinement of the models. Although real composites have random-packing arrays, the assumption of a regular array is a logical simplification if we are to have any hope of developing simple design equations. Such an assumption allows us to use simple relations among fiber size, spacing, and volume fraction. Hopkins and Chamis [3.11] have

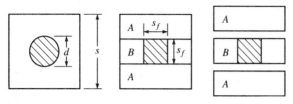

FIGURE 3.5
Division of representative volume element into subregions based on square fiber having equivalent fiber volume fraction.

developed a refined model for transverse and shear properties based on a square fiber-packing array and a method of dividing the RVE into subregions. The following derivation is adapted from Ref. [3.11].

A square array of fibers is shown in Fig. 3.1 and the RVE for such an array is shown in Fig. 3.5. The RVE is easily divided into subregions for more detailed analysis if we convert to a square fiber having the same area as the round fiber. The equivalent square fiber shown in Fig. 3.5 must then have the dimension

$$s_f = \sqrt{\frac{\pi}{4}} d \tag{3.43}$$

and from Eq. (3.10), the size of the RVE is

$$s = \sqrt{\frac{\pi}{4v_f}} d \tag{3.44}$$

The RVE is divided into subregions A and B, as shown in Fig. 3.5. In order to find the effective transverse modulus for the RVE, we first subject the series arrangement of fiber and matrix in subregion B to a transverse normal stress. Following the procedure of Sec. 3.2.2, the effective transverse modulus for this subregion, E_{B2}, is found to be

$$\frac{1}{E_{B2}} = \frac{1}{E_{f2}} \frac{s_f}{s} + \frac{1}{E_m} \frac{s_m}{s} \tag{3.45}$$

where the matrix dimension is $s_m = s - s_f$. From Eqs. (3.43) and (3.44), it is seen that

$$\frac{s_f}{s} = \sqrt{v_f} \quad \text{and} \quad \frac{s_m}{s} = 1 - \sqrt{v_f} \tag{3.46}$$

so that Eq. (3.45) now becomes

$$E_{B2} = \frac{E_m}{1 - \sqrt{v_f}(1 - E_m/E_{f2})} \tag{3.47}$$

The parallel combination of subregions A and B is now loaded by a transverse normal stress and the procedure of Sec. 3.2.1 is followed in order to find the effective transverse modulus of the RVE. The result, of course, is the rule of mixtures analogous to Eq. (3.20):

$$E_2 = E_{B2} \frac{s_f}{s} + E_m \frac{s_m}{s} \tag{3.48}$$

Substitution of Eqs. (3.46) and (3.47) in Eq. (3.48) then gives the final result

$$E_2 = E_m \left[(1 - \sqrt{v_f}) + \frac{\sqrt{v_f}}{1 - \sqrt{v_f}(1 - E_m/E_{f2})} \right] \tag{3.49}$$

A similar result may be found for G_{12}. The detailed derivation in Ref. [3.11] also includes the effect of a third phase, a fiber/matrix interphase material, which is assumed to be an annular volume surrounding the fiber. Such interphase regions exist in many metal matrix [3.11] and polymer matrix [3.12] composites. When the fiber diameter is equal to the interphase diameter, the equation for E_2 in Ref. [3.11] reduces to Eq. (3.49) above. The complete set of equations for effective moduli of the three-phase model are given in Ref. [3.11].

In separate publications Chamis [3.13, 3.14] presented the so-called "simplified micromechanics equations" (SME), which are based on this same method of subregions, except that only the terms for subregion B (see Fig. 3.5) are retained. Thus, the simplified micromechanics equation for E_2 would be the same as that for E_{B2} in Eq. (3.47), and similar equations for the other effective moduli are given in Refs. [3.13] and [3.14]. Also included in these references are tables of fiber and matrix properties to be used as input to the SME, and these tables are reproduced here in Tables 3.1 and 3.2. It is important to note that in such tables the transverse fiber modulus, E_{f2}, and the longitudinal fiber shear modulus, G_{f12}, are not actually measured but are inferred by substitution of measured composite properties and matrix properties in the SME. The inferred properties show that fibers such as graphite and aramid are highly anisotropic, whereas glass and boron are essentially isotropic. Similar back-calculations of anisotropic fiber properties using other analytical models have been reported by Kriz and Stinchcomb [3.15] and by Kowalski [3.16]. More recently, direct measurement of fiber transverse moduli has been reported by Kawabata [3.17]. Kawabata's measurements, based on transverse diametral compression of single graphite and aramid fibers, show even greater anisotropy than the inferred properties in Tables 3.1 and 3.2. However, Caruso and Chamis [3.18] have shown that the SME and the corresponding tables of properties give results which agree well with three-dimensional finite element models, as shown

TABLE 3.1
Fiber properties

Property	Units	Boron	HMS	AS	T300	KEV	S-G	E-G
Number of fibers per end	—	1	10 000	10 000	3000	580	204	204
Fiber diameter	in	0.0056	0.0003	0.0003	0.0003	0.00046	0.00036	0.00036
Density	lb/in^3	0.095	0.070	0.063	0.064	0.053	0.090	0.090
Longitudinal modulus	10^6 psi	58	55.0	31.0	32.0	22	12.4	10.6
Transverse modulus	10^6 psi	58	0.90	2.0	2.0	0.6	12.4	10.6
Longitudinal shear modulus	10^6 psi	24.2	1.1	2.0	1.3	0.42	5.17	4.37
Transverse shear modulus	10^6 psi	24.2	0.7	1.0	0.7	0.22	5.17	4.37
Longitudinal Poisson's ratio	—	0.20	0.20	0.20	0.20	0.35	0.20	0.22
Transverse Poisson's ratio	—	0.20	0.25	0.25	0.25	0.35	0.20	0.22
Heat capacity	Btu/lb/°F	0.31	0.20	0.20	0.22	0.25	0.17	0.17
Longitudinal heat conductivity	Btu/h/ft^2/°F/in	22	580	580	580	1.7	21	7.5
Transverse heat conductivity	Btu/h/ft^2/°F/in	22	58	58	58	1.7	21	7.5
Longitudinal thermal expansion coefficient	10^{-6} in/in/°F	2.8	−0.55	−0.55	−0.55	−2.2	2.8	2.8
Transverse thermal expansion coefficient	10^{-6} in/in/°F	2.8	5.6	5.6	5.6	30	2.8	2.8
Longitudinal tensile strength	ksi	600	250	350	350	400	600	400
Longitudinal compression strength	ksi	700	200	260	300	75	—	—
Shear strength	ksi	100	—	—	—	—	—	—

Transverse, shear, and compression properties are estimates inferred from corresponding composite properties.
Source: From Chamis [3.14].

(ΔT = -280°F) off/on

All matrices for alt prg

TABLE 3.2
Matrix properties

Name	Units	LM	IMLS	IMHS	HM	Polyimide	PMR
Density	lb/in^3	0.042	0.046	0.044	0.045	0.044	0.044
Modulus	10^6 psi	0.32	0.50	0.50	0.75	0.50	0.47
Shear modulus	10^6 psi	—	—	—	—	—	—
Poisson's ratio	—	0.43	0.41	0.35	0.35	0.35	0.36
Heat capacity	Btu/lb/°F	0.25	0.25	0.25	0.25	0.25	0.25
Heat conductivity	Btu/h/ft^2/°F/in	1.25	1.25	1.25	1.25	1.25	1.25
Thermal expansion coefficient	10^{-6} in/in/°F	57	57	36	40	20	28
Diffusivity	10^{-10} in^2/s	0.6	0.6	0.6	0.6	0.6	0.6
Moisture expansion coefficient	in/in/M	0.33	0.33	0.33	0.33	0.33	0.33
Tensile strength	ksi	8	7	15	20	15	8
Compression strength	ksi	15	21	35	50	30	16
Shear strength	ksi	8	7	13	15	13	8
Tensile fracture strain	in/in (%)	8.1	1.4	2.0	2.0	2.0	2.0
Compressive fracture strain	in/in (%)	15	4.2	5.0	5.0	4.0	3.5
Shear fracture strain	in/in (%)	10	3.2	3.5	4.0	3.5	5.0
Air heat conductivity	Btu/h/ft^2/°F/in	0.225	0.225	0.225	0.225	0.225	0.225
Glass transition temperature (dry)	°F	350	420	420	420	700	700

LM = low modulus; IMLS = intermediate modulus low strength; IMHS = intermediate modulus high strength; HM = high modulus. Thermal, hygral, compression, and shear properties are estimates only; $G_m = E_m/2(1 + \nu_m)$.
Source: From Chamis [3.14].

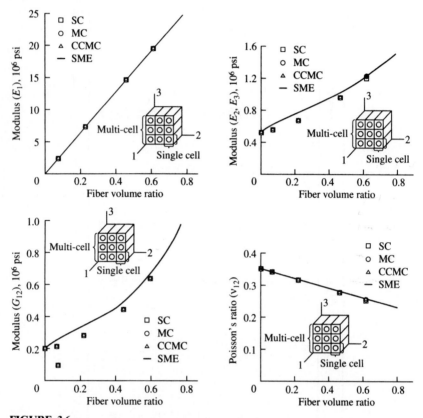

FIGURE 3.6

Comparison of three-dimensional finite element results for lamina elastic constants with predictions from simplified micromechanics equations for graphite/epoxy. (*From Caruso and Chamis* [3.18]. *Copyright ASTM. Reprinted with permission.*)

in Fig. 3.6. Since the SME for E_1 and ν_{12} are the same as Eqs. (3.22) and (3.40), respectively, this comparison provides further evidence of the validity of those equations.

Another set of equations for E_2 and G_{12} has been derived by Spencer [3.19], who used a square array model that included the effects of the strain concentration at points of minimum clearance between fibers in the RVE. Spencer's equation is

$$\frac{M_c}{M_m} = \frac{\Gamma - 1}{\Gamma} + \frac{1}{k}\left[\frac{\pi}{2} + \frac{2\Gamma}{\sqrt{\Gamma^2 - k^2}}\tan^{-1}\sqrt{\frac{\Gamma + k}{\Gamma - k}}\right] \qquad (3.50)$$

where $M_c = E_2$, $M_m = E_m$, and $k = 1 - E_m/E_{f2}$ for the transverse modulus equation and $M_c = G_{12}$, $M_m = G_m$, and $k = 1 - G_m/G_{f12}$ for the longitudinal shear modulus equation. The parameter $\Gamma = s/d$ in both equations. Spencer also suggests that Γ can be accurately approximated for a variety of packing geometries over the full range of fiber volume fractions, v_f, by the equation

$$\Gamma = \frac{1}{\sqrt{(1.1v_f^2 - 2.1v_f + 2.2)v_f}} \tag{3.51}$$

Spencer does not include a table of suggested properties for use with these equations. A comparison of the various models will be given later in Sec. 3.5 after the elasticity and semiempirical models have been discussed.

Example 3.5. For the composite in Example 3.3, compare transverse modulus values calculated by the inverse rule of mixtures [Eq. (3.35)], the method of subregions [Eq. (3.49)], and Spencer's equation [Eq. (3.50)].

Solution. The results are

Equation (3.35)   As previously calculated in Example 3.3,

$$E_2 = 0.82 \times 10^6 \, \text{psi} \quad (5.65 \, \text{GPa})$$

Equation (3.49) $E_2 = 0.9 \times 10^6 \, \text{psi} \quad (6.2 \, \text{GPa})$

Equation (3.50)   First estimate $\Gamma = 1.18$ from Eq. (3.51). Note that
the actual value of Γ from Example 3.2 is

$$\Gamma = s/d = 0.000623/0.0005 = 1.25$$

Using $\Gamma = 1.18$ in Eq. (3.50), we have

$$E_2 = 0.98 \times 10^6 \, \text{psi} \quad (6.76 \, \text{GPa})$$

As previously mentioned, the inverse rule of mixtures prediction for the transverse modulus is considerably lower than measured values. The higher values given by the method of subregions and Spencer's equation are more accurate. Further discussion on this will follow in the next section.

3.4 ELASTICITY MODELS

The theory of elasticity approach to micromechanical modeling begins in the same way as the mechanics of materials approach, by selecting the RVE and then subjecting the RVE to uniform stress or displacement at the boundary. The two approaches differ substantially in the solution of the resulting boundary value problem, however. The equations of elasticity must be satisfied at every point in the model, and no simplifying assumptions are made regarding the stress or strain distributions as in the mechanics of materials approach. Fiber-packing geometry is generally specified in the elasticity approach. A variety of closed-form and

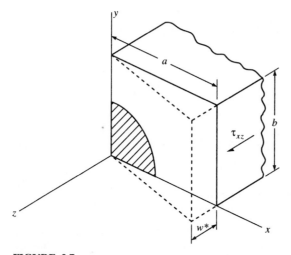

FIGURE 3.7
One quadrant of representative volume element from Adams and Doner elasticity solution for shear modulus G_{12}. (*From Adams and Doner* [3.20]. *Reprinted by permission of Technomic Publishing Co.*)

numerical solutions of the governing equations of elasticity have been reported in the literature [3.1–3.4], and a complete review of the work in this area is beyond the scope of this book. The objective here is to discuss several representative numerical and closed-form solutions in order to show what additional knowledge of micromechanical behavior can be obtained from the more rigorous elasticity approach.

Numerical solutions of the governing elasticity equations are often necessary for complex structural geometries such as those found in the RVEs used in micromechanics models. For example, Adams and Doner [3.20] used a finite difference solution to determine the shear modulus G_{12} for a rectangular array of fibers. A displacement boundary value problem was solved for one quadrant of the RVE, as shown in Fig. 3.7. Note that Adams and Doner use the z axis to define the fiber direction, whereas the x and y axes correspond to the transverse directions. The displacement components u, v, and w correspond to the x, y, and z axes, respectively. The imposed displacement w^* along $x = a$ causes a displacement field of the form

$$u = v = 0, \qquad w = w(x, y) \tag{3.52}$$

From the strain-displacement equations and Hooke's law, the only nonvanishing stress components are

$$\tau_{zx} = G \frac{\partial w}{\partial x} \qquad \text{and} \qquad \tau_{zy} = G \frac{\partial w}{\partial y} \tag{3.53}$$

where the shear modulus G may be either the fiber or matrix property, depending on the coordinates x and y. Isotropic behavior was assumed for both fiber and matrix materials. Substitution of Eqs. (3.53) in the only nontrivial equilibrium equation yielded the governing partial differential equation

$$G\left[\frac{\partial^2 w}{\partial x^2} + \frac{\partial^2 w}{\partial y^2}\right] = 0 \tag{3.54}$$

which was solved subject to the displacement boundary conditions

$$w(0, y) = 0, \qquad w(a, y) = w*$$
$$G\frac{\partial w}{\partial y} = 0 \qquad \text{along } y = 0 \text{ and } y = b \tag{3.55}$$

and continuity conditions at the fiber/matrix interface by using a finite difference scheme. The solution yielded the values of the displacements $w(x, y)$ at each node of the finite difference grid. Stresses were found by substituting these displacements in the finite difference forms of Eqs. (3.53), and the effective shear modulus was then determined from

$$G_{xz} = \frac{\bar{\tau}_{xz}}{w*/a} \tag{3.56}$$

where $\bar{\tau}_{xz}$ is the average shear stress along $x = a$. A similar boundary value problem for shear along $y = b$ yields the associated shear modulus G_{yz}. Typical results are shown in Fig. 3.8, where the ratio of the composite shear modulus to the matrix shear modulus is plotted vs. the shear modulus ratio G_f/G_m for various fiber volume fractions.

In a separate paper Adams and Doner [3.21] used a similar approach to determine the transverse modulus E_2, and typical results are shown in Fig. 3.9. It is seen in Figs. 3.8 and 3.9 that the reinforcement effect for both G_{12} and E_2 only becomes significant for fiber volume fractions above about 50 percent, but that combinations of high fiber stiffness and high fiber volume fractions can significantly increase G_{12} and E_2. Unfortunately, these same combinations also generate very high stress concentration factors at the fiber/matrix interfaces, as shown in the same papers [3.20, 3.21]. One of the advantages of the elasticity approach is that the complete stress and strain distributions in the RVE are generated, and the calculation of stress concentration factors is possible. One advantage of numerical solutions such as finite differences is the capability for analysis of complex geometries. For example, stiffness and stress concentration factors were also obtained for a variety of fiber cross-sectional shapes such as squares and ellipses in a rectangular array [3.20, 3.21].

The previously mentioned finite element analysis of Caruso and

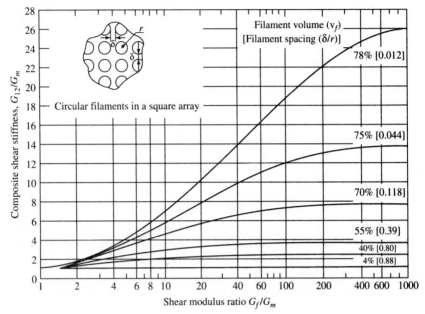

FIGURE 3.8
Normalized composite shear stiffness, G_{12}/G_m, vs. shear modulus ratio, G_f/G_m, for circular fibers in a square aray. (*From Adams and Doner* [3.20]. *Reprinted by permission of Technomic Publishing Co.*)

Chamis [3.18] and Caruso [3.22] is another example of a numerical elasticity solution. In this case a "single-cell" (SC) finite element model was developed from 192 three-dimensional isoparametric brick elements (Fig. 3.10). This "single-cell" model was then used as a building block for a "multicell" (MC) model consisting of nine single-cell models in a 3 × 3 array (Fig. 3.6). A third model (CCMC) used only the center cell in the nine-cell MC model for the calculations. Boundary and load conditions were consistent with those used for the previously discussed SME mechanics of materials solutions, so that the finite element results could be compared with the SME results. For example, equations such as Eq. (3.56) were used to determine stiffness from finite element results. Material properties for AS graphite fibers in an intermediate-modulus-high-strength (IMHS) epoxy matrix were used (Tables 3.1 and 3.2). Fibers were assumed to be orthotropic, whereas the matrix was assumed to be isotropic. As shown in Fig. 3.6, the finite element results show good agreement with SME results.

A two-dimensional finite element solution for the transverse modulus, E_2, of a square array of circular E-glass fibers in an epoxy matrix was obtained by Schroeder [3.23]. As shown in Fig. 3.11, the single RVE

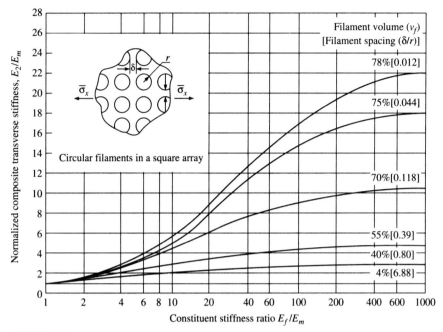

FIGURE 3.9
Normalized composite transverse stiffness, E_2/E_m, vs. modulus ratio, E_f/E_m, for circular fibers in a square array. (*From Adams and Doner* [3.21]. *Reprinted by permission of Technomic Publishing Co.*)

model (model 1) consisted of 456 two-dimensional isoparametric solid elements, whereas model 2 consisted of two rows of the single RVE models arranged in a square array. The elements were loaded in plane strain by imposing a displacement boundary condition along one vertical edge while fixing the opposite edge. The other two edges had "free edge" boundary conditions. Again, equations such as Eq. (3.56) were used to find stiffnesses from finite element results. Results from these finite element models will be compared with mechanics of materials models and a semiempirical solution in the next section.

As an example of a closed-form elasticity solution, Whitney and Riley [3.7] used axisymmetry Airy stress functions to solve for the stresses and strains in a so-called "self-consistent" model having a single isotropic fiber embedded in a concentric cylinder of isotropic matrix material. The cylindrical geometry of the "self-consistent" model is such that the model is not associated with any specific fiber-packing geometry. The resulting micromechanical stresses and strains were then used in energy balance equations similar to Eqs. (3.23) and (3.24) to solve for E_1 and E_2. The equation for E_1 reduces to the rule of mixtures when the

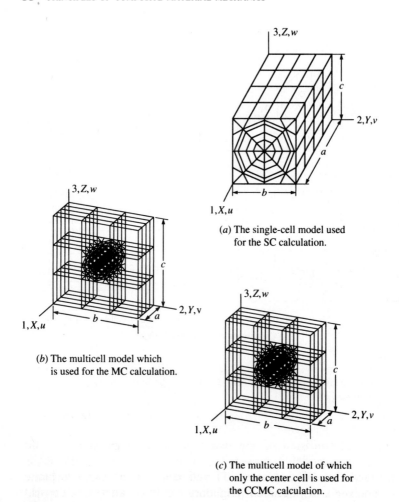

(a) The single-cell model used for the SC calculation.

(b) The multicell model which is used for the MC calculation.

(c) The multicell model of which only the center cell is used for the CCMC calculation.

FIGURE 3.10
Three-dimensional finite element models of representative volume elements. (*From Caruso and Chamis* [3.18]. *Copyright ASTM. Reprinted with permission.*)

Poisson's ratio of the fiber is equal to that of the matrix. The additional term is due to the mismatch in Poisson strains at the fiber/matrix interface [recall that this term was neglected in Eqs. (3.23) and (3.24)]. Predictions showed good agreement with experimental data for boron/epoxy. In a later paper Whitney extended the analysis to include anisotropic, transversely isotropic fibers [3.24].

Another closed-form micromechanical elasticity approach, the method of cells, was developed by Aboudi [3.25]. A representative cell consisting of a square fiber embedded in a square of matrix material was divided into four subcells. Equilibrium equations were then solved

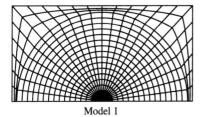

Model 1

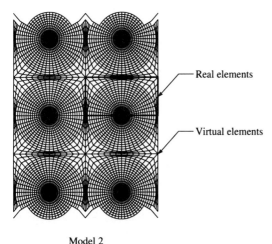

—— Real elements

—— Virtual elements

Model 2

FIGURE 3.11
Two-dimensional finite element models of representative volume elements. (*From Schroeder* [3.23].)

subject to continuity of displacements and tractions at the interfaces between the subcells and between neighboring cells on an average basis, along with the assumption of linear variations of displacements in each subcell. The equations are too lengthy to present here, but excellent agreement was observed with the experimental data on graphite/epoxy from Ref. [3.15]. One advantage of this approach is that it yields not only the in-plane lamina properties, but also the through-the-thickness properties such as G_{23} and ν_{23}.

Paul [3.26] obtained closed-form solutions for the bounds on the transverse modulus of a fiber composite (or the Young's modulus of an isotropic-particle-reinforced composite) by using a variational approach. By applying the theorem of minimum complementary energy to the situation where the composite is subjected to a uniaxial normal stress, Paul found the lower bound on E_2 to be the inverse rule of mixtures [see Eq. (3.35)]. The application of the theorem of minimum potential energy to the situation where the composite is subjected to a simple extensional strain gave the upper bound on E_2, which reduces to the rule of mixtures [Eq. (3.22)] when the Poisson's ratios of fiber and matrix materials are taken to be the same.

The bounds derived by Paul [3.26] are independent of packing geometry and are referred to as the elementary bounds. Thus, it should be no surprise that the bounds are very far apart, as shown in Fig. 3.4. Tighter bounds require the specification of packing geometry. For example, Hashin and Rosen [3.27] applied the principles of minimum potential and complementary energy to fiber composites with hexagonal and random arrays. Detailed summaries of these and other related results have been reported by Hashin [3.1] and Christensen [3.3]. More recently, Torquato [3.28] has reviewed advances in the calculation of improved bounds on the effective properties of random heterogeneous media. Such improved bounds are determined by using statistical correlation functions to model the random variations in the microstructure. Since the fiber-packing geometry in composites is of a random nature, such bounds should be more realistic than the bounds which are based on some idealized fiber-packing array.

3.5 SEMIEMPIRICAL MODELS

In Sec. 3.3 improved mechanics of materials models for prediction of E_2 and G_{12} were discussed. Another general approach to estimating these properties involves the use of semiempirical equations which are adjusted to match experimental results or elasticity results by the use of curve-fitting parameters. The equations are referred to as being "semiempirical" because, although they have terms containing curve-fitting parameters, they also have some basis in mechanics. The most widely used semiempirical equations were developed by Halpin and Tsai [3.29]. The Halpin–Tsai equation for the transverse modulus is

$$\frac{E_2}{E_m} = \frac{1 + \xi \eta v_f}{1 - \eta v_f} \tag{3.57}$$

where

$$\eta = \frac{(E_f/E_m) - 1}{(E_f/E_m) + \xi} \tag{3.58}$$

and ξ is the curve-fitting parameter which is also a measure of the degree of reinforcement of the matrix by the fibers. The corresponding equation for G_{12} is obtained by replacing the Young's moduli E_2, E_f, and E_m in the above equations by the shear moduli G_{12}, G_f, and G_m, respectively. Note that the values for the curve-fitting parameter may be different for E_2 and G_{12}. Halpin and Tsai found that the value $\xi = 2$ gave an excellent fit to the finite difference elasticity solution of Adams and Doner [3.21] for the transverse modulus of a square array of circular fibers having a fiber

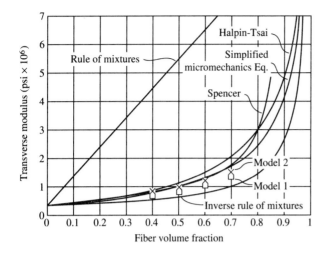

FIGURE 3.12
Comparison of predicted transverse modulus for E-glass/epoxy from two-dimensional finite element models with other predictions. (*From Schroeder* [3.23].)

volume fraction of 0.55. For the same material and fiber volume fraction a value of $\xi = 1$ gave excellent agreement with the Adams and Doner solution for G_{12} [3.20]. Schroeder [3.23] has compared the Halpin–Tsai equation with the Chamis SME equation [Eq. (3.47)], the Spencer equation [Eq. (3.50)], the Paul bounds [Eqs. (3.22) and (3.35)], and the Schroeder two-dimensional finite element analyses [3.23] for the case of the transverse modulus of E-glass/epoxy, and the results are shown in Fig. 3.12. Note that the SME equation and Spencer's equation show better agreement with the finite element results than the Halpin–Tsai equation, and that all results fall inside the predicted bounds. The suggested value of $\xi = 2$ was used in the Halpin–Tsai equation here, but perhaps a different value would produce better results for this case.

Jones [3.10] shows that when $\xi = 0$, the Halpin–Tsai equation reduces to the inverse rule of mixtures [Eq. (3.35)], whereas a value of $\xi = \infty$ yields the rule of mixtures [Eq. (3.22)]. Recall that Paul [3.26] proved that these equations also represent the bounds on E_2. Thus, the interpretation of the curve-fitting parameter, ξ, as a measure of the degree of fiber reinforcement has a theoretical basis. The use of the Halpin–Tsai equations in a variety of other applications and related empirical equations for estimating the curve-fitting parameter are discussed in more detail by Jones [3.10] and Halpin [3.4].

Tsai and Hahn [3.9] have proposed another semiempirical approach to calculating E_2 and G_{12}, which is based on the fact that the stresses in the fibers and the matrix are not equal under the corresponding loading conditions. Recall that the proof of such differences was demonstrated using a strain energy approach in Sec. 3.2.2. The method involves the use of empirical "stress-partitioning parameters" in derivations paralleling those used for the elementary mechanics of materials models. For example, the Tsai–Hahn equation for E_2 is found by introducing a stress-partitioning parameter, η_2, and using the relationship

$$\bar{\sigma}_{m2} = \eta_2 \bar{\sigma}_{f2} \tag{3.59}$$

in a derivation similar to that used for the elementary mechanics of materials model for E_2 in Sec. 3.2.2. The derivation was also based on the assumption that a rule of mixtures for stress similar to Eq. (3.18) also held for the transverse direction (2 direction). Although such an assumption is obviously not consistent with the RVE configuration and the loading condition shown in Fig. 3.3(c), it would be valid for a real composite with fiber packing such as that shown in Fig. 3.2. The result of this derivation is

$$\frac{1}{E_2} = \frac{1}{v_f + \eta_2 v_m} \left[\frac{v_f}{E_f} + \frac{\eta_2 v_m}{E_m} \right] \tag{3.60}$$

Note that Eq. (3.60) reduces to the inverse rule of mixtures [Eq. (3.35)] when the stress-partitioning parameter $\eta_2 = 1.0$. This is to be expected since Eq. (3.35) was based on the assumption that the stresses in the fiber and the matrix are the same. A similar equation can be derived for the shear modulus G_{12}, as shown in [3.9]. Figure 3.13 from Ref. [3.9] shows experimental data for the transverse modulus of a glass/epoxy composite

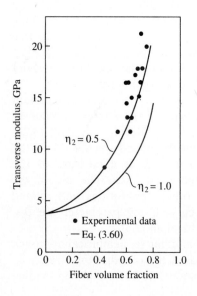

● Experimental data
— Eq. (3.60)

FIGURE 3.13

Transverse modulus for glass/epoxy according to Tsai–Hahn equation [Eq. (3.60)]. (*From Tsai and Hahn* [3.9]. *Reprinted by permission of Technomic Publishing Co.*)

compared with the predicted values from Eq. (3.60) for two different assumed values of the stress-partitioning parameter. The stress-partitioning parameter $\eta_2 = 0.5$ was found to yield accurate predictions of G_{12} based on comparisons with experimental data for the same glass/epoxy [3.9]. Formulas for estimating the stress-partitioning parameters from constituent material properties are also given in [3.9].

It would obviously be desirable to compare all the predictions discussed here with the experimental data shown in Fig. 3.13. Indeed, the results from Fig. 3.12 can be compared with those in Fig. 3.13 by correcting for the fact that the two figures were based on two different epoxy matrix materials but on the same glass fibers. Figure 3.12 is based on a low modulus (designated LM in Table 3.2) epoxy matrix with $E_m = 2.2\,\text{GPa}\ (0.32 \times 10^6\,\text{psi})$ and $v_m = 0.43$, whereas Fig. 3.13 is based on $E_m = 3.45\,\text{GPa}\ (0.5 \times 10^6\,\text{psi})$ and $v_m = 0.35$.

PROBLEMS

3.1. A rectangular array of elliptical fibers is shown in Fig. 3.14. Derive the relationship between the fiber volume fraction and the given geometrical parameters. What is the maximum possible fiber volume fraction for this packing geometry?

3.2. The fibers in a E-glass/epoxy composite are 0.0005 in (0.0127 mm) in diameter before coating with an epoxy-sizing 0.0001 in (0.00254 mm) thick. After the sizing has been applied, the fibers are bonded together with more epoxy of the same type. What is the maximum fiber volume fraction that can be achieved? Using the fiber and matrix moduli given in Eq. (3.28), determine the composite longitudinal modulus E_1 and the composite transverse modulus E_2 corresponding to the maximum fiber volume fraction.

3.3. A hybrid graphite–aramid/epoxy composite is made by randomly mixing continuous aligned fibers of the same diameter; so that there are two graphite fibers for each aramid fiber. The fibers are assumed to be arranged in a square array with the closest possible packing. The stress-strain curves

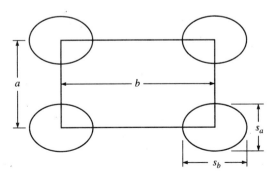

FIGURE 3.14
Rectangular array of elliptical fibers.

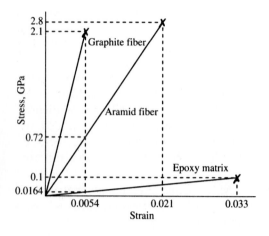

FIGURE 3.15
Stress-strain curves for fiber and matrix materials in a hybrid composite.

for longitudinal tensile loading of fiber and matrix materials are shown in Fig. 3.15. Determine the composite longitudinal modulus E_1.

3.4. Derive Eq. (3.40).

3.5. Derive Eq. (3.42).

3.6. Using an elementary mechanics of materials approach, find the micromechanics equation for predicting the minor Poisson's ratio, v_{21}, for a unidirectional fiber composite in terms of the corresponding fiber and matrix properties and volume fractions. Assume that the fibers are orthotropic, that the matrix is isotropic, and that all materials are linear elastic. This derivation should be independent of the one in Problem 3.4.

3.7. A composite shaft is fabricated by bonding an isotropic solid shaft having shear modulus G_1 and outside radius r_1 inside a hollow isotropic shaft having shear modulus G_2 and outside radius r_2. The composite shaft is to be loaded by a twisting moment which is distributed over the end of the shaft, as shown in Fig. 3.16. Using an elementary mechanics of materials approach, derive the equations for the stresses and deformations at any radius and the equation for the effective torsional shear modulus of the composite shaft in terms of the material and geometrical properties of shafts 1 and 2.

3.8. Using the method of subregions, derive an equation for the transverse modulus, E_2, for the RVE which includes a fiber/matrix interphase region,

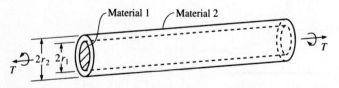

FIGURE 3.16
Composite shaft under torsional load.

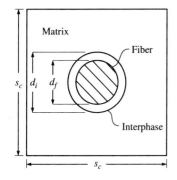

FIGURE 3.17
Representative volume element with fiber/matrix interphase region.

as shown in Fig. 3.17. *Hint*: The equation should reduce to Eq. (3.49) when the fiber diameter is the same as the interphase diameter.

3.9. Derive Eq. (3.54).

3.10. For a unidirectional composite with a rectangular fiber array [Fig. (3.7)], use the equations of elasticity to set up the displacement boundary value problem for determination of the transverse modulus, E_2. That is, find the governing partial differential equations for displacements u and v in the RVE, and specify the boundary and continuity conditions. Assume plane strain ($\epsilon_z = 0$). Do not attempt to solve the equations, but explain briefly how E_2 would be found. Assume that both fiber and matrix are isotropic.

3.11. Derive Eq. (3.60).

3.12. Show that a value of $\xi = 0$ reduces the Halpin–Tsai equation [Eq. (3.57)] to the inverse rule of mixtures [Eq. (3.35)], whereas a value $\xi = \infty$ reduces it to the rule of mixtures [Eq. (3.22)].

3.13. A representative volume element from a particle-reinforced composite is shown in Fig. 3.18. The particle has a cross-sectional area $A_p(x)$ which

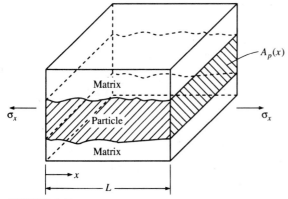

FIGURE 3.18
Representative volume element for a particle-reinforced composite with particle which has a varying cross-sectional area $A_p(x)$.

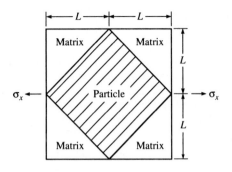

FIGURE 3.19

Representative volume element for a composite reinforced with a square particle and loaded along the diagonal of the particle.

varies with the distance x, and the stresses and strains in particle and matrix materials also vary with x. Find the expression for the effective Young's modulus of the composite, E_x, along the x direction. The answer should be left in terms of an integral involving the length, L; the particle modulus, E_p; the matrix modulus, E_m; and the particle area fraction $a_p(x) = A_p(x)/A_c$, where A_c is the total composite cross-sectional area. Assume that both the particle and matrix are isotropic.

3.14. Using the result from Problem 3.13, determine the effective Young's modulus, E_x, for the representative volume element shown in Fig. 3.19. In Fig. 3.19 the reinforcing particle has a square cross section and is oriented as shown. For a particle having a Young's modulus $E_p = 10 \times 10^6$ psi (68.95 GPa) and a matrix having a Young's modulus $E_m = 0.5 \times 10$ psi (3.45 GPa), determine the value of E_x and compare with the values from the rule of mixtures [Eq. (3.22)] and the inverse rule of mixtures [Eq. (3.35)]. Discuss your results in the context of the comments on theoretical bounds on the transverse modulus in Sec. 3.4.

3.15. A unidirectional composite is to be modeled by the representative volume element (RVE) shown in Fig. 3.20(a), where the fiber and matrix materials

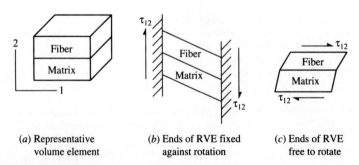

(a) Representative volume element

(b) Ends of RVE fixed against rotation

(c) Ends of RVE free to rotate

FIGURE 3.20

Representative volume element with two different loading and boundary conditions for Problem 3.15.

are assumed to be isotropic and perfectly bonded together. Using a mechanics of materials approach, derive the micromechanics equations for the effective in-plane shear modulus, G_{12}, for the following cases:

(a) The ends of the RVE are perfectly bonded to supports which are rigid against rotation, then subjected to the uniform in-plane shear stress, τ_{12}, by the nonrotating supports, as shown in Fig. 3.20(b), and

(b) The top and bottom surfaces of the RVE are subjected to the uniform in-plane shear stress, τ_{12}, and the ends of the RVE are free to rotate, as shown in Fig. 3.20(c).

REFERENCES

3.1. Hashin, Z., "Analysis of Composite Materials—A Survey," *Journal of Applied Mechanics*, **50**, 481–505 (September 1983).

3.2. Chamis, C. C. and Sendeckyj, G. P., "Critique on Theories Predicting Thermoelastic Properties of Fibrous Composites," *Journal of Composite Materials*, **2**(3), 332–358 (July 1968).

3.3. Christensen, R. M., *Mechanics of Composite Materials*, John Wiley & Sons, New York (1979).

3.4. Halpin, J. C., *Primer on Composite Materials: Analysis (Rev.)*, Technomic Publishing Co., Lancaster, PA (1984).

3.5. Gibson, R. F., *Elastic and Dissipative Properties of Fiber Reinforced Composite Materials in Flexural Vibration*, Ph.D. Dissertation, University of Minnesota (1975).

3.6. Adams, R. D., "Damping Properties Analysis of Composites," in T. J. Reinhart et al. (eds.), *Engineered Materials Handbook, Volume 1, Composites*, pp. 206–217, ASM International, Materials Park, OH (1987).

3.7. Whitney, J. M. and Riley, M. B., "Elastic Properties of Fiber Reinforced Composite Materials," *AIAA Journal* **4**(9), 1537–1542 (September 1966).

3.8. Gibson, R. F. and Plunkett, R., "Dynamic Mechanical Behavior of Fiber Reinforced Composites: Measurement and Analysis," *Journal of Composite Materials*, **10**, 325–341 (October 1976).

3.9. Tsai, S. W. and Hahn, H. T., *Introduction to Composite Materials*, Technomic Publishing Co., Lancaster, PA (1980).

3.10. Jones, R. M., *Mechanics of Composite Materials*, Hemisphere Publishing Co., New York (1975).

3.11. Hopkins, D. A. and Chamis, C. C., "A Unique Set of Micromechanics Equations for High Temperature Metal Matrix Composites," in P. R. DiGiovanni and N. R. Adsit (eds.), *Testing Technology of Metal Matrix Composites*, ASTM STP **964**, pp. 159–176, American Society for Testing and Materials, Philadelphia, PA (1988).

3.12. Drzal, L. T., Rich, M. J., Koenig, M. F., and Lloyd, P. F., "Adhesion of Graphite Fibers to Epoxy Matrices: II. The Effect of Fiber Finish," *Journal of Adhesion*, **16**, 133–152 (1983).

3.13. Chamis, C. C., "Simplified Composite Micromechanics Equations for Hygral, Thermal and Mechanical Properties," *SAMPE Quarterly*, **15**(3), 14–23 (April 1984).

3.14. Chamis, C. C., "Simplified Composite Micromechanics Equations for Mechanical, Thermal and Moisture-Related Properties," in J. W. Weeton, D. M. Peters, and K. L. Thomas (eds.), *Engineers' Guide to Composite Materials*, ASM International, Materials Park, OH (1987).

3.15. Kriz, R. D. and Stinchcomb, W. W., "Elastic Moduli of Transversely Isotropic Graphite Fibers and Their Composites," *Experimental Mechanics*, **19**, 41–49 (1979).

3.16. Kowalski, I. M., "Determining the Transverse Modulus of Carbon Fibers," *SAMPE Journal*, **22**(4), 38–42 (1986).

3.17. Kawabata, S., "Measurements of Anisotropic Mechanical Property and Thermal Conductivity of Single Fiber for Several High Performance Fibers," in *Proceedings of 4th Japan–U.S. Conference on Composite Materials*, pp. 253–262, Washington, DC (June 1988).

3.18. Caruso, J. J. and Chamis, C. C., "Assessment of Simplified Composite Micromechanics Using Three-Dimensional Finite Element Analysis," *Journal of Composites Technology and Research*, **8**(3), 77–83 (Fall 1986).

3.19. Spencer, A., "The Transverse Moduli of Fibre Composite Material," *Composites Science and Technology*, **27**, 93–109 (1986).

3.20. Adams, D. F. and Doner, D. R., "Longitudinal Shear Loading of a Unidirectional Composite," *Journal of Composite Materials*, **1**, 4–17 (1967).

3.21. Adams, D. F. and Doner, D. R., "Transverse Normal Loading of a Unidirectional Composite," *Journal of Composite Materials*, **1**, 152–164 (1967).

3.22. Caruso, J. J., "Application of Finite Element Substructuring to Composite Micromechanics," NASA TM 83729 (1984).

3.23. Schroeder, R., "Comparison of Transverse Modulus Determined by Finite Element Analysis with Theoretical and Empirical Formulas for E-Glass/Epoxy Composites," unpublished research, University of Idaho (1986).

3.24. Whitney, J. M., "Elastic Moduli of Unidirectional Composites with Anisotropic Fibers," *Journal of Composite Materials*, **1**, 188–193 (1967).

3.25. Aboudi, J., "Micromechanical Analysis of Composites by the Method of Cells," *Applied Mechanics Reviews*, **42**(7), 193–221 (1989).

3.26. Paul, B., "Prediction of Elastic Constants of Multi-Phase Materials," *Transactions of AIME*, **218**, 36–41, (February 1960).

3.27. Hashin, Z. and Rosen, B. W., "The Elastic Moduli of Fiber Reinforced Materials," *Journal of Applied Mechanics*, **31**, 223–232 (1964). Errrata March 1965, p. 219.

3.28. Torquato, S., "Random Heterogeneous Media: Microstructure and Improved Bounds on Effective Properties," *Applied Mechanics Reviews*, **44**(2), 37–76 (February 1991).

3.29. Halpin, J. C. and Tsai, S. W., "Effects of Environmental Factors on Composite Materials," AFML-TR-67-423 (June 1969).

CHAPTER
4

STRENGTH
OF A
CONTINUOUS
FIBER-
REINFORCED
LAMINA

$S_L = Long.\ strength$
$S_T = transverse\ strength$
$S_{LT} = In\ plane$

4.1 INTRODUCTION

The analysis of composite strength is more difficult than the analysis of elastic behavior which was discussed in Chaps. 2 and 3. As shown in Chap. 1, the strength of a composite is derived from the strength of the fibers, but this strength is highly directional in nature. For example, the longitudinal strength of the continuous fiber-reinforced lamina, s_L, is much greater than the transverse strength, s_T. In addition, the compressive strengths $s_L^{(-)}$ and $s_T^{(-)}$ associated with these directions may be different from the corresponding tensile strengths $s_L^{(+)}$ and $s_T^{(+)}$. The in-plane shear strength s_{LT} associated with the principal material axes is still another independent property. These five lamina strengths form the basis of a simplified lamina strength analysis, which will, in turn, be used later in a simplified laminate strength analysis. The relationships among these five lamina strengths and the allowable lamina strengths under off-axis or multiaxial loading are discussed in this chapter, as are several micromechanical models for predicting the lamina strengths. Interlaminar strengths will be discussed in Chaps. 7 and 9.

As shown in Chaps. 2 and 3, the linear elastic stress-strain relationships for the orthotropic lamina are simplified by the use of "effective moduli." The effective moduli, which relate the volume-averaged lamina stresses to the volume-averaged lamina strains [recall Eqs. (2.7) to (2.9)], are defined by simple uniaxial or shear stress conditions associated with the lamina principal material axes. Using a similar approach, the "effective strengths" of the lamina may be defined as ultimate values of the volume-averaged stresses which cause failure of the lamina under these same simple states of stress. Figure 4.1 shows the graphical interpretation of these simple states of stress, the effective strengths $s_L^{(+)}$, $s_L^{(-)}$, $s_T^{(+)}$, $s_T^{(-)}$, and s_{LT}, and the corresponding ultimate strains $e_L^{(+)}$, $e_L^{(-)}$, $e_T^{(+)}$, $e_T^{(-)}$, and e_{LT}. If we assume linear elastic behavior up to failure, the ultimate stresses are related to the ultimate strains by

$$s_L^{(+)} = E_1 e_L^{(+)}; \qquad s_T^{(+)} = E_2 e_T^{(+)}; \qquad s_{LT} = G_{12} e_{LT}$$
$$s_L^{(-)} = E_1 e_L^{(-)}; \qquad s_T^{(-)} = E_2 e_T^{(-)} \qquad (4.1)$$

Typical experimental values of the effective lamina strengths for selected composites are given in Table 4.1 from Ref. [4.1]. Note that the transverse tensile strength, $s_T^{(+)}$, is the lowest of all the strengths. As shown later, this condition is usually responsible for the so-called "first ply failure" in a laminate. It is also interesting to note in Table 4.1 that the compressive strengths are not necessarily equal to the corresponding tensile strengths; The transverse compressive strengths are generally greater than the transverse tensile strengths and the longitudinal compressive strengths are usually less than the longitudinal tensile strengths. The intrinsic compressive strength of composites has always been difficult to determine experimentally, however, and the validity of such compression test results is often questioned. Recent test results indicate that if the proper technique is used, the compression strength may be about the same as the tensile strength. Measurement of composite properties will be discussed in more detail later in Chap. 10.

In this section the lamina effective strengths under simple states of stress have been defined. In the next section we will discuss the use of these properties in several theories for predicting lamina strength under off-axis or multiaxial loading conditions. Elementary mechanics of materials models for micromechanical prediction of several of the lamina strengths will also be described in this chapter for illustrative purposes.

LONGITUDINAL UNIAXIAL LOADING

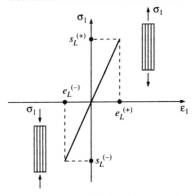

TRANSVERSE UNIAXIAL LOADING

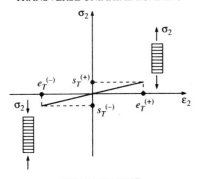

SHEAR LOADING

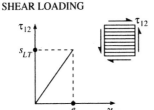

FIGURE 4.1
Stress-strain curves for uniaxial and shear loading showing lamina in-plane strengths and ultimate strains.

4.2 MULTIAXIAL STRENGTH CRITERIA

In the cases of off-axis or multiaxial loading we assume that lamina failure can be characterized by using a multiaxial strength criterion (or failure criterion) which incorporates the gross mechanical strengths described in the previous section. The objective of such a theory is to provide the designer with the capability to estimate quickly when lamina failure will occur under complex loading conditions other than simple

TABLE 4.1
Typical values of lamina strengths for several composites

Material	$s_L^{(+)}$ ksi (MPa)	$s_L^{(-)}$ ksi (MPa)	$s_T^{(+)}$ ksi (MPa)	$s_T^{(-)}$ ksi (MPa)	s_{LT} ksi (MPa)
Boron/5505 boron/epoxy $v_f = 0.5$†	230 (1586)	360 (2482)	9.1 (62.7)	35.0 (241)	12.0 (82.7)
AS/3501 graphite/epoxy $v_f = 0.6$†	210 (1448)	170 (1172)	7.0 (48.3)	36.0 (248)	9.0 (62.1)
T300/5208 graphite/epoxy $v_f = 0.6$†	210 (1448)	210 (1448)	6.5 (44.8)	36.0 (248)	9.0 (62.1)
Kevlar® 49/epoxy aramid/epoxy $v_f = 0.6$†	200 (1379)	40 (276)	4.0 (27.6)	9.4 (64.8)	8.7 (60.0)
Scotchply® 1002 E-glass/epoxy $v_f = 0.45$†	160 (1103)	90 (621)	4.0 (27.6)	20.0 (138)	12.0 (82.7)
E-glass/470-36 E-glass/vinylester $v_f = 0.30$‡	85 (584)	116 (803)	6.2 (43)	27.1 (187)	9.3 (64.0)

† From Chamis [4.1]. *Reprinted with permission of ASM International.*
‡ Courtesy of Ford Motor Company, Research Staff.
Kevlar® is a registered trademark of DuPont Company, Wilmington, Delaware, and Scotchply® is registered trademark of 3M Company, St. Paul, Minnesota.

uniaxial or shear stresses. In this semiempirical "mechanics of materials" approach we do not concern ourselves with the details of specific micromechanical failure modes such as fiber pullout, fiber breakage, fiber microbuckling, matrix cracking, and delamination. The actual failure process is complicated by the fact that these microfailure modes may occur in various combinations and sequences. Indeed, as pointed out by Hashin [4.2], our knowledge of the details of failure at the micromechanical level is so incomplete that "the failure process cannot be followed analytically." The existence and growth of cracks and other defects in the composite is also ignored with this approach. Studies of micromechanical failure modes generally require the use of more advanced approaches such as fracture mechanics and are the subjects of numerous journal publications. Additional discussion of such topics will be given in Section 4.3 and in Chap. 9.

Available multiaxial composite failure criteria have been discussed by Hashin [4.3], Wu [4.4], Sendeckyj [4.5], Chamis [4.6], Kaminski and

Lantz [4.7], and Franklin [4.8]. All the criteria are phenomenological, having evolved from attempts to develop analytical models to describe experimental observations of failure under combined stresses. As pointed out by Wu [4.4], a large experimental data base alone could form the basis for an empirical failure criterion, but the semiempirical mathematical model is preferable because it can reduce the number of required experiments and provide a more systematic approach to design. None of the available theories has been shown to accurately predict failure for all materials and loading conditions, however, and there is no universal agreement as to which theory is best.

Many of the failure criteria for anisotropic composites are based on generalizations of previously developed criteria for predicting the transition from elastic to plastic behavior in isotropic metallic materials. As such, they make use of the concept of a "failure surface" or "failure envelope" generated by plotting stress components in stress space. The coordinate axes for the stress space generally correspond to the stresses along the principal material axes. The theory predicts that those combinations of stresses whose loci fall inside the failure surface will not cause failure, whereas those combinations of stresses whose loci fall on or outside the surface will cause failure. *Thus, in the application of all the failure criteria, the first step is the transformation of calculated stresses to the principal material axes.* Since we are only dealing with two-dimensional stress states in a lamina at this point, the failure surface would be two dimensional. Failure surfaces for each of the criteria will be presented as they are discussed here.

4.2.1 Maximum Stress Criterion

The Maximum Stress Criterion for orthotropic laminae was apparently first suggested in 1920 by Jenkins [4.9] as an extension of the Maximum Normal Stress Theory (or Rankine's Theory) for isotropic materials, which is covered in elementary mechanics of materials courses [4.10]. This criterion predicts failure when any principal material axis stress component exceeds the corresponding strength. Thus, in order to avoid failure according to this criterion, the following set of inequalities must be satisfied:

$$-s_L^{(-)} < \sigma_1 < s_L^{(+)}$$
$$-s_T^{(-)} < \sigma_2 < s_T^{(+)} \qquad (4.2)$$
$$|\tau_{12}| < s_{LT}$$

where the numerical values of $s_L^{(-)}$ and $s_T^{(-)}$ are assumed to be positive. It is assumed that shear failure along the principal material axes is independent of the sign of the shear stress τ_{12}. Thus, only the magnitude of τ_{12} is important, as shown in the last of Eqs. (4.2). *As shown later,*

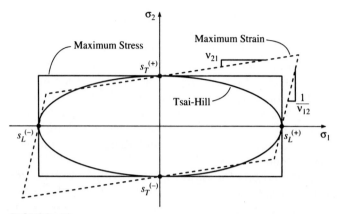

FIGURE 4.2
Maximum Stress, Maximum Strain, and Tsai-Hill failure surfaces in σ_1, σ_2 space.

however, the shear strength for off-axis loading may depend on the sign of the shear stress.

The failure surface for the Maximum Stress Criterion is σ_1-σ_2 space is a rectangle, as shown in Fig. 4.2. Note that this failure surface is independent of the shear stress τ_{12}, and that the criterion does not account for possible interaction between the stress components. That is, the predicted limiting value of a particular stress component is the same whether or not other stress components are present. Figure 4.3 shows a comparison of theoretical failure surfaces with experimental biaxial

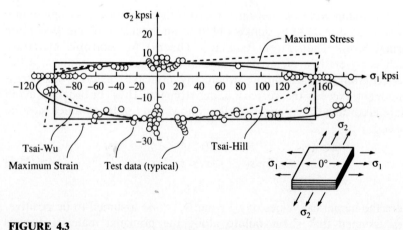

FIGURE 4.3
Comparison of predicted failure surfaces with experimental failure data for graphite/epoxy.
(*From Burk [4.11]. Copyright ©1983 AIAA. Reprinted with permission.*)

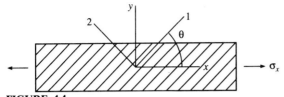

FIGURE 4.4
Off-axis uniaxial test of a unidirectional lamina specimen.

failure data for a unidirectional graphite/epoxy composite [4.11]. Since the strengths along the principal material directions provide the input to the criterion, we would expect the agreement to be good when the applied stress is uniaxial along those directions. Due to lack of stress interaction in the Maximum Stress Criterion, however, the agreement is not so good in biaxial stress situations. The scatter in the experimental data is unfortunately typical for composite strength tests.

Experimental biaxial failure data for comparison with predicted failure surfaces can be obtained by applying biaxial loading directly to the test specimens. Biaxial stress fields can also be generated indirectly by using off-axis uniaxial loading tests [4.12] or off-axis shear loading tests. According to Eqs. (2.31), the applied normal stress, σ_x, in the off-axis uniaxial loading test shown in Fig. 4.4. produces the following biaxial stress state along the principal material axes

$$\sigma_1 = \sigma_x \cos^2 \theta$$

$$\sigma_2 = \sigma_x \sin^2 \theta$$

$$\tau_{12} = -\sigma_x \sin \theta \cos \theta \tag{4.3}$$

where the applied normal stress, σ_x, may be positive or negative. The importance of the sign of the applied stress in the interpretation of the test results here is obvious. These stress components may then be substituted into equations similar to Eqs. (4.2) in order to generate failure surfaces. By plotting the predicted and measured values of σ_x at failure vs. lamina orientation, θ, the various failure criteria can be evaluated [4.13].

For the off-axis shear test described in Fig. 4.5, the applied shear stress, τ_{xy}, generates the following biaxial stress state along the principal material axes according to Eqs. (2.31)

$$\sigma_1 = 2\tau_{xy} \cos \theta \sin \theta$$

$$\sigma_2 = -2\tau_{xy} \cos \theta \sin \theta \tag{4.4}$$

$$\tau_{12} = \tau_{xy}(\cos^2 \theta - \sin^2 \theta)$$

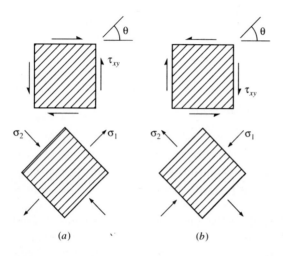

FIGURE 4.5
Off-axis shear test of a unidirectional lamina specimen. (a) Positive τ_{xy}; (b) negative τ_{xy}.

The importance of the sign of the applied shear stress in the interpretation of test results may not be so obvious here, and further discussion is warranted. For example, if the angle $\theta = 45°$, Eqs. (4.4) reduce to $\sigma_1 = \tau_{xy}$, $\sigma_2 = -\tau_{xy}$, $\tau_{12} = 0$. Thus, a positive applied shear stress, τ_{xy}, would produce longitudinal tension and transverse compression along the principal material axes, as shown in Fig. 4.5(a). On the other hand, a negative applied shear stress would produce longitudinal compression and transverse tension, as shown in Fig. 4.5(b). Given the fact that the transverse tensile strength is so much lower than the other strengths (see Table 4.1), the importance of the sign of the applied off-axis shear stress should now be obvious. It is easy to visualize a situation where a negative shear stress of a certain magnitude could cause a transverse tensile failure, whereas a positive shear stress of the same magnitude would not cause failure. A similar development for pure shear along the principal material axes shows that the sign of the shear stress makes no difference in that case. The importance of the sign of the shear stress extends beyond the interpretation of tests results as described here; it has implications for all phases of stress analysis in composite materials.

4.2.2 Maximum Strain Criterion *see ex p 113*

In 1967 Waddoups [4.14] proposed the Maximum Strain Criterion for orthotropic laminae as an extension of the Maximum Normal Strain Theory (or Saint Venant's Theory) for isotropic materials, which is also discussed in elementary mechanics of materials courses [4.10]. This criterion predicts failure when any principal material axis strain component exceeds the corresponding ultimate strain. In order to avoid failure according to this criterion, the following set of inequalities must be satisfied:

$$
\begin{aligned}
-e_L^{(-)} &< \epsilon_1 < e_L^{(+)} \\
-e_T^{(-)} &< \epsilon_2 < e_T^{(+)} \\
|\gamma_{12}| &< e_{LT}
\end{aligned}
\tag{4.5}
$$

Look @ page 100

where the numerical values of $e_L^{(-)}$ and $e_T^{(-)}$ are assumed to be positive and the ultimate strains are all *engineering strains* as defined by Eqs. (4.1). As with the Maximum Stress Criterion, it is assumed that shear failure along the principal material axes is independent of the sign of the shear strain γ_{12}.

Due to the similarity of Eqs. (4.5) and Eqs. (4.2), the failure surface for the Maximum Strain Criterion in ϵ_1-ϵ_2 space is a rectangle similar to that of the Maximum Stress Criterion in σ_1-σ_2 space. In σ_1-σ_2 space, however, the Maximum Strain Criterion failure surface is a skewed parallelogram, as shown in Figs. 4.2 and 4.3. The shape of the parallelogram can be deduced by combining the lamina stress-strain relationships in Eqs. (2.24) with the relationships given in Eqs. (4.1). For example, the limiting strain associated with the positive 1 direction is

$$
\epsilon_1 = \frac{s_L^{(+)}}{E_1} = \frac{\sigma_1}{E_1} - \frac{\nu_{12}\sigma_2}{E_1}
\tag{4.6}
$$

or

$$
\sigma_2 = \frac{\sigma_1 - s_L^{(+)}}{\nu_{12}}
\tag{4.7}
$$

which is the equation of a straight line having intercept $(s_L^{(+)}, 0)$ and slope $1/\nu_{12}$ (Fig. 4.2). A similar development using the limiting strain along the positive 2 direction yields the equation

$$
\sigma_2 = \nu_{21}\sigma_1 + s_T^{(+)}
\tag{4.8}
$$

which is the equation for a straight line having intercept $(0, s_T^{(+)})$ and slope ν_{21}. These lines form the right and top sides, respectively, of the parallelogram shown in Fig. 4.2, and similar consideration of the limiting strains in the negative 1 and 2 directions yields equations for the remaining two sides. It should be noted, however, that, depending on the magnitudes of the lamina strengths and stiffnesses, the intercepts of the Maximum Strain Criterion parallelogram may not be the same as those of the Maximum Stress Criterion rectangle in stress space. For some materials the lines defining the top and bottom of the Maximum Strain Criterion parallelogram intercept the horizontal axis at stresses less than the measured tensile and compressive longitudinal strengths, which contradicts experimental evidence [4.4, 4.7]. According to Wu [4.4], such contradictions develop as a result of an ambiguous conversion from strain space to stress space unless certain mathematical constraints on the properties are satisfied. Only for isotropic materials are the intercepts always the same for the Maximum Stress and Maximum Strain Criteria. As with the Maximum Stress Criterion, the Maximum Strain Criterion does not account for possible interaction between stress components, and the predicted failure surface does not show good agreement with experimental biaxial failure data for graphite/epoxy in Fig. 4.3. Off-axis uniaxial test data have led to similar conclusions [4.13], but both criteria are still used for orthotropic materials because the resulting equations are relatively simple.

4.2.3 Quadratic Interaction Criteria

The so-called Quadratic Interaction Criteria also evolved from early failure theories for isotropic materials, but they differ from the Maximum Stress and Maximum Strain Criteria in that they include terms to account for interaction between the stress components, and the quadratic forms of the equations for plane stress lead to elliptical failure surfaces. As shown in any mechanics of materials book, the Maximum Distortional Energy Criterion, or von Mises Criterion (circa early 1900s), is the most widely used quadratic interaction criteria for predicting the onset of yielding in isotropic metals [4.10]. In 1948 Hill [4.15] suggested that the von Mises Criterion could be modified to include the effects of induced anisotropic behavior in initially isotropic metals during large plastic deformations. For a general three-dimensional state of stress along the principal axes of anisotropy (the 123 axes) in such a material the failure surface (or yield surface) for the Hill Criterion in σ_1, σ_2, σ_3 space is described by the equation

$$A(\sigma_2 - \sigma_3)^2 + B(\sigma_3 - \sigma_1)^2 + C(\sigma_1 - \sigma_2)^2 + 2D\tau_{23}^2 + 2E\tau_{31}^2 + 2F\tau_{12}^2 = 1$$

$$(4.9)$$

where A, B, C, D, E, and F are determined from yield strengths in uniaxial or shear loading. In order to avoid failure, the left-hand side of Eq (4.9) must be <1, and failure is predicted if the left-hand side is ≥ 1. For a uniaxial test along the 1 direction with $\sigma_1 = Y_1$ and all other stresses equal to zero Eq. (4.9) reduces to

$$B + C = \frac{1}{Y_1^2} \qquad (4.10)$$

where Y_1 is the yield strength along 1 direction. Similarly, uniaxial tests along the 2 and 3 directions give the equations

$$A + C = \frac{1}{Y_2^2}; \qquad A + B = \frac{1}{Y_3^2} \qquad (4.11)$$

where Y_2 and Y_3 are the uniaxial yield strengths along the 2 and 3 directions, respectively. The yield strengths in tension and compression are assumed to be the same. Solving Eqs. (4.10) and (4.11) simultaneously for A, B, and C, we find that

$$2A = \frac{1}{Y_2^2} + \frac{1}{Y_3^2} - \frac{1}{Y_1^2}$$

$$2B = \frac{1}{Y_3^2} + \frac{1}{Y_1^2} - \frac{1}{Y_2^2} \qquad (4.12)$$

$$2C = \frac{1}{Y_1^2} + \frac{1}{Y_2^2} - \frac{1}{Y_3^2}$$

Similarly, for pure shear tests along the 23, 31, and 12 planes, Eq. (4.9) gives

$$2D = \frac{1}{Y_{23}^2}; \qquad 2E = \frac{1}{Y_{31}^2}; \qquad 2F = \frac{1}{Y_{12}^2} \qquad (4.13)$$

where Y_{12}, Y_{23}, and Y_{31} are the yield strengths in shear associated with the 12, 23, and 31 planes, respectively.

The extension of the Hill Criterion to prediction of failure in an orthotropic, transversely isotropic lamina was suggested by Azzi and Tsai [4.16] and Tsai [4.17], and the resulting equation is often referred to as the Tsai-Hill Criterion. If the 123 directions are assumed to be the principal material axes of the transversely isotropic lamina with the 1 direction being along the reinforcement direction, if plane stress is assumed ($\sigma_3 = \tau_{31} = \tau_{23} = 0$), and if Hill's anisotropic yield strengths are replaced by the corresponding effective lamina strengths, then $Y_1 = s_L$, $Y_2 = Y_3 = s_T$, $Y_{12} = s_{LT}$, and Eqs. (4.9), (4.12), and (4.13) reduce to the equation for the Tsai-Hill failure surface

Tsai-Hill

$$\frac{\sigma_1^2}{s_L^2} - \frac{\sigma_1 \sigma_2}{s_L^2} + \frac{\sigma_2^2}{s_T^2} + \frac{\tau_{12}^2}{s_{LT}^2} = 1 \qquad (4.14)$$

As with the Hill equation, failure is avoided if the left-hand side of Eq. (4.14) is <1, and failure is predicted if the left-hand side is ≥ 1. The failure surface generated by this equation is an ellipse, as shown in Fig. 4.2. The ellipse shown in Fig. 4.2 is symmetric about the origin because of the assumption of equal strengths in tension and compression. The Tsai-Hill equation can be used when tensile and compressive strengths are different by simply using the appropriate value of s_L and s_T for each quadrant of stress space. For example, if σ_1 is positive and σ_2 is negative, the values of $s_L^{(+)}$ and $s_T^{(-)}$ would be used in Eq. (4.14). The resulting failure surface is no longer symmetric about the origin, as shown for the case of graphite/epoxy in Fig. 4.3. Although such a procedure is inconsistent with the assumptions used in formulating the original von Mises and Hill Criteria, it has been successfully used for some composites [4.10, 4.17]. As shown in Fig. 4.3, the procedure seems to work reasonably well for the graphite/epoxy material except for the fourth quadrant of stress space. One way to account for different strengths in tension and compression is to include terms which are linear in the normal stresses σ_1, σ_2, and σ_3, as suggested by Hoffman [4.18].

In addition to the previously mentioned limitations of the quadratic interaction criteria based on the von Mises model, there is another problem. Since the von Mises and Hill Criteria are phenomenological theories for prediction of yielding in ductile metals, the equations are based on principal stress differences and the corresponding shear stresses and strains which drive slip and dislocation movement in metallic crystals. Experimental evidence suggests that a hydrostatic state of stress does not cause the slip and dislocation movement that are associated with yielding, and the Hill Criterion predicts that failure will never occur under a hydrostatic state of stress $\sigma_1 = \sigma_2 = \sigma_3$, $\tau_{12} = \tau_{23} = \tau_{31} = 0$. Due to shear coupling, however, a hydrostatic state of stress in an anisotropic material can produce shear strains and failure. Hoffman's equation [4.18], by virtue of its linear terms, could predict failure for the hydrostatic state of stress. However, all of these theories turn out to be special cases of a more general quadratic interaction criterion, which will be discussed next.

In 1971 Tsai and Wu [4.19] proposed an improved and simplified version of a tensor polynomial failure theory for anisotropic materials that had been suggested earlier by Gol'denblat and Kopnov [4.20]. In the Tsai-Wu general quadratic interaction criteria the failure surface in stress space is described by the tensor polynomial

$$F_i\sigma_i + F_{ij}\sigma_i\sigma_j = 1 \tag{4.15}$$

where the contracted notation $i, j = 1, 2, \ldots, 6$ is used and F_i and F_{ij} are experimentally determined strength tensors of the second and fourth rank, respectively. In order to avoid failure, the left-hand side of Eq. (4.15) must be <1, and failure is predicted when the left-hand side is ≥ 1. For the case of plane stress with $\sigma_3 = \sigma_{33} = 0$, $\sigma_4 = \tau_{23} = 0$, $\sigma_5 = \tau_{31} = 0$ Eq. (4.15) becomes

$$F_{11}\sigma_1^2 + F_{22}\sigma_2^2 + F_{66}\sigma_6^2 + F_1\sigma_1 + F_2\sigma_2 + 2F_{12}\sigma_1\sigma_2 = 1 \tag{4.16}$$

where the linear terms in the shear stress $\sigma_6 = \tau_{12}$ have been dropped because the shear strength along the principal material axes is not affected by the sign of the shear stress. Thus, only a quadratic term in the shear stress σ_6 remains. However, the linear terms in the normal stresses $\sigma_1 = \sigma_{11}$ and $\sigma_2 = \sigma_{22}$ are retained because they take into account the different strengths in tension and compression. In addition, the term $2F_{12}\sigma_1\sigma_2$ takes into account interaction between the normal stresses. With the exception of F_{12}, all the strength tensors in Eq. (4.16) can be expressed in terms of the uniaxial and shear strengths using the same approach that was used with the Hill Criterion. For example, for the tension and compression tests with uniaxial stresses $\sigma_1 = s_L^{(+)}$ and $\sigma_1 = -s_L^{(-)}$, respectively, simultaneous solution of the two equations resulting from Eq. (4.16) yields

$$F_{11} = \frac{1}{s_L^{(+)}s_L^{(-)}} \quad \text{and} \quad F_1 = \frac{1}{s_L^{(+)}} - \frac{1}{s_L^{(-)}} \tag{4.17}$$

where the numerical value of $s_L^{(-)}$ is assumed to be positive as in Table 4.1. From similar uniaxial and shear tests, it can be shown that

$$F_{22} = \frac{1}{s_T^{(+)}s_T^{(-)}}; \quad F_2 = \frac{1}{s_T^{(+)}} - \frac{1}{s_T^{(-)}}; \quad F_{66} = \frac{1}{s_{LT}^2} \tag{4.18}$$

where the numerical value of $s_T^{(-)}$ is assumed to be positive.

In order to find the interaction parameter, F_{12}, it is necessary to use a biaxial test involving both σ_1 and σ_2. For example, an expression for F_{12} can be obtained by substituting the biaxial stress conditions $\sigma_1 = \sigma_2 = P$, $\sigma_6 = 0$ into Eq. (4.16), where P is the biaxial failure stress [4.19]. Thus, in order to find F_{12} for this condition, we need to know P in addition to the previously defined uniaxial and shear failure stresses. There is no a priori reason that σ_1 must equal σ_2, however. Indeed, as pointed out by Hashin [4.2], F_{12} can have four different values because there are four different failure pairs σ_1, σ_2. Wu [4.4, 4.21] has suggested that in order to determine F_{12} accurately, the biaxial ratio $B = \sigma_1/\sigma_2$ must be optimized to account for the sensitivity of F_{12} to experimental scatter in the applied stresses. The optimization procedure is complicated, however, and the reader is referred to the articles by Wu [4.4, 4.21] for details. The Tsai-Wu failure surface for graphite/epoxy shown in Fig. 4.3 was based on such an optimization procedure for F_{12}. In Fig. 4.3 the agreement with experimental data seems to be much better for the Tsai-Wu failure surface than for the others, particularly in the fourth quadrant.

More recently, Tsai and Hahn [4.22] have proposed the equation

$$F_{12} = -\frac{(F_{11}F_{22})^{1/2}}{2} \tag{4.19}$$

which causes Eq. (4.16) to take on the form of a generalized von Mises Criteron for the yielding of isotropic materials. It is also interesting to note that Eq. (4.16) reduces to Eq. (4.14), the Tsai-Hill Criterion, when the tensile and compressive strengths are assumed to be equal and

$$F_{12} = -\frac{1}{2s_L^2} \tag{4.20}$$

The development of improved multiaxial strength criteria for composites continues to be the subject of numerous publications. For example, Hashin [4.2, 4.3] has suggested that for a given composite each failure mode and its contributing stresses should be identified, and that each of these failure modes should be modeled separately by a quadratic criterion. Tennyson et al. [4.23] have extended the tensor polynomial criterion to include cubic terms. Obviously, the evaluation of the strength parameters in such an equation is a formidable task. It was shown, however, that in the particular case of failure in laminated tubes under internal pressure loading the cubic criterion is more accurate than the quadratic criterion. Although considerable progress has been made, there is still a need for systematic experimental verification of the various theories for a variety of stress conditions. Finally, the theories discussed in this section are based on the macromechanical behavior of the composite without regard for the micromechanical behavior of fiber and matrix materials. In the next section several micromechanical models for predicting composite strength will be presented.

Example 4.1. The filament wound pressure vessel described in Example 2.1 is fabricated from E-glass/epoxy having the lamina strengths listed in Table 4.1. Determine the internal pressure p which would cause failure of the vessel according to (*a*) the Maximum Stress Criterion and (*b*) the Tsai-Hill Criterion.

Solution. The first step in the application of both theories is to determine the stresses along the principal material axes. From the results of Example 2.1, $\sigma_1 = 20.5p$, $\sigma_2 = 17.0p$, and $\tau_{12} = 6.0p$ (all in MPa). Note that both normal stresses are positive, so that the tensile strengths should be used in the failure theories.

(*a*) For the Maximum Stress Criterion the three possible values of p at failure are found as follows:

$$\sigma_1 = 20.5p = s_L^{(+)} = 1103 \text{ MPa}; \qquad \text{therefore,} \qquad p = 53.8 \text{ MPa}$$
$$\sigma_2 = 17.0p = s_T^{(+)} = 27.6 \text{ MPa}; \qquad \text{therefore,} \qquad p = 1.62 \text{ MPa}$$
$$\tau_{12} = 6.0p = s_{LT} = 82.7 \text{ MPa}; \qquad \text{therefore,} \qquad p = 13.78 \text{ MPa}$$

Thus, the transverse tensile failure governs, and failure occurs first at $p = 1.62$ MPa.

(*b*) For the Tsai-Hill Criterion Eq. (4.14) yields

$$\left(\frac{20.5p}{1103}\right)^2 - \frac{(10.5p)(17.0p)}{(1103)^2} + \left(\frac{17.0p}{27.6}\right)^2 + \left(\frac{6.0p}{82.7}\right)^2 = 1$$

Solving for p, we find that $p = 1.61$ MPa. Thus, for this case the two criteria yield approximately the same result. This is not always true, however.

Example 4.2. Using the Maximum Strain Criterion, determine the uniaxial failure stress, σ_x, for off-axis loading of the unidirectional lamina in Fig. 4.4 if the material is AS/3501 graphite/epoxy and the angle $\theta = 30°$.

Solution. First, the strains along the principal material axes must be found in terms of the applied stress, σ_x. Upon substituting the stress transformation Eqs. (4.3) in the lamina stress-strain Eqs. (2.24) we find that

$$\epsilon_1 = \frac{1}{E_1}(\cos^2\theta - \nu_{12}\sin^2\theta)\sigma_x$$

$$\epsilon_2 = \frac{1}{E_2}(\sin^2\theta - \nu_{21}\cos^2\theta)\sigma_x$$

and

$$\gamma_{12} = -\frac{1}{G_{12}}(\sin\theta\cos\theta)\sigma_x$$

Assuming linear elastic behavior up to failure and using the stress-strain relations in Eqs. (4.1), the Maximum Strain Criterion [Eq. (4.5)] becomes

$$\sigma_x < \frac{s_L^{(+)}}{\cos^2\theta - \nu_{12}\sin^2\theta}$$

$$\sigma_x < \frac{s_T^{(+)}}{\sin^2\theta - \nu_{21}\cos^2\theta}$$

$$\sigma_x < \frac{s_{LT}}{\sin\theta\cos\theta}$$

where only the tensile strengths have been used because σ_x is positive. Using the AS/3501 data in Tables 2.2 and 4.1, we find that in order to avoid longitudinal tensile failure,

$$\sigma_x < \frac{1448}{(0.866)^2 - 0.3(0.5)^2}\text{MPa} \qquad \text{or} \qquad \sigma_x < 2145 \text{ MPa}$$

In order to avoid transverse tensile failure,

$$\sigma_x < \frac{48.3}{(0.5)^2 - 0.0195(0.866)^2}\text{MPa} \qquad \text{or} \qquad \sigma_x < 205 \text{ MPa}$$

and in order to avoid shear failure,

$$\sigma_x < \frac{62.1}{0.866(0.5)}\text{MPa} \qquad \text{or} \qquad \sigma_x < 143 \text{ MPa}$$

Thus, according to the Maximum Strain Criterion, the mode of failure is shear, and the applied stress at failure is $\sigma_x = 143$ MPa. The reader is encouraged to check that for compressive loading or other loading angles both the mode of failure and the failure stress may be different. The off-axis tensile test has been used to check the validity of the various failure criteria [4.12, 4.13].

4.3 MICROMECHANICS MODELS FOR LAMINA STRENGTH

In this section the use of elementary mechanics of materials approaches to micromechanical modeling of lamina strength will be described. We should not expect such simple models for strength to be as accurate as those for stiffness because strength is affected more than stiffness by material and geometric nonhomogeneity and the resulting local perturbations in the stress and strain distributions. As shown in Chap. 3, the effects of such local stress and strain perturbations on stiffness are reduced due to the smoothing effect of integration in the effective modulus theories. On the other hand, material failure is often initiated at the sites of such stress and strain concentrations, so the effect on strength is much greater. For example, as shown in Fig. 4.6 from Ref. [4.24], the variability of strength in reinforcing fibers alone may be quite significant, and statistical methods must be used for accurate analysis. In addition, differences between tensile and compressive modes of failure make it necessary to develop different micromechanics models for tensile strengths and compressive strengths.

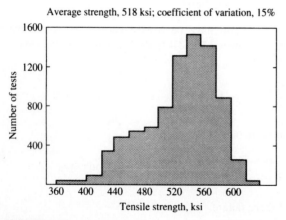

FIGURE 4.6
Statistical distribution of tensile strength for boron filaments (*From Weeton, et al.* [4.24]. *Reprinted by permission of ASM International.*)

4.3.1 Longitudinal Strength

Simple micromechanics models for composite longitudinal tensile strength can be developed from the rule of mixtures for longitudinal stress, Eq. (3.18), and the representative stress-strain curves for fiber, matrix, and composite materials shown in Figs. 4.7(a) and 4.7(b). In Fig. 4.7(a) the matrix failure strain, $e_{m1}^{(+)}$, is assumed to be greater than the fiber failure strain, $e_{f1}^{(+)}$, which is typical for many polymer matrix composites. A model for this case by Kelly and Davies [4.25] will be summarized here. Figure 4.7(b) shows the case where the fiber failure strain is greater than the matrix failure strain, which is typical for

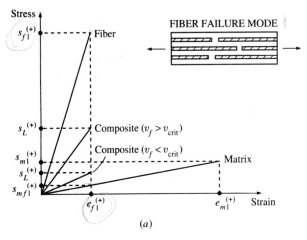

(a)

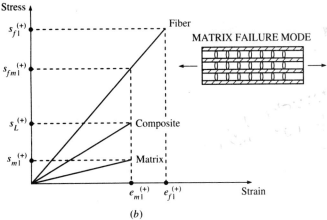

(b)

FIGURE 4.7
Representative stress-strain curves for typical fiber, matrix, and composite materials. (a) Matrix failure strain greater than fiber failure strain; (b) fiber failure strain greater than matrix failure strain.

ceramic matrix composites. A model based on the one proposed by Hull [4.26] will be described for this case. For both cases shown in Fig. 4.7 the analyses will be developed on the assumptions of (1) equal strengths in all fibers, (2) linear elastic behavior up to failure, and (3) equal longitudinal strains in composite, fiber, and matrix [recall Eqs. (3.21)].

For the case described in Fig. 4.7(a) the composite must fail at a strain level corresponding to the fiber tensile failure strain, $e_{f1}^{(+)} = s_{f1}^{(+)}/E_{f1}$. Theoretically, if the matrix could support the full applied load after fiber failure, the strain could be increased to the matrix failure strain. For all practical purposes, however, fiber failure means composite failure. Thus, when the fiber longitudinal stress reaches the fiber tensile strength, $s_{f1}^{(+)}$, the matrix longitudinal stress reaches a value $s_{mf1} = E_m e_{f1}^{(+)}$, the composite longitudinal stress reaches the composite tensile strength, $s_L^{(+)}$, and Eq. (3.18) becomes

$$s_L^{(+)} = s_{f1}^{(+)} v_f + s_{mf1} v_m = s_{f1}^{(+)} v_f + s_{mf1}(1 - v_f) \qquad (4.21)$$

However, Eq. (4.21) only has meaning if the fiber volume fraction is sufficiently large. As shown in Figs. 4.7(a) and 4.8(a), if the fiber volume

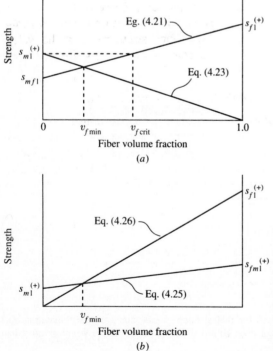

(a)

(b)

FIGURE 4.8
Variation of composite longitudinal tensile strength with fiber volume fraction for composite having: (a) Matrix failure strain greater than fiber failure strain; (b) fiber failure strain greater than matrix failure strain.

fraction $v_f < v_{f\text{crit}}$, the composite strength from Eq. (4.21) is less than the matrix strength, where

$$v_{f\text{crit}} = \frac{s_{m1}^{(+)} - s_{mf1}}{s_{f1}^{(+)} - s_{mf1}} \tag{4.22}$$

Once the fibers fail in composites having $v_f < v_{f\text{crit}}$, however, the remaining cross-sectional area of matrix that can support the load is such that

$$s_L^{(+)} = s_{m1}^{(+)} v_m = s_{m1}^{(+)}(1 - v_f) \tag{4.23}$$

As shown in Fig. 4.8(a), Eqs. (4.21) and 4.23) intersect at

$$v_{f\text{min}} = \frac{s_{m1}^{(+)} - s_{mf1}}{s_{f1}^{(+)} - s_{mf1} + s_{m1}^{(+)}} \tag{4.24}$$

For practical composites, however, $v_{f\text{crit}}$ is generally less than 5 percent. Since $v_{f\text{min}} < v_{f\text{crit}}$, both of these values must be much smaller than the actual fiber volume fraction of the composite, and the composite longitudinal strength for the case of Fig. 4.7(a) would therefore be given by Eq. (4.21).

For the case described in Fig. 4.7(b) composite failure may be defined in two ways, depending on whether we choose to use fiber failure or matrix failure as the criterion. If matrix failure is the criterion, composite failure will occur at the strain level corresponding to the matrix tensile failure strain, $e_{m1}^{(+)}$. Thus, when the matrix stress reaches the matrix tensile strength, $s_{m1}^{(+)}$, the fiber stress reaches the value $s_{fm1} = E_{f1}e_{m1}^{(+)}$, the composite stress reaches the composite strength, $s_L^{(+)}$, and Eq. (3.18) becomes

Longitudinal
$$s_L^{(+)} = s_{fm1}v_f + s_{m1}^{(+)}(1 - v_f) \tag{4.25}$$

As with Eq. (4.21), this equation only has physical meaning for a certain range of fiber volume fractions. As shown in Fig. 4.7(b), if the fibers could still withstand additional loading after matrix failure, the fiber strain may reach the fiber failure strain, $e_{f1}^{(+)}$. Due to the matrix failure, however, the remaining load-bearing area of fibers is such that the composite strength is now given by

$$s_L^{(+)} = s_{f1}^{(+)} v_f \tag{4.26}$$

As shown in Fig. 4.8(b), Eqs. (4.25) and (4.26) intersect at

$$v_{f\text{min}} = \frac{s_{m1}^{(+)}}{s_{f1}^{(+)} - s_{fm1} + s_{m1}^{(+)}} \tag{4.27}$$

Thus, for $v_f < v_{f\text{min}}$, the composite strength would be given by Eq. (4.25), and for $v_f > v_{f\text{min}}$ the composite strength would be given by Eq. (4.26). For practical composites, however, $v_{f\text{min}}$ is much smaller than the actual fiber volume fraction, so the composite longitudinal strength for the case of Fig. 4.7(*b*) would be given by Eq. (4.26). As shown in Figs. 4.7(*a*) and 4.8(*a*), the composite strength for this case is greater than the matrix strength for all fiber volume fractions.

Of the three assumptions made at the beginning of this section, the weakest one is that all fibers in the composite have the same strength, $s_{f1}^{(+)}$. As shown in Fig. 4.6, fiber strength is not uniform, and some fibers fail at stresses well below the ultimate composite strength. In addition, fiber strength decreases with increasing fiber length due to the increased probability of imperfections in the fiber. Various statistical models have been proposed for the sequence of events beginning with the first fiber failure and culminating with overall composite failure [4.27, 4.28]. Such analyses are beyond the scope of this book, however, and the reader is referred to the article by Rosen [4.29] for a review of the various models.

While the assumption regarding linear elastic behavior up to failure is not valid for many ductile matrix materials, the errors generated by this assumption are believed to be small. For example, the contribution of the matrix strength to the composite strength in Eq. (4.21) is small and the matrix strength does not appear at all in Eq. (4.26). If the matrix has yielded before or during fiber failure, the term s_{mf1} in Eqs. (4.21), (4.22), and (4.24) can be replaced by the matrix yield strength, s_y. Excellent agreement has been reported between Eq. (4.21) modified in this way and experimental results for a tungsten fiber/copper matrix system over a wide range of fiber volume fractions [4.30].

It has long been assumed that the models for longitudinal strength cannot be used for compressive strength because the modes of failure are different. This assumption has been supported by observed differences in measured tensile and compressive strengths, as shown in Table 4.1. Accurate measurement of the intrinsic compressive strength has proved to be very difficult, however, and test results to date typically depend on specimen geometry and/or test method. Whitney [4.31] has pointed out that the failure mode is the key issue because different compression test methods may produce different failure modes. Whether the failure mode in the test is the same as the failure mode in the composite structure being designed is another question. There appears to be three basic longitudinal compression failure modes, which are shown schematically in Fig. 4.9:

1. Microbuckling of fibers in either shear or extensional mode
2. Transverse tensile rupture due to Poisson strains
3. Shear failure of fibers without buckling

Variations on these basic mechanisms have also been observed. For example, the shear mode of fiber microbuckling (Fig. 4.9) often leads to

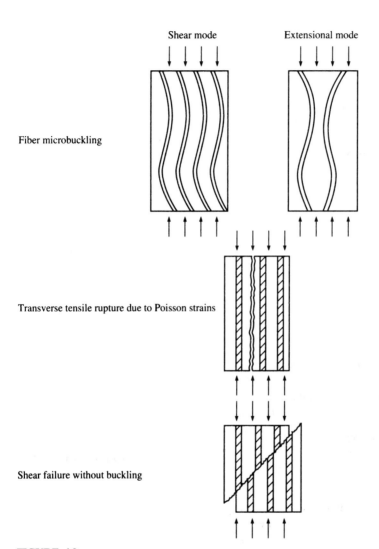

FIGURE 4.9
Three possible failure modes for longitudinal compressive loading of a unidirectional composite.

"shear crippling" due to kink band formation [4.26, 4.32]. Although these problems make it difficult to assess the accuracy of various micromechanics models for compressive strength, several representative models will be summarized.

Mechanics of materials models for local buckling, or microbuckling, of fibers in the matrix have been developed by Rosen [4.33] and Schuerch [4.34]. It is assumed that fiber buckling occurs in either the extensional mode, where fibers buckle in an out-of-phase pattern and the matrix is extended or compressed, or the shear mode, where fibers buckle in an in-phase pattern and the matrix is sheared (Fig. 4.9). Two-dimensional models were used, with the fibers represented as plates separated by matrix blocks. By the energy method, the work done by external forces during deformation, W, is equal to the corresponding change in the strain energy of the fibers, ΔU_f, plus the change in the strain energy of the matrix, ΔU_m:

$$\Delta U_f + \Delta U_m = W \qquad (4.28)$$

Assuming a sinusoidally buckled shape, the buckling stress (or compressive strength) for the extensional, or out-of-phase mode, was found to be

$$s_L^{(-)} = 2v_f[v_f E_m E_f/3(1 - v_f)]^{1/2} \qquad (4.29)$$

whereas the buckling stress for the shear, or in-phase mode, was found to be

$$s_L^{(-)} = G_m/(1 - v_f) \qquad (4.30)$$

The extensional mode turns out to be important only for very low fiber volume fractions, where it predicts the lowest buckling stress, and is not important for practical composites (Fig. 4.10). While the shear mode

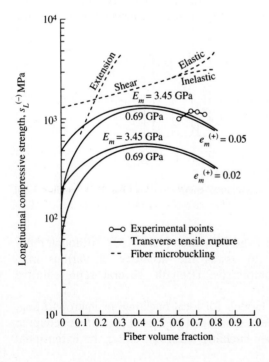

FIGURE 4.10
Variation of predicted compressive strength of glass/epoxy with fiber volume fraction for fiber microbuckling and transverse tensile rupture modes of failure. (*From Agarwal and Broutman* [4.36]. *Copyright ©1990 John Wiley & Sons, Inc. Reprinted by permission of John Wiley & Sons, Inc.*)

gives the lowest buckling stress over the range of practical fiber volume fractions, it overpredicts considerably by comparison with test results. One way to reduce the buckling stress predicted by Eq. (4.30) is to take into account possible inelastic deformation of the matrix material by using a reduced value of the matrix shear modulus, G_m, but predictions still tend to be too high. The nonlinear model of Hahn and Williams [4.32] includes the effects of initial fiber curvature and material non-linearity, and reasonable predictions of compressive strength for graphite/epoxy were reported. Greszczuk [4.35] has shown that if the matrix shear modulus is high enough, the mode of failure shifts from microbuckling to compressive failure of the reinforcement. Since advanced composites tend to have high modulus matrix materials, this may explain why attempts to apply microbuckling failure theories to these materials have not succeeded. For such cases Greszczuk recommended that Eq. (4.21) be used with $s_{f1}^{(-)}$ in place of $s_{f1}^{(+)}$. It should be added, however, that this conclusion was based on tests of laminates consisting of aluminum strips bonded together with urethane or epoxy resins. The difficulty in measurement of fiber compressive strength, $s_{f1}^{(-)}$, may preclude the use of this model for fiber composites.

A model for transverse tensile rupture due to Poisson strains (Fig. 4.9) has been presented by Agarwal and Broutman [4.36]. The model is based on the application of the Maximum Strain Criterion to the tensile transverse Poisson strain under longitudinal compressive loading. Under the applied longitudinal stress, σ_1, the resulting transverse Poisson strain is

$$\epsilon_2 = \nu_{12}\epsilon_1 = -\nu_{12}\frac{\sigma_1}{E_1} \tag{4.31}$$

Thus, when the Poisson strain $\epsilon_2 = e_T^{(+)}$, the corresponding longitudinal stress is $\sigma_1 = -s_L^{(-)}$, and the compressive strength is

$$s_L^{(-)} = \frac{E_1 e_T^{(+)}}{\nu_{12}} \tag{4.32}$$

As shown in Fig. 4.10, Eq. (4.32) shows better agreement with measured compressive strengths of glass/epoxy than the microbuckling theories do [4.36].

Failure of the fibers in direct shear due to the maximum shear stress $\tau_{max} = s_L^{(-)}/2$ at an angle of 45° to the loading axis is a third possible failure mode under longitudinal compression (Fig. 4.9). Hull [4.26] reports good agreement with experimental data for graphite/epoxy when the maximum shear stress is given by a rule of mixtures, so that the compressive strength is

$$s_L^{(-)} = 2(s_{f12}v_f + s_{m12}v_m) \tag{4.33}$$

where $s_{f12}$and s_{m12} are the shear strengths of fiber and matrix, respectively. The direct shear mode of failure for graphite/epoxy has been reported in several other publications as well [4.37, 4.38]. For example, Crasto and Kim [4.38] have used a novel minisandwich beam to attain shear failure of the fibers in the composite facing without buckling—the resulting compressive strengths are much higher than those obtained with conventional compression testing.

A number of other factors have been shown to affect longitudinal compressive strength, and this continues to be a very active research area. For example, although the fiber/matrix interfacial strength does not appear in any of the equations presented here, it would appear to be very important in the case of transverse tensile rupture due to Poisson strains. The experiments of Madhukar and Drzal [4.39] have shown that the compressive strength of graphite/epoxy is strongly related to the interfacial shear strength, and that fiber surface treatments which improve the interfacial shear strength also improve the compressive strength.

4.3.2   Transverse Strength

Since failure of the lamina under transverse tension occurs at such low stresses (Table 4.1), this mode of failure is generally the first to occur. In laminates the so-called "first ply failure" is generally due to transverse tension. The low value of the transverse tensile strength, $s_T^{(+)}$, and the corresponding transverse failure strain, $e_T^{(+)}$, are due to strain concentration in the matrix around the fibers, as shown in Eq. (4.34):

$$e_T^{(+)} = \frac{e_{m2}^{(+)}}{F} = \frac{e_m^{(+)}}{F} \qquad (4.34)$$

where $e_{m2}^{(+)} = e_m^{(+)}$, the matrix tensile failure strain (matrix is assumed
       to be isotropic)
   F = strain concentration factor ($F > 1$).

Thus, the strain concentration causes the composite transverse tensile failure strain to be less than the matrix failure strain. The strain concentration factor is more appropriate than the stress concentration factor here because the stress-strain relationships for transverse loading are often nonlinear, reflecting the nonlinear behavior of many matrix materials. However, if linear behavior to failure can be assumed, the corresponding transverse strength is

$$s_T^{(+)} = \frac{E_2 s_m^{(+)}}{E_m F} \qquad (4.35)$$

see 4.38

It is assumed here that the fiber is perfectly bonded to the matrix, so in composites having poor interfacial strength the composite transverse strength would be less than predicted by Eq. (4.35).

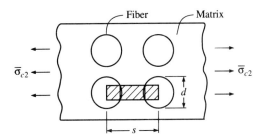

FIGURE 4.11(a)
Mechanics of materials model for
strain concentration under transverse loading.

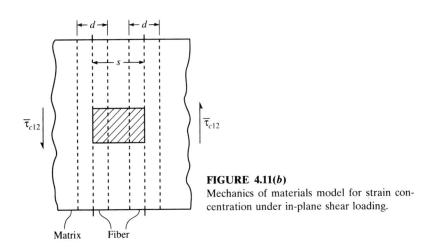

FIGURE 4.11(b)
Mechanics of materials model for strain concentration under in-plane shear loading.

A mechanics of materials approximation for the strain concentration factor has been developed by Kies [4.40], who considered an element
in a transversely loaded lamina, as shown in Fig. 4.11(a). For the shaded
strip shown in Fig. 4.11(a) the total elongation is found by summing the
deformations in the fiber and matrix:

$$\bar{\delta}_{c2} = \bar{\delta}_{f2} + \bar{\delta}_{m2} = s\bar{\epsilon}_{c2} = d\bar{\epsilon}_{f2} + (s - d)\bar{\epsilon}_{m2} \qquad (4.36)$$

where the symbols are defined in Sec. 3.2.2 and Fig. 4.11(a). For the
series arrangement of fiber and matrix materials in the shaded strip it is
assumed that the stresses in composite, matrix, and fiber are equal and
that each material satisfies Hooke's law [Eqs. (3.33)], as in Sec. 3.2.2.
Equation (4.36) can then be written as

$$s\bar{\epsilon}_{c2} = \left(d\frac{E_{m2}}{E_{f2}} + s - d \right)\bar{\epsilon}_{m2} \qquad (4.37)$$

which can be rearranged to give the expression for the strain concentration factor

Assume square.

$$F = \frac{\bar{\epsilon}_m}{\bar{\epsilon}_{c2}} = \frac{1}{\dfrac{d}{s}\left[\dfrac{E_m}{E_{f2}} - 1\right] + 1} \qquad (4.38)$$

for S use G's not E's

where the subscript "2" for matrix properties has been dropped due to the assumption of isotropy. Recall from Eqs. (3.10) and (3.12) that the ratio of fiber diameter to fiber spacing, d/s, is related to the fiber volume fraction, v_f. For example, from substitution of the properties listed in Eqs. (3.28) in Eqs. (3.12) and (4.38), the strain concentration factor for a triangular array of E-glass fibers in an epoxy matrix ($v_f = 0.45$) is $F = 3.00$. This value is in good agreement with experimentally determined values based on the ratio of matrix failure strain to transverse composite failure strain, or the ratio of matrix yield strain to transverse composite yield strain [4.41]. A slightly higher value is predicted by a finite difference solution of the theory of elasticity model [4.42].

It is important to note that according to Eq. (4.38), the strain concentration factor increases with increasing v_f and increasing E_{f2}. For example, the variation of F with fiber volume fraction is shown in Fig. 4.12 for $E_m/E_{f2} \ll 1$ [4.43]. The sharp rise in F for $v_f > 0.6$ is particularly noteworthy. *Thus, as we strive to improve the composite longitudinal properties by using higher fiber volume fractions and higher modulus fibers, we pay the penalty of lower composite transverse strength!*

The same method outlined above can be used to estimate the transverse compressive strength, $s_T^{(-)}$, by replacing the tensile strains or

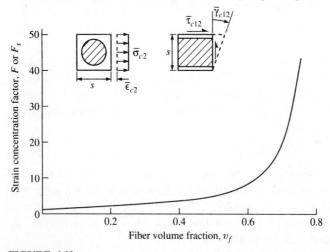

FIGURE 4.12
Variation of strain concentration factor F or F_s with fiber volume fraction. Valid for F when $E_m/E_{f2} \ll 1$ and for F_s when $G_m/G_{f12} \ll 1$. (From Chamis [4.43].)

strengths with the corresponding compressive strains or strengths. Alternatively, the corresponding matrix strength can be used as an upper bound on the composite strength, but the actual composite strength would be lower because of fiber/matrix interfacial bond failure, strain concentrations around fibers and/or voids, or longitudinal fiber splitting [4.44].

4.3.3 In-Plane Shear Strength

The in-plane shear strength, s_{LT}, can also be estimated using the procedure outline in the previous section. For the shaded element shown in Fig. 4.11(b) the expression analogous to Eq. (4.36) for the in-plane shear strain is

$$s\bar{\gamma}_{c12} = d\bar{\gamma}_{f12} + (s - d)\bar{\gamma}_{m12} \qquad (4.39)$$

and the in-plane shear strain concentration factor is

$$F_s = \frac{\bar{\gamma}_{m12}}{\bar{\gamma}_{c12}} = \frac{1}{\dfrac{d}{s}\left[\dfrac{G_{m12}}{G_{f12}} - 1\right] + 1} \qquad S + 4.38 \; \text{modified.} \qquad (4.40)$$

where γ_{m12} = average matrix in-plane shear strain
   γ_{c12} = average composite in-plane shear strain
   γ_{f12} = average fiber in-plane shear strain
   G_{f12} = fiber in-plane shear modulus

Note that this equaton has the same form as Eq. (4.38). Thus, Fig. 4.12 also gives the variation of F_s with v_f when $G_m/G_{f12} \ll 1$ and the previous comments regarding the effect of v_f on F are also valid for F_s. The other necessary equations are obtained by replacing the tensile strains or strengths in Eqs. (4.34) and (4.35) with the corresponding in-plane shear strains or strengths. Again, the matrix shear strength can be used as an upper bound on the composite shear strength, as the actual composite strength would be lower for the same reasons mentioned in the previous section.

   In conclusion, only the basics of micromechanical strength prediction have been discussed here. More detailed micromechanics analyses of strength under other types of loading such as interlaminar shear and flexure as well as micromechanical effects of voids and residual stresses on strength have been summarized by Chamis [4.43].

   Example 4.3. Determine the longitudinal and transverse tensile strength of the graphite/epoxy material described in Examples 3.1 to 3.3 if the tensile strengths of fiber and matrix materials are 2413 MPa and 103 MPa, respectively.

Solution. The fiber tensile failure strain is

$$e_{f1}^{(+)} = \frac{s_{f1}^{(+)}}{E_{f1}} = \frac{2.413}{220} = 0.011$$

The matrix tensile failure strain is

$$e_m^{(+)} = \frac{s_m^{(+)}}{E_m} = \frac{0.103}{3.45} = 0.03$$

Thus, the material fails as described in Fig. 4.7(a) at a strain level of $e_{f1}^{(+)} = 0.011$. Since $v_f = 0.506$ and $v_m = 0.482$ from Example 3.1, the composite longitudinal tensile strength is given by Eq. (4.21):

$$
\begin{aligned}
s_L^{(+)} &= s_{f1}^{(+)}v_f + s_{mf1}v_m \\
&= s_{f1}^{(+)}v_f + E_m e_{f1}^{(+)}v_m \\
&= 2413(0.506) + 3450(0.011)(0.482) \\
&= 1239 \text{ MPa} \quad (180{,}000 \text{ psi})
\end{aligned}
$$

Note that the matrix contribution here is only 18.3 MPa out of 1239 MPa, or about 1.5 percent.

The strain concentration factor for calculation of the transverse tensile strength is given by Eq. (4.38):

$$F = \frac{1}{\dfrac{d}{s}\left[\dfrac{E_m}{E_{f2}} - 1\right] + 1} = \frac{1}{\dfrac{0.0127}{0.0158}\left[\dfrac{3.45}{13.79} - 1\right] + 1} = 2.52$$

If linear elastic behavior to failure can be assumed, the transverse tensile strength is given by Eq. (4.35):

$$s_T^{(+)} = \frac{E_2 s_m^{(+)}}{E_m F} = \frac{5.65(103)}{3.45(2.52)} = 66.9 \text{ MPa} \quad (9703 \text{ psi})$$

PROBLEMS

4.1. An orthotropic lamina has the following properties:

$$
\begin{array}{ll}
E_1 = 160 \text{ GPa} & s_L^{(+)} = 1800 \text{ MPa} \\
E_2 = 10 \text{ GPa} & s_L^{(-)} = 1400 \text{ MPa} \\
v_{12} = 0.3 & s_T^{(+)} = 40 \text{ MPa} \\
G_{12} = 7 \text{ GPa} & s_T^{(-)} = 230 \text{ MPa} \\
& s_{LT} = 100 \text{ MPa}
\end{array}
$$

Construct the failure surfaces in the σ_1-σ_2 stress space for this material according to the (a) Maximum Stress Criterion, (b) Maximum Strain Criterion, and (c) Tsai–Hill Criterion.

4.2. Using the material properties from Problem 4.1 and assuming that the stiffnesses are the same in tension and compression, determine the

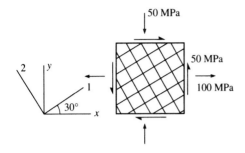

FIGURE 4.13
Stresses acting on element of balanced orthotropic lamina.

allowable off-axis shear stress, τ_{xy}, at $\theta = 45°$ (refer to Fig. 4.5) according to the (a) Maximum Stress Criterion, (b) Maximum Strain Criterion, and (c) Tsai-Hill Criterion. Compare and discuss the results and check both positive and negative values of τ_{xy}.

4.3. An element of a balanced orthotropic lamina is under the state of stress shown in Fig. 4.13. The properties of the lamina are

$$E_1 = E_2 = 70\,\text{GPa} \qquad s_L^{(+)} = s_L^{(-)} = s_T^{(+)} = s_T^{(-)} = 560\,\text{MPa}$$

$$\nu_{12} = \nu_{21} = 0.25 \qquad s_{LT} = 25\,\text{MPa}$$

$$G_{12} = 5\,\text{GPa}$$

Using the Maximum Strain Criterion, determine whether or not failure will occur.

4.4. If some of the compliances and strengths of an orthotropic lamina satisfy certain conditions, the Maximum Strain Criterion failure surface will intercept the horizontal axis at a point like $(\sigma_1, 0)$ instead of at $(s_L^{(-)}, 0)$, as shown in Fig. 4.14. Express these conditions in terms of an inequality.

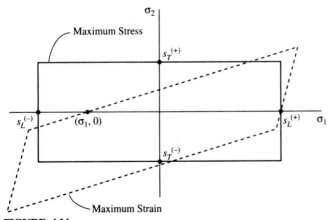

FIGURE 4.14
Example showing that intercepts for Maximum Strain Criterion failure surface are not always the same as those for Maximum Stress Criterion.

4.5. An element of an orthotropic lamina having the properties given in Problem 4.1 is subjected to an off-axis tensile test, as shown in Fig. 4.4. Using the Maximum Strain Criterion, determine the values of σ_x at failure and the mode of failure for each of the following values of the angle θ: (a) 2°, (b) 30°, (c) 75°.

4.6. Repeat Problem 4.5 for an off-axis *compression* test.

4.7. A material having the properties given in Problem 4.1 is subjected to a biaxial tension test and the biaxial failure stress is found to be $\sigma_1 = \sigma_2 = 35$ MPa. Determine the Tsai-Wu interaction parameter F_{12}, then use the Tsai-Wu Criterion to determine whether or not failure will occur for the stress condition $\sigma_1 = 100$ MPa, $\sigma_2 = -50$ MPa, $\tau_{12} = 90$ MPa.

4.8. The Tsai-Wu interaction parameter F_{12} is determined from biaxial failure stress data. One way to generate a biaxial state of stress is by using a uniaxial 45° off-axis tension test. Derive the expression for F_{12} based on such a test, assuming that all the uniaxial and shear strengths are known.

4.9. Determine the longitudinal tensile strength of the hybrid graphite–aramid/epoxy composite described in Problem 3.3 and Fig. 3.15.

4.10. Compare and discuss the estimated longitudinal compressive strengths of Scotchply 1002 E-glass/epoxy based on (a) fiber microbuckling and (b) transverse tensile rupture. Assume linear elastic behavior to failure. For the epoxy matrix, assume $E_m = 3.79$ GPa, $\nu_m = 0.35$.

4.11. An element of an orthotropic lamina is subjected to an off-axis shear stress, τ_{xy}, as shown in Fig. 4.5(a). Using the Tsai-Hill Criterion and assuming that the lamina strengths are the same in tension and compression, develop an equation relating the allowable value of τ_{xy} to the lamina strengths, s_L, s_T, and s_{LT}, and the fiber orientation θ.

REFERENCES

4.1. Chamis, C. C., "Simplified Composite Micromechanics Equations for Mechanical, Thermal and Moisture-Related Properties," in J. W. Weeton, D. M. Peters, and K. L. Thomas (eds.), *Engineers' Guide to Composite Materials*, 3-8–3-24, ASM International, Materials Park, OH (1987).

4.2. Hashin, Z., "Analysis of Composite Materials—A Survey," *Journal of Applied Mechanics*, **50**, 481–505 (1983).

4.3. Hashin, Z. "Failure Criteria for Unidirectional Fiber Composites," *Journal of Applied Mechanics*, **47**, 329–334 (1980).

4.4. Wu, E. M., "Phenomenological Anisotropic Failure Criterion," in G. P. Sendeckyj (ed.), *Composite Materials Volume 2: 353–431, Mechanics of Composite Materials*, Academic Press, New York (1974).

4.5. Sendeckyj, G. P., "A Brief Survey of Empirical Multiaxial Strength Criteria for Composites," *Composite Materials: Testing and Design (Second Conference)*, ASTM STP **497**, 41–51, American Society for Testing and Materials, Philadelphia, PA (1972).

4.6. Chamis, C. C., "Failure Criteria for Filamentary Composites," *Composite Materials: Testing and Design,* ASTM STP **460,** 336–351, American Society for Testing and Materials, Philadelphia, PA (1969).

4.7. Kaminski, B. E. and Lantz, R. B., "Strength Theories of Failure for Anisotropic Materials," *Composite Materials: Testing and Design,* ASTM STP **460,** 160–169, American Society for Testing and Materials, Philadelphia, PA (1969).

4.8. Franklin, H. G., "Classical Theories of Failure of Anisotropic Materials," *Fiber Science and Technology,* **1**(2), 137–150 (1968).

4.9. Jenkins, C. F., "Report on Materials of Construction Used in Aircraft and Aircraft Engines," Great Britain Aeronautical Research Committee (1920).

4.10. Higdon, A., Ohlsen, E. H., Stiles, W. B., Weese, J. A., and Riley, W. F., *Mechanics of Materials,* 3d ed., John Wiley & Sons, New York (1976).

4.11. Burk, R. C., "Standard Failure Criteria Needed for Advanced Composites," *Astronautics and Aeronautics,* **21**(6), 58–62 (1983).

4.12. Pipes, R. B. and Cole, B. W., "On the Off-Axis Strength Test for Anisotropic Materials," *Journal of Composite Materials,* **7,** 246–256 (1973).

4.13. Jones, R. M., *Mechanics of Composite Materials,* Hemisphere Publishing Co. New York (1975).

4.14. Waddoups, M. E., "Advanced Composite Material Mechanics for the Design and Stress Analyst," General Dynamics, Fort Worth Division Report FZM-4763, Fort Worth, TX (1967).

4.15. Hill, R., "A Theory of the Yielding and Plastic Flow of Anisotropic Metals," in *Proceedings of the Royal Society of London,* Series A, **193,** 281–297 (1948).

4.16. Azzi, V. D. and Tsai, S. W., "Anisotropic Strength of Composites," in *Proceedings of the Society for Experimental Stress Analysis,* **XXII**(2), 283–288 (1965).

4.17. Tsai, S. W., "Strength Theories of Filamentary Structures," in R. T. Schwartz and H. S. Schwartz (eds.), *Fundamental Aspects of Fiber Reinforced Plastic Composites,* 3–11, Chap. 1, Wiley Interscience, New York (1968).

4.18. Hoffman, O., "The Brittle Strength of Orthotropic Materials," *Journal of Composite Materials,* **1,** 200–206 (1967).

4.19. Tsai, S. W. and Wu, E. M., "A General Theory of Strength for Anisotropic Materials," *Journal of Composite Materials,* **5,** 58–80 (1971).

4.20. Gol'denblat, I. and Kopnov, V. A., "Strength of Glass Reinforced Plastics in the Complex Stress State," *Mekhanika Polimerov,* **1,** 70–78 (1965). English translation: *Polymer Mechanics,* **1,** 54–60 (1966).

4.21. Wu, E. M., "Optimal Experimental Measurements of Anisotropic Failure Tensors," *Journal of Composite Materials,* **6,** 472–489 (1972).

4.22. Tsai, S. W. and Hahn, H. T., *Introduction to Composite Materials,* Technomic Publishing Co., Lancaster, PA (1980).

4.23. Tennyson, R. C., MacDonald, D., and Nanyaro, A. P., "Evaluation of the Tensor Polynomial Failure Criterion for Composite Materials," *Journal of Composite Materials,* **12,** 63–75 (1978).

4.24. Weeton, J. W., Peters, D. M., and Thomas, K. L. (eds.), *Engineers' Guide to Composite Materials,* ASM International, Materials Park, OH (1987).

4.25. Kelly, A. and Davies, G. J., "The Principles of the Fibre Reinforcement of Metals," *Metallurgical Review,* **10,** 1–77 (1965).

4.26. Hull, D., *An Introduction to Composite Materials,* Cambridge University Press, Cambridge, MA (1981).

4.27. Rosen, B. W., "Tensile Failure of Fibrous Composites," *AIAA Journal,* **2,** 1985–1991 (1964).

4.28. Zweben, C. and Rosen, B. W., "A Statistical Theory of Material Strength with Applications to Composite Materials," *Journal of Mechanics and Physics of Solids,* **18,** 180–206 (1970).

4.29. Rosen, B. W., "Composite Materials Analysis and Design," in T. J. Reinhart (ed.), *Engineered Materials Handbook, Volume 1 Composites*, Sec. 4, ASM International, Materials Park, OH (1987).

4.30. Holister, G. S. and Thomas, C., *Fibre Reinforced Materials*, Elsevier Publishing Co., Ltd., New York (1966).

4.31. Whitney, J. M., "Failure Modes in Compression Testing of Composite Materials," *How Concept Becomes Reality—Proceedings of 36th International SAMPE Symposium*, **36**, 1069–1078, Society for Advancement of Material and Process Engineering, Covina, CA (1991).

4.32. Hahn, H. T. and Williams, J. G., "Compression Failure Mechanisms in Unidirectional Composites," J. M. Whitney (ed.), in *Composite Materials: Testing and Design (Seventh Conference)*, ASTM STP **893**, 115–139, American Society for Testing and Materials, Philadelphia, PA (1986).

4.33. Rosen, B. W., 'Mechanics of Composite Strengthening," *Fiber Composite Materials*, American Society for Metals, Metals Park, OH (1965).

4.34. Schuerch, H., "Prediction of Compressive Strength in Uniaxial Boron Fiber Metal Matrix Composites," *AIAA Journal*, **4**, 102–106 (1966).

4.35. Greszczuk, L. B., "Microbuckling of Lamina-Reinforced Composites," in C. A. Berg et al. (eds.), *Composite Materials: Testing and Design (Third Conference)*, ASTM STP **546**, 5–29, American Society for Testing and Materials, Philadelphia, PA (1974).

4.36. Agarwal, B. D. and Broutman, L. J., *Analysis and Performance of Fiber Composites*, 2d ed., John Wiley & Sons, Inc., New York (1990).

4.37. Hancox, N. L., "The Compression Strength of Unidirectional Carbon Fibre Reinforced Plastics," *Journal of Materials Science*, **10**(2), 234–242 (1975).

4.38. Crasto, A. S. and Kim, R. Y., "Compression Strengths of Advanced Composites from a Novel Mini-Sandwich Beam," *SAMPE Quarterly*, **22**(3), 29–39 (1991).

4.39. Madhukar, M. S. and Drzal, L. T., "Effect of Fiber-Matrix Adhesion on Longitudinal Compressive Properties of Graphite/Epoxy Composites," in *Proceedings of the American Society for Composites Fifth Technical Conference*, 849–858, Technomic Publishing Co., Lancaster, PA (1990).

4.40. Kies, J. A., "Maximum Strains in the Resin of Fiber Glass Composites," U.S. Naval Research Laboratory Report No. 5752 (1962).

4.41. Gibson, R. F., *Elastic and Dissipative Properties of Fiber Reinforced Composite Materials in Flexural Vibration*, Ph.D. Dissertation, University of Minnesota (1975).

4.42. Adams, D. F. and Doner, D. R., "Transverse Normal Loading of a Unidirectional Composite," *Journal of Composite Materials*, **1**, 152–164 (1967).

4.43. Chamis, C. C., "Micromechanics Strength Theories," in L. J. Broutman (ed.), *Composite Materials, Volume 5: Fracture and Fatigue*, Chap. 3, Academic Press, New York (1974).

4.44. Piggott, M. R., *Load Bearing Fibre Composites*, Pergamon Press Ltd., Oxford, England (1980).

CHAPTER
5

ANALYSIS
OF
LAMINA
HYGROTHERMAL
BEHAVIOR

5.1 INTRODUCTION

The analytical models for composite mechanical behavior presented up to now have been based on the assumption of constant environmental conditions. Composites are usually subjected to changing environmental conditions during both initial fabrication and final use, however, and it is important to be able to include the effects of such changes in the analysis. Among the many environmental conditions that may influence composite mechanical behavior, changes in temperature and moisture content are singled out for discussion here because of the particularly important effects they have on polymer matrix materials and those properties of polymer composites which are matrix-dominated. Effects of temperature are usually referred to as "thermal" effects, whereas those of moisture are often referred to as "hygroscopic" effects. The word "hygrothermal" has evolved as a way of describing the combined effects of temperature and moisture.

There are two principal effects of changes in the hygrothermal environment on mechanical behavior of polymer composites:

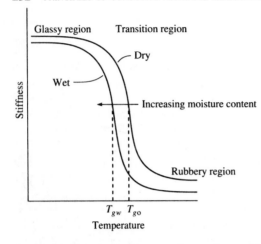

FIGURE 5.1
Variation of stiffness with temperature for a typical polymer showing the glass transition temperature, T_g, and the effect of absorbed moisture on T_g. Note: T_{g0} = "dry" T_g, and T_{gw} = "wet" T_g.

1. Matrix-dominated properties such as stiffness and strength under transverse, off-axis, or shear loading are altered. Increased temperature causes a gradual softening of the polymer matrix material up to a point. If the temperature is increased beyond the so-called "glass transition" region (indicating a transition from glassy behavior to rubbery behavior), however, the polymer becomes too soft for use as a structural material (Fig. 5.1). Plasticization of the polymer by absorbed moisture causes a reduction in the glass transition temperature, T_g, and a corresponding degradation of composite properties (Fig. 5.1).

2. Hygrothermal expansions or contractions change the stress and strain distributions in the composite. Increased temperature and/or moisture content causes swelling of the polymer matrix, whereas reduced temperature and/or moisture content causes contraction. Since the fibers are not affected as much by hygrothermal conditions, the swelling or contraction of the matrix is resisted by the fibers and residual stresses develop in the composite. A similar effect at the laminate level is due to differential expansions or contractions of constituent laminae.

This chapter is therefore concerned with analytical modeling of hygrothermal degradation of matrix-dominated properties and modification of the lamina stress-strain relationships to include hygrothermal effects. Micromechanical analysis of mechanical and thermophysical properties will also be discussed because of its importance in the analytical modeling of both effects.

5.2 HYGROTHERMAL DEGRADATION OF PROPERTIES

As evidence of hygrothermal degradation of properties, consider the data of Browning et al. [5.1], who tested graphite/epoxy composites and their epoxy matrix materials under various hygrothermal conditions. Figure 5.2 shows the stress-strain curves for a typical epoxy matrix material under the various combinations of temperature and absorbed moisture. The corresponding stress-strain curves for the graphite/epoxy composite under transverse loading are shown in Fig. 5.3. Note that the imposed hygrothermal conditions cause substantial reductions of both strength and stiffness in both cases, with the so-called "hot-wet" conditions (combined high temperature and high moisture content) generating the most severe degradation. Similar degradation was observed in the case of in-plane shear loading of the composite since the behavior is matrix-dominated in both cases. On the other hand, the corresponding stress-strain curves for the composite under longitudinal loading showed little effect because longitudinal strength and stiffness are fiber-dominated.

Another example of the hygrothermal sensitivity of matrix-dominated composite properties is the data of Gibson et al. [5.2], who used a vibrating beam method to measure the flexural moduli of several E-glass/polyester sheet-molding compounds after soaking at various times in a water bath. Table 5.1 gives a description of the materials, Figure 5.4 shows the percent weight gain due to moisture pickup, and Fig. 5.5 shows the variation in modulus with soaking time. Composites having some continuous fibers and high fiber contents absorbed little moisture and showed negligible change in modulus with soaking time. On the other hand, composites having matrix-dominated behavior (those with

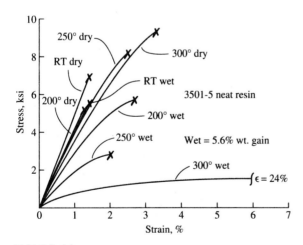

FIGURE 5.2
Stress-strain curves for 3501-5 epoxy resin at different temperatures and moisture contents. (*From Browning et al. [5.1]. Copyright ©ASTM. Reprinted with permission.*)

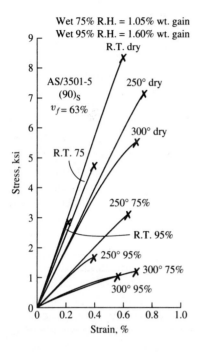

FIGURE 5.3

Stress-strain curves for AS/3501-5 graphite/epoxy composite under transverse loading at different temperatures and moisture contents. (*From Browning et al.* [5.1]. *Copyright* ©*ASTM. Reprinted with permission.*)

chopped fibers only and low fiber contents) exhibited the most moisture pickup and the greatest reduction in modulus.

Considerable insight into the physics of temperature and moisture distribution in a material is gained from the analysis of Shen and Springer [5.3], who considered the one-dimensional distribution of temperature, T, and moisture concentration, c, in a plate of thickness, h, which is

TABLE 5.1
Description of composite materials for Figs. 5.4 and 5.5

	Weight of percentages of constituents		
Material	Chopped E-glass fibers	Continuous E-glass fibers	Polyester resin, fillers, etc.
PPG SMC-R25†	25	0	75
PPG SMC-R65	65	0	35
PPG XMC-3	25	50 (±7.5°, X-pattern)	25
OCF SMC-R25‡	25	0	75
OCF C20/R30	30	20 (aligned)	50

† Manufactured by PPG Industries, Fiber Glass Division, Pittsburgh, PA 15222.

‡ Manufactured by Owens-Corning Fiberglas Corporation, Toledo, OH 43659.

Source: From Gibson et al. [5.2].

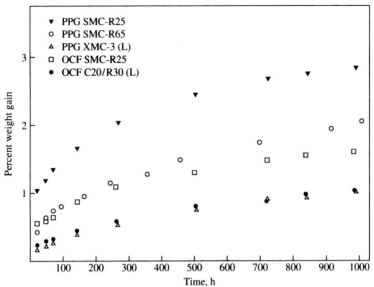

FIGURE 5.4
Percent weight gain due to moisture pickup vs. soaking time for several E-glass/polyester sheet-molding compounds. Materials described in Table 5.1. (*From Gibson et al.* [5.2]. *Reprinted by permission of Technomic Publishing Co.*)

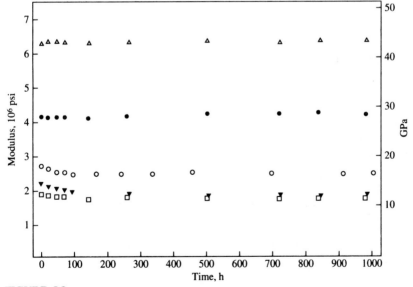

FIGURE 5.5
Variation of flexural modulus of several E-glass/polyester sheet-molding compounds with soaking time in distilled water at 21–24°C. Materials described in Table 5.1 and in Fig. 5.4. (*From Gibson et al.* [5.2]. *Reprinted by permission of Technomic Publishing Co.*)

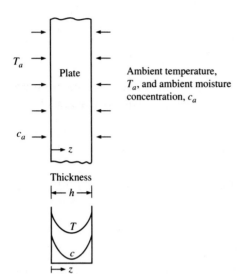

T_a

Plate

Ambient temperature, T_a, and ambient moisture concentration, c_a

c_a

z

Thickness

$\leftarrow h \rightarrow$

T

c

z

FIGURE 5.6
Schematic representation of temperature and moisture distributions through the thickness of a plate which is exposed to an environment of temperature, T_a, and moisture concentration, c_a, on both sides.

suddenly exposed on both sides to an environment of temperature, T_a, and moisture concentration, c_a (Fig. 5.6). The temperature and moisture concentration are assumed to vary only through the thickness along the z direction and the initial temperature, T_i, and initial moisture concentration, c_i, are assumed to be uniform. The temperature distribution is governed by the Fourier heat conduction equation

$$\rho C \frac{\partial T}{\partial t} = \frac{\partial}{\partial z} K_z \frac{\partial T}{\partial z} \qquad (5.1)$$

whereas the moisture is governed by Fick's second law

$$\frac{\partial c}{\partial t} = \frac{\partial}{\partial z} D_z \frac{\partial c}{\partial z} \qquad (5.2)$$

where ρ = density of material
  C = specific heat of material
  K_z = thermal conductivity of material along the z direction
  D_z = mass diffusivity along the z direction
  t = time

These equations are solved subject to the initial and boundary conditions

$$\left.\begin{array}{c} T = T_i \\ c = c_i \end{array}\right\} \quad 0 < z < h, \quad t \le 0$$

$$\left.\begin{array}{c} T = T_a \\ c = c_a \end{array}\right\} \quad z = 0; \quad z = h; \quad t > 0$$

Shen and Springer [5.3] point out that, due to the numerical values of the thermophysical properties C, K_z, D_z, and ρ for typical composites, the temperature approaches equilibrium about one million times faster than the moisture concentration. Thus, the material temperature can usually be assumed to be the same as the ambient temperature, but the moisture distribution requires further analysis. If the diffusivity is assumed to be constant, Fick's second law becomes

$$\frac{\partial c}{\partial t} = D_z \frac{\partial^2 c}{\partial^2 z} \tag{5.3}$$

The solution to this equation subject to the previously stated initial and boundary conditions is [5.3, 5.4]

$$\frac{c - c_i}{c_m - c_i} = 1 - \frac{4}{\pi} \sum_{j=0}^{\infty} \frac{1}{(2j+1)} \sin \frac{(2j+1)\pi z}{h}$$
$$\times \exp\left[-\frac{(2j+1)^2 \pi^2 D_z t}{h^2} \right] \tag{5.4}$$

where the moisture concentration at the surface of the material, c_m, is related to the moisture content of the environment, c_a. Browning et al. [5.1] used Eq. (5.4) to predict moisture profiles for a graphite/epoxy plate after drying out for various periods of time, as shown in Fig. 5.7. While Eq. (5.4) gives the *local* moisture concentration, $c(z, t)$, we normally measure the *total* amount of moisture averaged over the sample. The

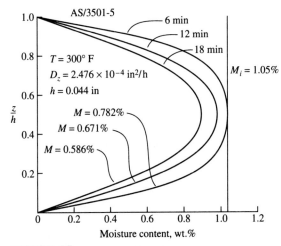

FIGURE 5.7
Predicted moisture profiles through the thickness of a graphite/epoxy plate after drying out for various periods of time. (*From Browning et al. [5.1]. Copyright ©ASTM. Reprinted with permission.*)

M = % moisture in part as % of part's weight (handwritten annotation)

average concentration is given by [5.4]

$$\bar{c} = \frac{1}{h} \int_0^h c(z, t)\, dz = (c_m - c_i)$$

$$\times \left[1 - \frac{8}{\pi^2} \sum_{j=0}^{\infty} \frac{\exp\left[-(2j+1)^2 \pi^2 (D_z t / h^2)\right]}{(2j+1)^2} \right] + c_i \quad (5.5)$$

The weight percent moisture, M, is the quantity that is normally measured, and since $\bar{c}$ is linearly related to M, we can write [5.3]

$$G = \frac{M - M_i}{M_m - M_i} = 1 - \frac{8}{\pi^2} \sum_{j=0}^{\infty} \frac{\exp\left[-(2j+1)^2 \pi^2 (D_z t / h^2)\right]}{(2j+1)^2} \quad (5.6)$$

(handwritten: *keg* above)

where M_i is the initial weight percent of moisture in the material and M_m is the weight percent of moisture in the material when the material reaches fully saturated equilibrium with the ambient conditions. Thus, the parameter G describes the moisture weight gain as a function of time. Such data can be obtained experimentally by weighing the specimen at various times during exposure to a moist environment. Figure 5.8 from Ref. [5.3] shows a comparison of measured and predicted values of G as a function of the dimensionless ratio $D_z t / h^2$ for graphite/epoxy. The agreement is seen to be excellent. Thus, the moisture diffusion process in these composites seems to follow Fick's law. Non-Fickian diffusion has also been observed in some cases where microcracking is developed as a result of the hygrothermal degradation [5.5].

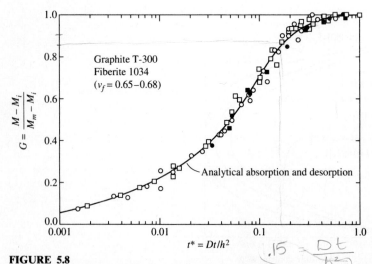

Graphite T-300
Fiberite 1034
($v_f = 0.65$–0.68)

Analytical absorption and desorption

$G = \dfrac{M - M_i}{M_m - M_i}$

$t^* = Dt/h^2$

(handwritten: $.15 = \dfrac{Dt}{h^2}$)

FIGURE 5.8

Comparison of predicted [Eq. (5.6)] and measured moisture absorption and desorption of T300/1034 graphite/epoxy composites. Open symbols represent measured absorption and dark symbols represent measured desorption. (*From Shen and Springer* [5.3]. *Reprinted by permission of Technomic Publishing Co.*)

The hygrothermal degradation of composite strength and/or stiffness can be estimated by using an empirical equation to degrade the corresponding matrix property, then by using the degraded matrix property in the appropriate micromechanics equation. Chamis and Sinclair [5.6] and Chamis [5.7] have demonstrated such a procedure based on the equation

$$F_m = \frac{P}{P_o} = \left[\frac{T_{gw} - T}{T_{go} - T_o}\right]^{1/2} \quad \text{(in deg°F)} \tag{5.7}$$

where F_m = matrix mechanical property retention ratio
P = matrix strength or stiffness after hygrothermal degradation
P_o = reference matrix strength or stiffness before degradation
T = temperature at which P is to be predicted (°F)
T_{go} = glass transition temperature for reference dry condition (°F)
T_{gw} = glass transition temperature for wet matrix material at moisture content corresponding to property P (°F) (Fig. 5.1)
T_o = test temperature at which P_o was measured (°F)
(All temperatures are degrees Fahrenheit.)

The form of Eq. (5.7) is based on the experimental observation that degradation is gradual until the temperature T approaches T_{gw}, whereupon the degradation accelerates. The value of T_{gw} can be obtained from experimental data on the glass transition temperature of the matrix resin as a function of absorbed moisture. For example, the data of DeIasi and Whiteside [5.8] show the reduction in T_{gw} with increasing moisture content for six epoxy resins (Fig. 5.9). Chamis [5.7] suggests that T_{gw} can be estimated by using the following empirical equation:

$$T_{gw} = (0.005M_r^2 - 0.10M_r + 1.0)T_{go} \quad \text{wet matrix mat'l} \tag{5.8}$$

where M_r is the weight percent of moisture in the matrix resin and values of T_{go} for several matrix materials are found in Table 3.2. DeIasi and Whiteside [5.8] and Browning et al. [5.1] also found that data such as that in Fig. 5.9 are in good agreement with predictions from the theory of polymer plasticization by diluents.

Once the mechanical property retention ratio is found from Eq. (5.7), it is then used to degrade the matrix property in the appropriate micromechanics equation. For example, the rule of mixtures for the longitudinal modulus [Eq. (3.22)] now becomes

$$E_1 = E_{f1}v_f + F_m E_{mo}v_m \tag{5.9}$$

where E_{mo} is the reference value of the matrix modulus in the dry condition. It is assumed that Poisson's ratio is not hygrothermally degraded [5.7].

Reasonably accurate predictions are also obtained when Eq. (5.7)

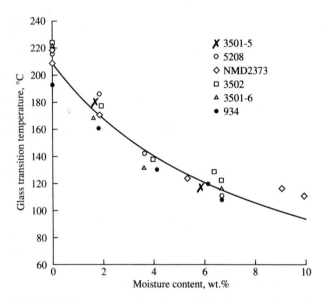

FIGURE 5.9
Variation of glass transition temperature with equilibrium moisture content for several epoxy resins. (*From Delasi and Whiteside* [5.8]. *Copyright* ©*ASTM. Reprinted with permission.*)

is applied directly to matrix-dominated composite properties (i.e., P and P_o are taken to be matrix-dominated composite properties instead of matrix properties). For example, Chamis and Sinclair [5.6] found good agreement between the predictions of Eq. (5.7) and experimental data on hygrothermal degradation of transverse compression, transverse tension, and intralaminar shear strengths of boron/epoxy and graphite/epoxy composites (Fig. 5.10). Thus, the hygrothermally degraded composite property may be estimated by applying Eq. (5.7) directly to the matrix-dominated composite property measured under reference conditions, or by applying Eq. (5.7) to matrix data measured under reference conditions, then substituting the result into the appropriate micromechanics equation.

Empirical equations such as Eqs. (5.7) and (5.8) should always be used with caution. Curve-fitting parameters such as the exponent of 1/2 in Eq. (5.7) and the coefficients of M_r in Eq. (5.8) are based on experimental data for epoxy matrix materials. While the equations may be suitable for other composites as well, the user should check predictions against available experimental data where possible.

The procedure just outlined is based on the combined effects of temperature and moisture, and the two effects were seen to be coupled by the lowering of the glass transition temperature by absorbed

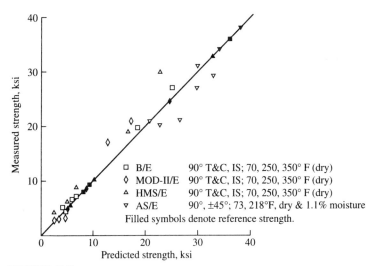

FIGURE 5.10
Comparison of predicted [Eq. (5.7)] and measured strengths of several hygrothermally degraded composites. (*From Chamis and Sinclair* [5.6]. *Copyright* ©*ASTM. Reprinted with permission.*)

moisture. There is another important connection between the two effects. Moisture absorption occurs by diffusion, which is known to be a thermally activated process. The diffusivity D which appears in Fick's law is related to temperature by the Arrhenius relationship:

$$D = D_o \exp\left(-E_a/RT\right) \qquad (5.10)$$

where D_o = material constant
E_a = activation energy for diffusion
R = universal gas constant
T = absolute temperature

Proof that this relationship holds for composites is given by the fact that plots of experimentally determined values of $\log D$ vs. $1/T$ fall on a straight line, as shown in Fig. 5.11 from Loos and Springer [5.9]. The result of this relationship is that increased temperature causes an increase in the rate of moisture absorption, as shown in Fig. 5.12.

The applied stress also has an effect on moisture absorption in polymers and polymer composites [5.10, 5.11]. For example, Fahmy and Hurt [5.10] have shown that the diffusivity of a polymer is increased under tensile stress and decreased under compressive stress. Thus, in a composite the residual stresses due to differential thermal expansion of fiber and matrix materials may cause increased moisture absorption along a path running through a tensile stress field.

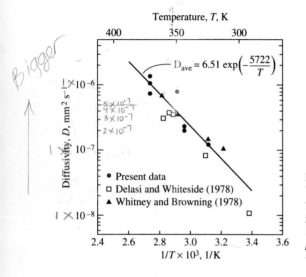

FIGURE 5.11
Variation of transverse diffusivity with temperature for AS/3501-5 graphite/epoxy composite. (*From Loos and Springer* [5.9]. *Reprinted by permission of Technomic Publishing Co.*)

Example 5.1. An epoxy resin sample has a thickness $h = 5$ mm and a diffusivity $D = 3 \times 10^{-8}$ mm²/s. Determine the moisture absorption of an initially dry sample after a time $t = 100$ days.

Solution. The moisture absorption is predicted by Eq. (5.6), which involves an infinite series. In order to examine the convergence characteristics of the series, we will look at the first four terms. Each term in the series contains

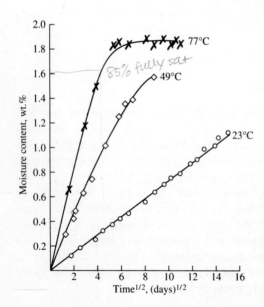

FIGURE 5.12
Effect of temperature on rate of moisture absorption in AS/3501-5 graphite/epoxy composite. (*From Delasi and Whiteside* [5.8]. *Copyright ©ASTM. Reprinted with permission.*)

the dimensionless ratio $\pi^2 Dt/h^2$, which has the numerical value

$$\frac{\pi^2 Dt}{h^2} = \frac{\pi^2 (3 \times 10^{-8})(100)(60)(60)(24)}{(5)^2} = 0.102$$

Since the sample was initially dry, the initial weight of moisture in the material is $M_i = 0$. Thus, Eq. (5.6) reduces to the ratio M/M_m, which is the ratio of the weight percent of moisture at time t to the weight percent of moisture in the fully saturated equilibrium condition. The first four terms are

$$\frac{M}{M_m} = 1 - \frac{8}{\pi^2} \left[\exp(-0.102) + \frac{\exp(-9(0.102))}{9} \right.$$
$$\left. + \frac{\exp(-25(0.102))}{25} + \frac{\exp(-49(0.102))}{49} + \cdots \right]$$

The values of M/M_m corresponding to the different number of terms are

	One term	Two terms	Three terms	Four terms
$\dfrac{M}{M_m} =$	0.267	0.230	0.228	0.228

Thus, the series has converged after three terms. Rapid convergence is a characteristic of this solution, and in many cases only one term is sufficient [5.4].

Example 5.2. The composite described in Examples 3.1 to 3.3 is to be used in a "hot-wet" environment with temperature $T = 200°F$ (93°C) and resin moisture content $M_r = 3\%$. If the glass transition temperature of the dry matrix resin is 350°F (177°C) and if the properties given in Examples 3.1 to 3.3 are for a temperature of 70°F (21°C), determine the hygrothermally degraded values of the longitudinal and transverse moduli.

Solution. From Eq. (5.8), the glass transition temperature in the wet condition is

$$T_{gw} = (0.005(3)^2 - 0.1(3) + 1.0)350 = 261°F \quad (127°C)$$

From Eq. (5.7), the hygrothermally degraded Young's modulus of the epoxy resin is

$$E_m = ((261 - 200)/(350 - 70))^{1/2}(0.5)(10^6)$$
$$= 0.233(10^6) \text{ psi} \quad (1.61 \text{ GPa})$$

From Eq. (5.9), the hygrothermally degraded longitudinal modulus is

$$E_1 = 32(10^6)(0.506) + 0.233(10^6)(0.482) = 16.3(10^6) \text{ psi} \quad (112 \text{ GPa})$$

Thus, the hygrothermally degraded value of E_1 is 99.2 percent of the reference value calculated in Example 3.3. As stated earlier, the fiber-dominated properties are not affected much by temperature and moisture.

Similarly, using the degraded value of E_m in Eq. (3.35), we find that the hygrothermally degraded transverse modulus is estimated to be

$E_2 = 0.434(10^6)$ psi (3.0 GPa), which is only 53 percent of the reference value calculated in Example 3.3. Thus, the matrix-dominated properties such as the transverse modulus are strongly affected by hygrothermal conditions.

5.3 LAMINA STRESS-STRAIN RELATIONSHIPS INCLUDING HYGROTHERMAL EFFECTS

In Chap. 2 the lamina stress-strain relationships were developed for linear elastic behavior and constant environmental conditions. The thermal expansion or contraction of structural materials due to temperature change is a well-known phenomenon, however, and the thermal strains for an isotropic material are usually described by an equation of the form

$$\epsilon_i^T = \begin{cases} \alpha \, \Delta T & \text{if } i = 1, 2, 3 \\ 0 & \text{if } i = 4, 5, 6 \end{cases} \tag{5.11}$$

where $i = 1, 2, \ldots, 6$ (recall contracted notation)
ΔT = temperature change $(T - T_o)$
T = final temperature
T_o = initial temperature where $\epsilon_i^T = 0$ for all i
α = coefficient of thermal expansion (CTE)

This relationship is based on the experimental observation that a temperature change in an unrestrained isotropic material induces an equal expansion or contraction in all directions with no distortion due to shear deformation. In this case $\alpha > 0$ because an increase in temperature causes an increase in strain. As we will see later, however, some anisotropic fiber materials have *negative* CTEs along the fiber axis and positive CTEs along the transverse direction. In general, the strain-temperature relationship is nonlinear, but the assumption of linearity is valid over a sufficiently narrow temperature range. Typical thermal expansion data for an epoxy resin is shown in Fig. 5.13. If operation over

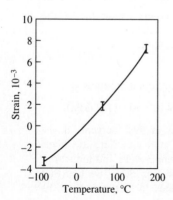

FIGURE 5.13
Thermal expansion vs. temperature for 3501-6 epoxy resin. (*From Cairns and Adams* [5.12]. *Reprinted by permission of Technomic Publishing Co.*)

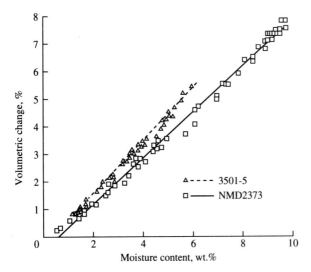

FIGURE 5.14
Hygroscopic expansion vs. moisture content for two epoxy resins. (*From DeIasi and Whiteside* [5.8]. *Copyright ©ASTM. Reprinted with permission.*)

a wide temperature range is expected, the reader is referred to data such as that of Cairns and Adams [5.12], who have developed cubic polynomial expressions to fit experimental thermal expansion data for epoxy, glass/epoxy, and graphite/epoxy from −73°C to 175°C. A procedure for estimating the hygrothermal degradation of matrix-dominated thermal properties will be discussed in Sec. 5.4.

In polymeric materials moisture has been shown to cause hygroscopic expansions or contractions analogous to thermal strains. For example, the experimentally determined moisture-induced swelling of several epoxy resins is shown in Fig. 5.14. The experimental observation is that the moisture-induced strains in isotropic materials can be expressed as

$$\epsilon_i^M = \begin{cases} (\beta)c & \text{if } i = 1, 2, 3 \\ 0 & \text{if } i = 4, 5, 6 \end{cases} \tag{5.12}$$

where c = moisture concentration

$$= \frac{\text{mass of moisture in a unit volume}}{\text{mass of dry material in a unit volume}}$$

β = coefficient of hygroscopic expansion (CHE)

The reference condition is assumed to be the moisture-free state $c = 0$, where $\epsilon_i^M = 0$. Hygroscopic strains are generally nonlinear functions of moisture content [5.12], but the linear relationship in Eq. (5.12) is valid if the range of moisture contents is not too wide. Thus, in an isotropic material the total hygrothermal strain can be written as

$$\epsilon_i^H = \epsilon_i^T + \epsilon_i^M \tag{5.13}$$

Because fibers usually have CTEs and CHEs that are quite different from those of matrix materials (see Tables 3.1 and 3.2), the hygrothermal strains in a composite lamina are different in longitudinal and transverse directions. Notice also in Table 3.1 that the longitudinal CTEs of some fibers are negative, whereas the transverse CTEs are positive. As shown later, this leads to the interesting possibility of designing a composite with a CTE of near zero. Thus, subscripts are needed for α and β, and the hygrothermal strains associated with the 12 principal material axes in the specially orthotropic lamina should be expressed as

$$\epsilon_i^H = \begin{cases} \alpha_i \, \Delta T + \beta_i c & \text{if } i = 1, 2, 3 \\ 0 & \text{if } i = 4, 5, 6 \end{cases} \tag{5.14}$$

If the material is transversely isotropic, $\alpha_2 = \alpha_3$ and $\beta_2 = \beta_3$. Typical values of α_i and β_i for several composites are given in Table 5.2 from Ref. [5.13]. Notice that the negative longitudinal CTE of graphite fibers leads to a very small longitudinal CTE for the lamina. Notice also the large differences between longitudinal and transverse hygrothermal coefficients.

The total strains along the principal material axes in the specially orthotropic lamina are found by summing the mechanical strains due to applied stresses [Eq. (2.24)] and the hygrothermal strains [Eq. (5.14)]:

$$\begin{Bmatrix} \epsilon_1 \\ \epsilon_2 \\ \gamma_{12} \end{Bmatrix} = \underbrace{\begin{bmatrix} S_{11} & S_{12} & 0 \\ S_{21} & S_{22} & 0 \\ 0 & 0 & S_{66} \end{bmatrix} \begin{Bmatrix} \sigma_1 \\ \sigma_2 \\ \tau_{12} \end{Bmatrix}}_{mech} + \underbrace{\begin{Bmatrix} \alpha_1 \\ \alpha_2 \\ 0 \end{Bmatrix} \Delta T}_{Thermal} + \underbrace{\begin{Bmatrix} \beta_1 \\ \beta_2 \\ 0 \end{Bmatrix} c}_{Moisture} \tag{5.15}$$

TABLE 5.2
Typical thermal and hygroscopic expansion properties

Material	Thermal expansion coefficients $(10^{-6}\,\text{m/m})/°C$		Hygroscopic expansion coefficients, m/m	
	α_1	α_2	β_1	β_2
AS graphite/epoxy	0.88	31.0	0.09	0.30
E-glass/epoxy	6.3	20.0	0.014	0.29
AF-126-2 adhesive	29.0	29.0	0.20	0.20
1020 steel	12.0	12.0	—	—

Source: From Graves and Adams [5.13].

$$\begin{Bmatrix} \epsilon_1 \\ \epsilon_2 \\ \gamma_{12} \end{Bmatrix} = \underbrace{\{S\}[\sigma]}_{mech} + \underbrace{\{\alpha\}\Delta T}_{Thermal} + \underbrace{\{\beta\}c}_{Moisture}$$

or, in more concise matrix notation

$$\{\epsilon\} = [S]\{\sigma\} + \{\alpha\}\Delta T + \{\beta\}c \qquad (5.16)$$

whereupon the stresses are given by

$$\{\sigma\} = [S]^{-1}(\{\epsilon\} - \{\alpha\}\Delta T - \{\beta\}c) \qquad (5.17)$$

Note that if the material is unrestrained during the hygrothermal exposure, there are no stresses generated and the strains are given by

$$\{\epsilon\} = \{\alpha\}\Delta T + \{\beta\}c \qquad (5.18)$$

If the material is completely restrained during hygrothermal exposure, however, the total strain must be zero. Thus,

$$\{\epsilon\} = 0 = [S]\{\sigma\} + \{\alpha\}\Delta T + \{\beta\}c \qquad (5.19)$$

and the resulting hygrothermal stresses are given by

$$\{\sigma\} = [S]^{-1}(-\{\alpha\}\Delta T - \{\beta\}c) \qquad (5.20)$$

Note that there are no hygrothermal shear strains or shear stresses along the principal material axes. This is not true for the generally orthotropic (off-axis) case, however. For an arbitrary set of axes xy oriented at an angle θ to the 12 axes, the stress-strain relationships can be transformed as in Chap. 2. The complete stress-strain relations for the generally orthotropic lamina are

$$\begin{Bmatrix} \epsilon_x \\ \epsilon_y \\ \gamma_{xy} \end{Bmatrix} = \begin{bmatrix} \bar{S}_{11} & \bar{S}_{12} & \bar{S}_{16} \\ \bar{S}_{12} & \bar{S}_{22} & \bar{S}_{26} \\ \bar{S}_{16} & \bar{S}_{26} & \bar{S}_{66} \end{bmatrix} \begin{Bmatrix} \sigma_x \\ \sigma_y \\ \tau_{xy} \end{Bmatrix} + \begin{Bmatrix} \alpha_x \\ \alpha_y \\ \alpha_{xy} \end{Bmatrix}\Delta T + \begin{Bmatrix} \beta_x \\ \beta_y \\ \beta_{xy} \end{Bmatrix}c$$

$$(5.21)$$

In the transformations it must be remembered that the CTEs and the CHEs transform like *tensor strains* [recall Eqs. (2.33)], so that

$$\begin{Bmatrix} \alpha_x \\ \alpha_y \\ \alpha_{xy}/2 \end{Bmatrix} = [T]^{-1}\begin{Bmatrix} \alpha_1 \\ \alpha_2 \\ 0 \end{Bmatrix} \qquad (5.22)$$

*inverse
see page 50
eg 2.30*

and a similar equation is used for the CHEs. Notice that the hygrothermal effects *do* induce shear strains in the off-axis case due to α_{xy} and β_{xy}, the shear coefficients of thermal and hygroscopic expansion, respectively. This is quite different from the case of isotropic materials, where hygrothermal effects do not cause shear strains along any axes. The variations of α_x, α_y, and α_{xy} with lamina orientation according to Eqs.

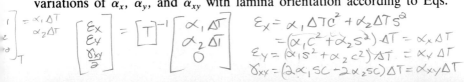

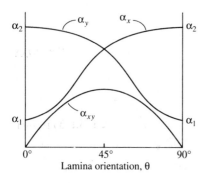

FIGURE 5.15
Variation of lamina thermal expansion coefficients with lamina orientation for a lamina having $\alpha_2 > \alpha_1 > 0$.

(5.22) are shown in Fig. 5.15. The same curves could also be used for β_x, β_y, and β_{xy}. The hygrothermal shear coefficients α_{xy} and β_{xy} have their maximum values at $\theta = 45°$ and are proportional to the differences $(\alpha_1 - \alpha_2)$ and $(\beta_1 - \beta_2)$, respectively. Thus, the greater the degree of anisotropy (i.e., the larger the ratio α_1/α_2 or β_1/β_2), the greater the hygrothermally induced shear strains. It is important to note that if $\alpha_1 < 0$ and $\alpha_2 > 0$, then it is possible to find an angle θ where $\alpha_x = 0$. Thus, we can design a laminate consisting of plies of such a material, so that the CTE along a particular direction is zero.

> **Example 5.3.** An orthotropic lamina forms one layer of a laminate which is initially at temperature T_o. Assuming that the lamina is initially stress-free, that adjacent lamina are rigid, that the properties do not not change as a result of the temperature change, and that the lamina picks up no moisture, determine the maximum temperature that the lamina can withstand according to the Maximum Stress Criterion.

see eq 5.19 & solve for {σ}

> **Solution.** Due to the assumption that adjacent laminae are rigid, deformation is prevented and the total strains must all be zero. The resulting hygrothermal stresses are therefore given by Eq. (5.20) with $c = 0$,
>
> $$\{\sigma\} = -[S]^{-1}\{\alpha\}(T - T_o) = -[Q]\{\alpha\}(T - T_o)$$
>
> Thus, for the Maximum Stress Criterion it is necessary to check each of the following conditions:
> For tensile stresses

NO shear Because $\alpha_{12} = 0$

> $$-(Q_{11}\alpha_1 + Q_{12}\alpha_2)(T - T_o) = s_L^{(+)}$$
> $$-(Q_{12}\alpha_1 + Q_{22}\alpha_2)(T - T_o) = s_T^{(+)}$$
>
> For compressive stresses
>
> $$-(Q_{11}\alpha_1 + Q_{12}\alpha_2)(T - T_o) = s_L^{(-)}$$
> $$-(Q_{12}\alpha_1 + Q_{22}\alpha_2)(T - T_o) = s_T^{(-)}$$
>
> (*Note*: There are no hygrothermal shear stresses along the 12 axes.)

select lowest T

After substituting numerical values for the initial temperature, T_o, the lamina stiffness, Q_{ij}, the coefficients of thermal expansion, α_i, the strengths $s_L^{(+)}$, etc. in the above equations, the equation which yields the lowest temperature T would be the condition governing failure. It is worthwhile to note that adjacent laminae are not really rigid, but we will need to use laminate theory later to consider deformations of adjacent laminae. It is also worthwhile to note that if hygrothermal degradation of properties is to be taken into account, Eq. (5.7) could be used to express the hygrothermally degraded lamina strengths and stiffnesses in terms of the temperature, T. In this case T would appear on both sides of the above equations and the problem would be more difficult to solve.

5.4 MICROMECHANICS MODELS FOR HYGROTHERMAL PROPERTIES

We have seen in Chaps. 3 and 4 that the mechanical properties of a composite lamina can be estimated from the corresponding properties of the constituent materials using micromechanics models. Similarly, micromechanics equations for the thermophysical properties that appear in hygrothermal analysis can be developed. Various theoretical approaches ranging from elementary mechanics of materials to energy methods have been proposed.

An equation for the longitudinal coefficient of thermal expansion, α_1, can be developed using the elementary mechanics of materials approach from Chap. 3. Recall that in the derivation of the rule of mixtures for the longitudinal modulus [Eq. (3.22)] the one-dimensional forms of the stress-strain relationships along the 1 direction for the lamina, fiber, and matrix materials [Eq. (3.19)] were substituted in the rule of mixtures for longitudinal stress, Eq. (3.18). The corresponding one-dimensional form of the stress-strain relationship including the thermal effect is

$$\epsilon_1 = \frac{\sigma_1}{E_1} + \alpha_1 \Delta T \qquad (5.23)$$

or
$$\sigma_1 = E_1(\epsilon_1 - \alpha_1 \Delta T) \qquad (5.24)$$

If we now substitute equations similar to Eq. (5.24) for composite, fiber, and matrix, respectively, into Eq. (3.18), the result is

$$E_1(\bar{\epsilon}_1 - \alpha_1 \Delta T) = E_{f1}(\bar{\epsilon}_{f1} - \alpha_{f1} \Delta T)v_f + E_{m1}(\bar{\epsilon}_{m1} - \alpha_{m1} \Delta T)v_m \qquad (5.25)$$

where α_{f1} and α_{m1} are the longitudinal CTEs of fiber and matrix materials, respectively (see Tables 3.1 and 3.2), and the remaining terms are defined in Chap. 3. By combining Eqs. (5.25), (3.21), and (3.22), we

In 1 dir $\varepsilon_{1c} = \varepsilon_{1f} = \varepsilon_{1m} = \varepsilon$ $\Delta T, \sigma \leftarrow \boxed{} \rightarrow \sigma, \Delta T$ *Assume perfect Bond*

$(2.1)_c = \frac{\sigma_c}{E_c} + \alpha_c \Delta T$

$\sigma_c = \sigma_f V_f + \sigma_m V_m$

$E_c [\varepsilon - \alpha_c \Delta T] = V_f E_f [\varepsilon - \alpha_f \Delta T] + V_m E_m [\varepsilon - \alpha_m \Delta T]$ *Vol frac* $\sigma_c = E_c[\varepsilon_{1c} - \alpha_c]$

$E_c = E_m V_m + E_f V_f$

$-E_c \alpha_c \Delta T = -\alpha \Delta T V_f E_f - \alpha_m \Delta T V_m E_m$ *solve for* α_c

get a modified rule of mixtures for the longitudinal CTE:

α_1 *in general*
$$\alpha_1 = \frac{E_{f1}\alpha_{f1}v_f + E_{m1}\alpha_{m1}v_m}{E_{1c}} = \frac{E_{f1}\alpha_{f1}v_f + E_{m1}\alpha_{m1}v_m}{E_{f1}v_f + E_{m1}v_m} \tag{5.26}$$

For the case of isotropic constituents the above equation becomes

α_1 *Isotropic*
$$\alpha_1 = \frac{E_f\alpha_f v_f + E_m\alpha_m v_m}{E_f v_f + E_m v_m} \tag{5.27}$$

This equation derived by a mechanics of materials approach turns out to be the same as the result obtained by Schapery [5.14], who used a more rigorous energy method. Hashin [5.15] derived a more complicated expression for the case of orthotropic constituents. Schapery [5.14] derived the following expression for the transverse CTE of a composite with isotropic constituents:

α_2 *in general* *vol frac*
$$\alpha_2 = (1 + v_m)\alpha_m v_m + (1 + v_f)\alpha_f v_f - \alpha_1 v_{12} \tag{5.28}$$

where α_1 is the longitudinal CTE given by Eq. (5.27) and v_{12} is the major Poisson's ratio given by Eq. (3.40). The variations of α_1 and α_2 with fiber volume fraction for a typical graphite/epoxy composite are shown in Fig. 5.16. Rosen [5.16] has observed that for such composites having high fiber volume fractions, the predicted α_1 is practically zero. Measurements of the CTEs for such materials by Ishikawa et al. [5.17] have confirmed that α_1 is so small as to fluctuate between positive and negative values due to small changes in temperature or fiber volume fraction. Over the range of practical fiber volume fractions α_2 is much greater than α_1. It is also interesting to note that at low fiber volume fractions α_2 can be greater than α_m.

By substituting the one-dimensional forms of the stress-strain

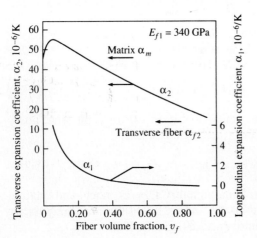

FIGURE 5.16
Variation of predicted longitudinal and transverse coefficients of thermal expansion with fiber volume fraction for typical unidirectional graphite/epoxy composite. (*From Rosen* [5.16]. *Reprinted by permission of ASM International.*)

$\Delta L_c = \Delta L_f + \Delta L_m \rightarrow \dfrac{\Delta L_f}{L_c} = \varepsilon_f \rightarrow \Delta L_f = \varepsilon_f L_f$

$\varepsilon_c = \dfrac{\varepsilon_f L_f}{L_c} + \dfrac{\varepsilon_m L_m}{L_c} \rightarrow \dfrac{L_f}{L_c}\left(\dfrac{A}{A}\right) \dfrac{V_f = V_f}{V}$

$\rightarrow \varepsilon_c = \varepsilon_f V_f + \varepsilon_m V_m \quad \text{if } \sigma = E\varepsilon$

$\dfrac{\sigma_c}{E_c} = \dfrac{\sigma_f}{E_f} V_f + \dfrac{\sigma_m}{E_m} V_m$

relationships with hygroscopic effects into Eq. (3.18) and following the same procedure outlined above in the derivation of Eq. (5.26), a similar relationship is found for the longitudinal CHE:

$$\beta_1 = \frac{E_{f1}\beta_{f1}v_f + E_{m1}\beta_{m1}v_m}{E_{f1}v_f + E_{m1}v_m} \tag{5.29}$$

In polymer matrix composites the amount of moisture absorbed by the fibers is usually negligible in comparison with the moisture absorbed by the matrix, so that the term involving β_{f1} can be ignored. For isotropic constituents the equation for β_1 would be analogous to Eq. (5.27). According to Ashton et al. [2.4], the equations derived by Schapery [i.e. Eqs. (5.27) and (5.28)] can be used for any expansional coefficients such as the CTE or the CHE. Thus, the transverse CHE would be given by

$$\beta_2 = (1 + v_m)\beta_m v_m + (1 + v_f)\beta_f v_f - \beta_1 v_{12} \tag{5.30}$$

where β_1 is given by the isotropic form of Eq. (5.29).

Recall that in the equations governing the temperature and moisture distributions [Eqs. (5.1) and (5.2)], thermophysical properties such as specific heat, thermal conductivity, and diffusivity appeared. According to Chamis [5.7], the composite specific heat is given by

$$C_c = \frac{1}{\rho_c}(\rho_f C_f v_f + \rho_m C_m v_m) \tag{5.31}$$

where C_f and C_m are the specific heats of fiber and matrix, respectively, the composite density, ρ_c, is given by Eq. (3.6), and the remaining terms are defined in Chap. 3. Ashton et al. [2.4] and Shen and Springer [5.3] have observed that the rule of mixtures formulations

$$K_1 = K_f v_f + K_m v_m \tag{5.32}$$

and
$$D_1 = D_f v_f + D_m v_m \tag{5.33}$$

can be used to find the longitudinal thermal conductivity and mass diffusivity, respectively, as well as other transport properties. Equations for the transverse thermal conductivity and diffusivity based on the method of subregions (see Sec. 3.3) have been presented by Hopkins and Chamis [3.11] and Chamis [5.7]. These equations can be formed by substituting the appropriate properties (thermal conductivities or diffusivities instead of transverse moduli) in an equation of the form shown in Eq. (3.49). Ashton et al. [2.4] have suggested that the Halpin-Tsai equations (see Sec. 3.5) can also be used for transverse transport properties such as thermal conductivity and mass diffusivity. Off-axis properties can be found by recognizing that thermal conductivity and diffusivity are both second-order tensor quantities which transform according to the form shown in Eqs. (2.30).

Finally, a procedure for estimating hygrothermal degradation of matrix properties such as α, β, K, and C has been proposed by Chamis [5.7]. Based on the observation that the effect of increased temperature on these properties is opposite from the corresponding effect on strength and stiffness, Chamis suggests that the matrix hygrothermal property retention ratio can be approximated by

$$F_h = \frac{R}{R_o} = \left[\frac{T_{go} - T_o}{T_{gw} - T} \right]^{1/2} \tag{5.34}$$

where R = matrix hygrothermal property after hygorthermal degradation

R_o = reference matrix hygrothermal property before degradation

Following a procedure similar to that outlined in Sec. 5.2, the matrix hygrothermal property is degraded according to Eq. (5.34). Then the degraded matrix property is used in a micromechanics equation such as Eqs. (5.26) through (5.33) to estimate the hygrothermally degraded composite property.

Example 5.4. A composite lamina is to be designed to have a specified coefficient of thermal expansion along a given direction. Outline a procedure to be used in the design.

Solution. First, it is necessary to use micromechanics equations like Eqs. (5.27) and (5.28) to find a combination of fiber and matrix materials having consitituent CTEs and moduli and volume fractions, so that the specified CTE lies between the values of α_1 and α_2. As shown by Eqs. (5.22) and Fig. 5.15, the value of the specified α_x along the direction defined by the angle θ must lie between the values of α_1 and α_2. The required angle θ is then found by setting α_x equal to the specified value and solving the first of Eqs. (5.22). In a practical design problem other constraints would have to be considered as well.

PROBLEMS

5.1. Using Eq. (5.6) for moisture diffusion, derive an equation for the time required for an initially dry material to reach 99.9% of its fully saturated equilibrium moisture content. The series in Eq. (5.6) converges rapidly, so for the purposes of this problem it is necessary only to consider the first term. The answer should be expressed in terms of the thickness, h, and the diffusivity, D_z.

5.2. The dependence of the transverse (through-the-thickness) diffusivity of unidirectional AS/3501-5 graphite/epoxy composite on temperature is given in Fig. 5.11. For a temperature of 77°C and a thickness of 2.54 mm, use the results from Fig. 5.11 and Problem 5.1 to estimate the time required for this material to reach 99.9% of its fully saturated equilibrium moisture content from an initially dry condition. Compare your estimate with the experimental data in Fig. 5.12. Does the estimate seem to be reasonable?

5.3. For the material described in Problems 5.1 and 5.2 at a temperature of 77°C, determine the time required for drying the material from 99.9 to 50% of its fully saturated equilibrium moisture content.

5.4. Using only the linear part of the moisture absorption curve for a temperature of 77°C in Fig. 5.12, and assuming a thickness of 2.54 mm, estimate the diffusivity D_z. Compare this value with the estimate from Fig. 5.11. 5.17×10^{-7} ~

5.5. For the composite properties and environmental conditions described in Examples 3.3, 4.3, and 5.2, determine the hygrothermally degraded values of the longitudinal and transverse tensile strengths. Compare with the reference values of these strengths from Example 4.3.

5.6. For the composite properties and environmental conditions described in Examples 3.3, 4.3, and 5.2, compare the reference and hygrothermally degraded values of the longitudinal compressive strength. Assume $v_{12} = 0.3$. Compare and discuss the different effects that hygrothermal conditions have on longitudinal tensile and compressive strengths.

5.7. The filament wound E-glass/epoxy pressure vessel described in Example 4.1 is to be used in a hot-wet environment with temperature $T = 100°F$ (38°C) and moisture content $M_m = 4\%$. The glass transition temperature of the dry epoxy resin is 250°F (121°C), and the lamina strengths listed in Table 4.1 are for a temperature of 70°F (21°C) and a moisture content of zero. Determine the internal pressure p which would cause failure of the vessel according to the Maximum Stress Criterion. Compare with the result from Example 4.1.

5.8. A graphite/epoxy lamina is clamped between rigid plates in a mold (Fig. 5.17) while curing at a temperature of 125°C. After curing the lamina/mold assembly (still clamped together) is cooled from 125°C to 25°C. The cooling process occurs in moist air and the lamina absorbs 0.5% of its weight in moisture. The lamina has the following properties:

$$E_1 = 140 \text{ GPa} \qquad \alpha_1 = -0.3 \times 10^{-6}/K$$
$$E_2 = 10 \text{ GPa} \qquad \alpha_2 = 28 \times 10^{-6}/K$$
$$v_{12} = 0.3 \qquad \beta_1 = 0$$
$$G_{12} = 7 \text{ GPa} \qquad \beta_2 = 0.44$$

Assuming that the lamina properties do not change over this temperature

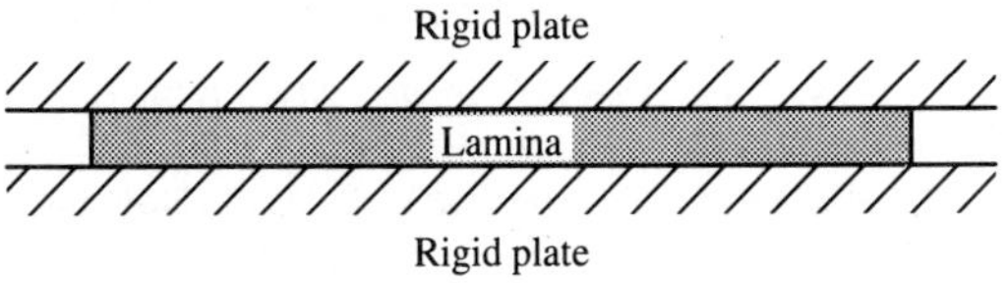

Rigid plate

Lamina

Rigid plate

FIGURE 5.17
Lamina clamped between rigid plates in a mold.

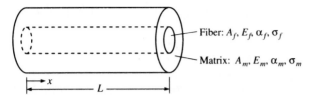

FIGURE 5.18
Representative volume element for Problem 5.11.

range and that the lamina is initially dry and stress free, determine the residual hygrothermal stresses in the lamina at 25°C for angles $\theta = 0°$ and 45°.

5.9. A unidirectional continuous fiber composite is to be made from T300 graphite fibers in a high modulus (HM) epoxy matrix, and the composite is to have a longitudinal coefficient of thermal expansion of zero. Using the fiber and matrix properties in Tables 3.1 and 3.2, determine the required fiber volume fraction. Is this a practical composite? Sketch a graph showing the longitudinal CTE of the composite vs. the fiber volume fraction, and show the range of fiber volume fractions over which the longitudinal CTE would be negative.

5.10. A unidirectional graphite/epoxy lamina having the properties described in Problem 5.8 is to be designed to have a coefficient of thermal expansion of zero along a particular axis. Determine the required lamina orientation for such a design. $\alpha = 0$ $\alpha_x = [cos^2\theta \quad sin^2\theta] \begin{bmatrix} \alpha_1 \\ \alpha_2 \end{bmatrix}$

5.11. A representative volume element (RVE) consisting of a cylindrical iso-tropic fiber embedded and perfectly bonded in a cylinder of isotropic matrix material is shown in Fig. 5.18. If the ends of the RVE at $x = 0$ and $x = L$ and the outer surface of the RVE are stress-free and the RVE is subjected to a uniform temperature change ΔT, determine the fiber stress, σ_f, and the matrix stress, σ_m, along the fiber direction at the midpoint of the RVE (at $x = L/2$). Use a mechanics of materials approach and express answers in terms of the coefficients of thermal expansion α_f and α_m, the cross-sectional areas A_f and A_m, the Young's moduli E_f and E_m, and the temperature change, ΔT, where the subscripts f and m refer to fiber and matrix, respectively.

REFERENCES

5.1. Browning, C. E., Husman, G. E., and Whitney, J. M., "Moisture Effects in Epoxy Matrix Composites," *Composite Materials: Testing and Design: Fourth Conference,* ASTM STP **617**, 481–496, American Society for Testing and Materials, Philadelphia, PA (1977).

5.2. Gibson, R. F., Yau, A. Mende, E. W., and Osborn, W. E., "The Influence of Environmental Conditions on the Vibration Characteristics of Chopped Fiber Reinforced Composite Materials," *Journal of Reinforced Plastics and Composites,* **1**(3), 225–241 (1982).

5.3. Shen, C. H. and Springer, G. S., "Moisture Absorption and Desorption of Composite Materials," *Journal of Composite Materials,* **10,** 2–20 (1976).

5.4. Jost, W., *Diffusion in Solids, Liquids, Gases,* Academic Press, Inc., New York (1952).

5.5. Loos, A. C., Springer, G. S., Sanders, B. A., and Tung, R. W., "Moisture Absorption of Polyester-E Glass Composites," in G. S. Springer (ed.), *Environmental Effects on Composite Materials,* 51–62, Technomic Publishing Co., Lancaster, PA (1981).

5.6. Chamis, C. C. and Sinclair, J. H., "Durability/Life of Fiber Composites in Hygrothermomechanical Environments," in I. M. Daniel (ed.), *Composite Materials: Testing and Design (Sixth Conference), ASTM STP* **787,** 498–512, American Society for Testing and Materials, Philadelphia, PA (1982).

5.7. Chamis, C. C., "Simplified Composite Micromechanics Equations for Mechanical, Thermal, and Moisture-Related Properties," in J. W. Weeton et al. (eds.), *Engineers' Guide to Composite Materials,* 3-8–3-24, ASM International, Materials Park, OH (1987).

5.8. Delasi, R. and Whiteside, J. B., "Effect of Moisture on Epoxy Resins and Composites," in J. R. Vinson (ed.), *Advanced Composite Materials—Environmental Effects,* ASTM STP **658,** 2–20, American Society for Testing and Materials, Philadelphia, PA (1987).

5.9. Loos, A. C. and Springer, G. S., "Moisture Absorption of Graphite/Epoxy Composites Immersed in Liquids and in Humid Air," in G. S. Springer (ed.), *Environmental Effects on Composite Materials,* 34–50, Technomic Publishing Co., Lancaster, PA (1981).

5.10. Fahmy, A. A. and Hurt, J. C., "Stress Dependence of Water Diffusion in Epoxy Resins," *Polymer Composites,* **1**(2), 77–80 (1980).

5.11. Marom, G. and Broutman, L. J., "Moisture Penetration into Composites under External Stress," *Polymer Composites,* **2**(3), 132–136 (1981).

5.12. Cairns, D. S. and Adams, D. F., "Moisture and Thermal Expansion Properties of Unidirectional Composite Materials and the Epoxy Matrix," in G. S. Springer (ed.), *Environmental Effects on Composite Materials: Volume 2,* 300–316, Technomic Publishing Co., Lancaster, PA (1984).

5.13. Graves, S. R. and Adams, D. F., "Analysis of a Bonded Joint in a Composite Tube Subjected to Torsion," *Journal of Composite Materials,* **15,** 211–224 (1981).

5.14. Schapery, R. A., "Thermal Expansion Coefficients of Composite Materials Based on Energy Principles," *Journal of Composite Materials,* **2**(3), 380–404 (1968).

5.15. Hashin, Z., "Analysis of Properties of Fiber Composites with Anisotropic Constituents," *Journal of Applied Mechanics,* **46,** 543–550 (1979).

5.16. Rosen, B. W., "Composite Materials Analysis and Design," in T. J. Reinhart (ed.), *Engineered Materials Handbook, Volume 1 Composites,* Sec. 4, ASM International, Materials Park, OH (1987).

5.17. Ishikawa, T., Koyama, K., and Kobayashi, S., "Thermal Expansion Coefficients of Unidirectional Composites," *Journal of Composite Materials,* **12,** 153–168 (1978).

CHAPTER
6

ANALYSIS OF A DISCONTINUOUS FIBER-REINFORCED LAMINA

6.1 INTRODUCTION

In Chaps. 2 to 5 we have discussed the analysis of continuous fiber-reinforced composites. The effects of fiber discontinuity or fiber length on composite mechanical behavior were not taken into account in these analyses since it was assumed that the fibers extended from one end of the lamina to the other end. This chapter is concerned with the mechanical behavior of laminae having discontinuous fiber, or short fiber reinforcement.

Short fiber-reinforced composites are not as strong or as stiff as continuous fiber-reinforced composites and are not likely to be used in critical structural applications such as aircraft primary structures. However, short fiber composites do have several attractive characteristics that make them worthy of consideration for other applications. For example, in components having complex geometrical contours, continuous fibers may not be practical because they may not conform to the desired shape without being damaged or distorted from the desired pattern. On the other hand, short fibers can be easily mixed with the liquid matrix resin,

156

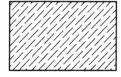 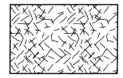

(a) Aligned discontinuous   (b) Off-axis aligned   (c) Randomly oriented
    fibers   discontinuous fibers   discontinuous fibers

FIGURE 6.1
Types of discontinuous fiber reinforcement.

and the resin/fiber mixture can be injection or compression molded to produce parts having complex shapes. Such processing methods are also fast and inexpensive, which makes them very attractive for high-volume applications. Composites having randomly oriented short fiber reinforcement are nearly isotropic, whereas unidirectional continuous fiber composites are highly anisotropic. In many applications the advantages of low cost, ease of fabricating complex parts, and isotropic behavior are enough to make short fiber composites the material of choice.

Short fiber composites with three types of fiber reinforcement will be considered here, as shown in Fig. 6.1: aligned discontinuous fibers, off-axis aligned discontinuous fibers, and randomly oriented discontinuous fibers. Although the randomly oriented short fiber composites are the most widely used of the three types, the development of the analytical models logically begins with the simplest case—aligned short fibers.

6.2 ALIGNED DISCONTINUOUS FIBERS

The analysis of the specially orthotropic aligned discontinuous fiber composite in Fig. 6.1(a) begins with the selection of a representative volume element (RVE) consisting of a short fiber embedded in a cylinder of matrix material, as shown in Fig. 6.2. Several models are based on the simplified RVE in Fig. 6.2(b), which does not include matrix material at

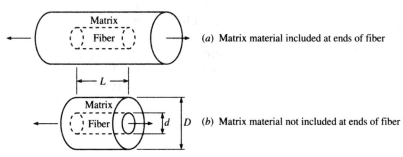

(a) Matrix material included at ends of fiber

(b) Matrix material not included at ends of fiber

FIGURE 6.2
Representative volume elements for aligned discontinuous fiber composite.

BEFORE DEFORMATION

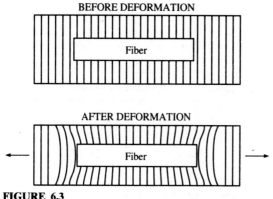

FIGURE 6.3
Schematic representation of matrix shear deformation in a short fiber composite.

the ends of the fiber as the model in Figure 6.2(a) does. Before beginning the analysis, however, it is instructive to consider the geometry of deformation in the RVE of Fig. 6.2(a). As shown by the grid lines before and after deformation in Fig. 6.3, the stiffness mismatch between fiber and matrix $(E_f \gg E_m)$ leads to large shear deformations near the fiber ends but no shear deformation at the middle of the fiber. As we will see later, the stress transfer between matrix and fiber occurs primarily through interfacial shear, which is the greatest near the fiber ends. On the other hand, the normal stress in the fiber builds from a minimum at the fiber ends to a maximum at the middle of the fiber.

The above observations based on the geometry of deformation will now be confirmed by considering the free-body diagram of a differential element of the fiber from the RVE, as shown in Fig. 6.4. For static equilibrium of the forces along the x direction

$$\sum F_x = (\sigma_f + d\sigma_f)\frac{\pi d^2}{4} - \sigma_f \frac{\pi d^2}{4} - \tau(\pi d)\, dx = 0 \qquad (6.1)$$

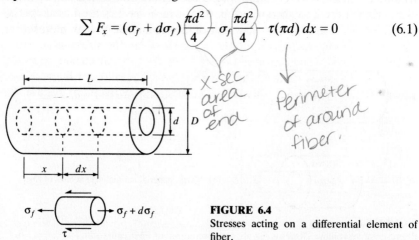

FIGURE 6.4
Stresses acting on a differential element of fiber.

where σ_f = fiber normal stress along the x direction at a distance x from end of fiber

F_x = force along the x direction

τ = interfacial shear stress at a distance x from end of fiber

d = fiber diameter, a constant

dx = length of differential element

$d\sigma_f$ = differential change in stress σ_f

Simplifying and rearranging the above equation, we get the differential equation relating the rate of change of the fiber normal stress along the x direction to the interfacial shear stress:

$$\frac{d\sigma_f}{dx} = \frac{4\tau}{d} \tag{6.2}$$

Separating variables and integrating, we find that

σ₀ is @ end of fiber

$$\int_{\sigma_0}^{\sigma_f} d\sigma_f = \frac{4}{d} \int_0^x \tau \, dx \tag{6.3}$$

Don't take into account end effects

It is commonly assumed that essentially all of the stress transfer from matrix to fiber occurs by interfacial shear around the periphery of the fiber, and that the fiber normal stress, σ_0, which is transferred across the ends of the fiber, is negligible. With this assumption, Eq. (6.3) becomes

not @ the ends.

$$\sigma_f = \frac{4}{d} \int_0^x \tau \, dx \tag{6.4}$$

Thus, if we want to determine the fiber stress, σ_f, we must know the interfacial shear stress, τ, as a function of the distance x. Two basic approaches have been proposed, both of which are based on assumptions regarding the behavior of the matrix material. Kelly and Tyson [6.1] assumed that the matrix is rigid-plastic, as shown in the stress-strain curve in Fig. 6.5(*a*). Cox [6.2] assumed that the matrix is linear elastic, as shown in Fig. 6.5(*b*). Both models are based on the assumption of linear elastic fibers. We will consider both models, but it is convenient to use the

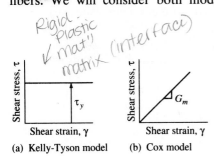

Rigid-Plastic mat'l matrix (interface)

(a) Kelly-Tyson model

Shear stress, τ

τ_y

Shear strain, γ

(b) Cox model

Shear stress, τ

G_m

Shear strain, γ

FIGURE 6.5

Assumed stress-strain curves for matrix material in Kelly-Tyson and Cox models.

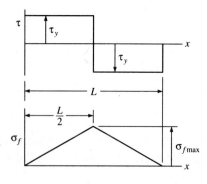

FIGURE 6.6
Variation of interfacial shear stress, τ, and fiber normal stress, σ_f, with distance along the fiber according to Kelly-Tyson model.

Kelly-Tyson model for illustrative purposes at this point. The Kelly-Tyson model is much simpler than the Cox model because the interfacial shear stress, τ, is everywhere equal to the matrix yield stress in shear, τ_y. Thus, for the Kelly-Tyson model the resulting fiber stress from Eq. (6.4) is now

$$\sigma_f = \frac{4}{d}\tau_y x \tag{6.5}$$

This equation tells us that the fiber stress varies linearly with the distance from the fiber end, but we also know that the fiber stress distribution must be symmetric about $x = L/2$. Since it has been assumed that $\sigma_f = \sigma_0 = 0$ at $x = 0$ and, by symmetry, at $x = L$, the fiber stress distribution and the corresponding shear stress distribution must be as shown in Fig. 6.6. The stress distributions in Fig. 6.6 are actually valid only for fibers having lengths less than a certain value, as we will see later. The maximum fiber stress for such a fiber occurs at $x = L/2$ and is given by

$$\sigma_{f\max} = \frac{4}{d}\tau_y\frac{L}{2} = \frac{2\tau_y L}{d} \tag{6.6}$$

The maximum fiber stress cannot keep increasing indefinitely as the fiber length L is increased, however. If the fiber is assumed to be elastic, $\sigma_{f\max}$ cannot exceed the value $E_{f1}\sigma_{c1}/E_1$, which is the fiber stress in a continuous fiber composite under longitudinal composite stress, σ_{c1} (recall Sec. 3.2.1). Thus, as $\sigma_{f\max}$ approaches the limiting value $E_{f1}\sigma_{c1}/E_1$, the fiber length, L, approaches a value L_i, which has been referred to as the "ineffective length" [6.3], or the "load transfer length" [6.4]. The equation for L_i is therefore

$$L_i = \frac{dE_{f1}\sigma_{c1}}{2\tau_y E_1} \tag{6.7}$$

The effect of increasing fiber length on the fiber stress and shear stress

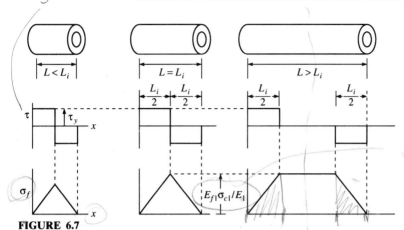

FIGURE 6.7
Effect of fiber length on stress distributions along fiber according to Kelly-Tyson model.

distributions is shown graphically in Fig. 6.7. Note that no matter how long the fiber is the load transfer between fiber and matrix (by virtue of the interfacial shear stress, τ) only occurs over the length, L_i. The length L_i has been referred to as the "ineffective length" because the fiber stress is less than its maximum value for this portion of the fiber. The term "load transfer length" comes from the fact that the load transfer between fiber and matrix only occurs over this portion of the fiber. Although these results are for the Kelly-Tyson model, similar results are obtained from the Cox model.

Another limiting value of the fiber stress, $\sigma_{f\,\text{max}}$, is the fiber tensile strength, $s_{f1}^{(+)}$, where it is assumed that $s_{f1}^{(+)} \geq E_{f1}\sigma_{c1}/E_1$. From Eq. (6.6), the fiber length corresponding to this maximum fiber stress is

$$L_c = \frac{ds_{f1}^{(+)}}{2\tau_y} \qquad (6.8)$$

which has been referred to as the "critical length."

Equation (6.8) has important implications for the calculation of composite strength. Recall that in Chap. 4 the equations for estimating the longitudinal tensile strength of a unidirectional composite were developed on the basis of either initial fiber failure or initial matrix failure. We now see that if the fiber length $L < L_c$, matrix failure will occur first and it is appropriate to use Eq. (4.26) to estimate the composite strength. If $L > L_c$, however, fiber failure occurs first and Eq. (4.21) should be used to estimate composite strength.

Alternatively, Eq. (6.8) can be rearranged to give the interfacial shear strength, τ_y, corresponding to the critical length

$$\tau_y = \frac{ds_{f1}^{(+)}}{2L_c} \qquad (6.9)$$

This equation has been used by Drzal et al. [6.5, 6.6] and others to determine the interfacial shear strength from measurements of critical length. In such an experiment, a specimen consisting of a single fiber embedded in a strip of translucent matrix material is mounted under a microscope and then subjected to an increasing tensile load. Once the fiber stress reaches $s_{f1}^{(+)}$, the fiber breaks up into segments having a statistical distribution about the critical length, L_c, and the corresponding statistical parameters describing the interfacial shear strength are calculated using Eq. (6.9).

Expressions for the longitudinal modulus of the aligned discontinuous fiber composite can be found using either the Kelly-Tyson model or the Cox model, but only the derivation of the Cox model, extended further by Kelly [6.7], will be discussed here. A similar model, which is often referred to as a "shear lag" model, was developed by Rosen [6.8]. For the RVE of Fig. 6.2(b), recall from Eq. (6.2) that the rate of change of the axial load in the fiber with respect to distance along the fiber is a linear function of the interfacial shear stress. Cox further assumed that the interfacial shear stress is proportional to the difference between u and v, where u is the axial displacement at a point in the fiber and v is the axial displacement the matrix would have at the same point in the RVE with no fiber present. Thus, the rate of change of the fiber axial load P is given by

$$\frac{dP}{dx} = H(u - v) \tag{6.10}$$

where H is a proportionality constant to be determined from geometrical and material property data. Differentiating Eq. (6.10) once, we find that

$$\frac{d^2P}{dx^2} = H\left(\frac{du}{dx} - \frac{dv}{dx}\right) = H\left(\frac{P}{A_f E_{f1}} - e\right) \tag{6.11}$$

where the expression $\dfrac{du}{dx} = \dfrac{P}{A_f E_{f1}}$ is taken from elementary mechanics of materials and $\dfrac{dv}{dx} = e$ is the matrix strain with no fiber present.

Equation (6.11) can be rearranged in the standard form of a second-order differential equation with constant coefficients as

$$\frac{d^2P}{dx^2} - \beta^2 P = -He \tag{6.12}$$

where $\beta^2 = \dfrac{H}{A_f E_{f1}}$. The solution to Eq. (6.12) is of the form

$$P = P_p + P_h \tag{6.13}$$

where P_p = particular solution = $A_f E_{f1} e$
P_h = homogeneous solution = $R \sinh \beta x + S \cosh \beta x$

The coefficients R and S must be determined from the boundary conditions $P = 0$ at $x = 0$ and $x = L$. After using trigonometric identities and further manipulation, the resulting fiber stress is

$$\sigma_f = \frac{P}{A_f} = E_{f1} e \left[1 - \frac{\cosh \beta (0.5L - x)}{\cosh (0.5\beta L)} \right] \tag{6.14}$$

The average fiber stress is then

$$\bar{\sigma}_f = \frac{\displaystyle\int_0^{L/2} \sigma_f \, dx}{L/2} = E_{f1} e \left[1 - \frac{\tanh (\beta L/2)}{\beta L/2} \right] \tag{6.15}$$

From equilibrium of the composite for longitudinal loading, recall the rule of mixtures for stress [Eq. (3.18)], which is also valid for the RVE of Fig. 6.2(b):

$$\bar{\sigma}_{c1} = \bar{\sigma}_{f1} v_f + \bar{\sigma}_m v_m \tag{6.16}$$

Substituting Eq. (6.15) in Eq. (6.16), dividing Eq. (6.16) by e; assuming that the applied composite stress produces a strain, e, in composite, fiber, and matrix; and using Hooke's law for composite and matrix, we find the equation for the longitudinal modulus of the Cox model:

[handwritten: ∝ (call it ∝ for now)]

[handwritten: effect of disc. fibers close to 1 when L ≈ 1mm]

$$E_{C1} = E_{f1} \left[1 - \frac{\tanh (\beta L/2)}{\beta L/2} \right] v_f + E_m v_m \tag{6.17}$$

Note that the assumption of equal strains in fiber and matrix here does not violate the original assumptions about u and v being different because v is the displacement in a piece of unreinforced matrix material. The term inside the brackets represents the effect of fiber length on the composite modulus.

The parameter β in the above equations and the interfacial shear stress, τ, can be determined by considering the shear strain in the matrix, as shown by Kelly [6.7]. The results are

[handwritten: H]

$$\beta^2 = \frac{2\pi G_m}{A_f E_{f1} \ln (D/d)} \tag{6.18}$$

and

$$\tau = \frac{d E_{f1} e \beta}{4} \left[\frac{\sinh [\beta (0.5L - x)]}{\cosh (0.5\beta L)} \right] \tag{6.19}$$

where G_m is the matrix shear modulus and D is the outside diameter of the RVE, as shown in Fig. 6.2. The predicted variations of the fiber stress and the interfacial shear stress from the Cox model when the fiber length

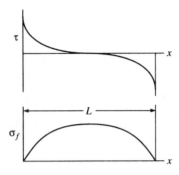

FIGURE 6.8
Variation of interfacial shear stress, τ, and fiber normal stress, σ_f, with distance along the fiber according to Cox model.

$L < L_i$ are shown schematically in Fig. 6.8. Notice the difference between these stress distributions and the ones from the Kelly-Tyson model in Fig. 6.6. For the Cox stresses evaluated at the midpoint of the fiber ($x = L/2$), as $L \to L_i$ the term in brackets in Eq. (6.14) approaches the value 1.0, whereas the term in brackets in Eq. (6.19) approaches zero.

Another variation on the Cox model was developed by Gibson et al. [6.9], who used the Cox stresses, σ_f and τ, in a strain energy method similar to that outlined in Eqs. (3.23) and (3.24). The longitudinal modulus calculated by the energy method was found to agree closely with Eq. (6.17), and the predicted variation of E_{C1} with fiber aspect ratio, L/d, is shown for several composites in Fig. 6.9. Notice that as the fiber length $L \to \infty$, $E_{C1} \to E_{f1}v_f + E_m v_m$, and that as $L \to 0$, $E_{C1} \to E_m v_m$. It is also interesting to see that the fiber length does not have to be very large relative to the fiber diameter in order to bring the modulus E_{C1} very close to the limiting value given by the rule of mixtures.

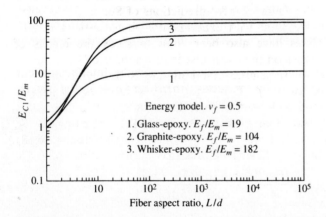

FIGURE 6.9
Variation of modulus ratio, E_{C1}/E_m, with fiber aspect ratio, L/d, for several composites. (*From Gibson et al.* [6.9]. *Reprinted by Permission of Chapman & Hall.*)

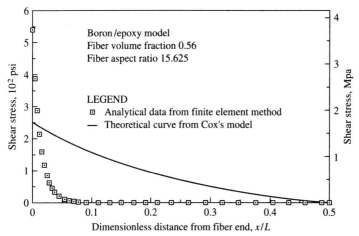

FIGURE 6.10
Predicted shear stress distributions along fiber from finite element analysis and Cox model. (*From Hwang* [6.10].)

Although the Kelly-Tyson model and the Cox model both provide valuable insight into the concepts of load transfer and fiber length effects, neither model accurately predicts the stress distributions. For example, more recent results from finite element analyses [6.10, 6.11] and experimental photoelasticity [6.7, 6.12, 6.13] indicate that both the magnitude and the rate of change of the interfacial shear stresses near the end of the fiber are much higher than those predicted by the Kelly-Tyson or Cox models. A typical comparison of predicted shear stress distributions along the fiber from finite element analysis and from the Cox model is shown in Fig. 6.10. The finite element predictions of Sun and Wu [6.11] also showed good agreement with experimental photoelasticity results. Finite element analyses have also been used to study the effects of different fiber end shapes on the stress distributions [6.10, 6.11].

It is important to remember that both the Kelly-Tyson and Cox models were derived for the RVE in Fig. 6.2(*b*), which does not include matrix material at the ends of the fiber. One result is that the actual modulus values are lower than predicted by Eq. (6.17). For example, the experimental results of Suarez et al. [6.14] on aligned discontinuous graphite/epoxy composites having various fiber aspect ratios, L/d, are shown in Fig. 6.11. The measured moduli are seen to be well below the predicted curve from the Cox model. In order to shift the predicted curve to match the experimental results better, Suarez et al. introduced the concept of an "effective fiber aspect ratio," $(L/d)_{\text{eff}}$, which would account for the fact that the reinforcement was not a single fiber but, rather, a bundle of fibers having an aspect ratio lower than that of a single fiber.

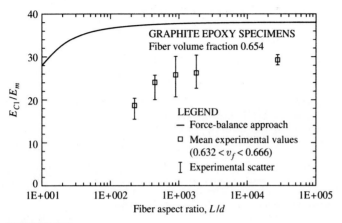

FIGURE 6.11
Comparison of measured and predicted (Cox model) longitudinal moduli of aligned discontinuous fiber graphite/epoxy for various fiber aspect ratios. $(L/d)_{\text{eff}} = L/d$. (*From Suarez et al.* [6.14].)

The effective fiber aspect ratio is defined as

$$(L/d)_{\text{eff}} = Z(L/d) \qquad (6.20)$$

where Z is a curve-fitting parameter which accomplishes a horizontal shift of the curve of E_{C1} vs. L/d. Before the horizontal shift the predicted curve was shifted vertically by using a reduced fiber modulus to account for possible degradation of fiber properties or fiber misalignment during fabrication. The results of vertical and horizontal shifting of the graphite/epoxy curve of Fig. 6.11 are shown in Fig. 6.12, and the agreement is very good. Similar results were reported for aramid/epoxy and boron/epoxy. This approach did not take into account the matrix material between the fiber ends, however.

Hwang and Gibson [6.15] studied the effect of the fiber end gap on the composite modulus by using both finite element analysis and a modified Cox model. The modified Cox model consists of the Cox model [Fig. 6.2(*b*)] with one piece of matrix material attached on each end, as shown schematically in Fig. 6.13. Following the development of Eq. (3.35) for the series arrangement of elements under longitudinal stress with the assumption of equal stresses in each element, the modified Cox modulus is

$$\frac{1}{E_{MC1}} = \frac{v_{C1}}{E_{C1}} + \frac{v_m}{E_m} = \frac{L/(L+e)}{E_{C1}} + \frac{e/(L+e)}{E_m} \qquad (6.21)$$

where E_{MC1} = longitudinal modulus of modified Cox model
v_{C1} = volume fraction of the Cox model in modified Cox model

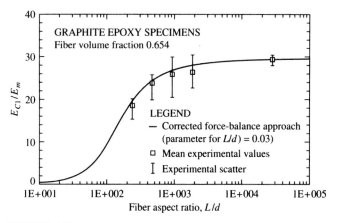

FIGURE 6.12
Comparison of measured and predicted (Cox model corrected for fiber aspect ratio) longitudinal moduli of aligned discontinuous fiber graphite/epoxy for various fiber aspect ratios. $(L/d)_{\text{eff}} = 0.03L/d$. (*From Suarez et al.* [6.14].)

$$L = \text{length of Cox model}$$
$$e = \text{distance between fiber ends in modified Cox model}$$
$$L + e = \text{length of modified Cox model}$$

Figure 6.14 shows a comparison of predictions from a finite element model and the modified Cox model with experimental data for boron/epoxy. The modified Cox model shows good agreement with both the finite element analysis and experimental data.

Halpin [6.16] has proposed a modification of the Halpin-Tsai equations (recall Sec. 3.5) as another approach to estimating the

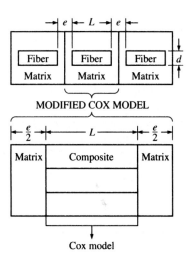

FIGURE 6.13
Modified Cox model which includes matrix material at ends of fiber. (*From Hwang and Gibson* [6.15]. *Reprinted by permission of ASME.*)

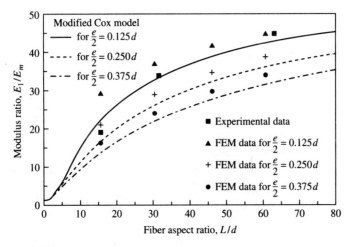

FIGURE 6.14
Comparison of predictions from modified Cox model and finite element analysis with experimental data for boron/epoxy aligned discontinuous fiber composite at different fiber aspect ratios. (*From Hwang and Gibson* [6.15]. *Reprinted by permission of ASME.*)

longitudinal modulus of the aligned discontinuous fiber composite. The proposed equations are

$$\frac{E_1}{E_m} = \frac{1 + \xi \eta v_f}{1 - \eta v_f} \tag{6.22}$$

where

$$\eta = \frac{(E_{f1}/E_m) - 1}{(E_{f1}/E_m) + \xi} \tag{6.23}$$

and the suggested value of the curve-fitting parameter is $\xi = 2L/d$. Figure 6.15 shows that the predictions from these equations give good agreement with experimental data. Halpin also concluded that E_2, G_{12}, and v_{12} are

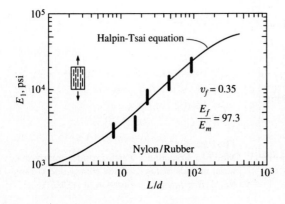

FIGURE 6.15
Dependence of longitudinal modulus on fiber aspect ratio for aligned discontinuous fiber nylon/rubber composite. Predictions from Halpin-Tsai equations are compared with experimental results. (*From Halpin* [6.16]. *Reprinted by permission of Technomic Publishing Co.*)

not significantly affected by the fiber length [6.16]. Thus, Eqs. (3.57) and (3.58) for E_2 in the continuous fiber case can also be used for the discontinuous fiber case. Similar equations can be used for G_{12}, as described in Sec. 3.5, and Eq. (3.40) can be used for v_{12}.

Example 6.1. An aligned short fiber graphite/epoxy composite is to be fabricated so that it behaves as a continuous fiber composite with a composite modulus of $E_1 = 80$ GPa. The 0.01-mm-diameter fibers have a modulus of elasticity $E_{f1} = 240$ GPa and a tensile strength $s_{f1}^{(+)} = 2.5$ GPa. The epoxy matrix can be assumed to be a rigid-plastic materal with a yield strength of 20 MPa in shear. Determine (a) the fiber length necessary to just reach the "continuous fiber stress" at the midpoint for a composite stress of 50 MPa and (b) the fiber length and the composite stress necessary to develop the ultimate tensile strength in the fiber.

Solution. (a) The "continuous fiber stress" is

$$\sigma_{f\,\max} = E_{f1}\sigma_{c1}/E_1 = 240(50)/80 = 150 \text{ MPa}$$

and the corresponding fiber length from Eq. (6.6) is

$$L = d\sigma_{f\,\max}/2\tau_y = 0.01(150)/2(20) = 0.0375 \text{ mm}$$

(b) The fiber length corresponding to a fiber stress $s_{f1}^{(+)}$ is found from Eq. (6.8):

$$L_c = d\sigma_{f1}^{(+)}/2\tau_y = 0.01(2500)/2(20) = 0.625 \text{ mm}$$

and the corresponding composite stress is

$$\sigma_{c1} = E_1\sigma_{f1}^{(+)}/E_{f1} = 80(2.5)/240 = 0.833 \text{ GPa} = 833 \text{ MPa}$$

6.3 OFF-AXIS ALIGNED DISCONTINUOUS FIBERS

 OMIT

The generally orthotropic aligned discontinuous fiber composite can be conveniently analyzed by using the RVE shown in Fig. 6.16, where the short fiber is oriented at an angle α with the loading axis. Chon and Sun [6.17] used this RVE to develop a generalized shear-lag analysis of the off-axis short fiber composite. Only the key results will be summarized here as the equations are quite lengthy. The predicted variations of the interfacial shear stress and the fiber stress with the distance along the fiber for various angles α are shown in Fig. 6.17. Note that the results

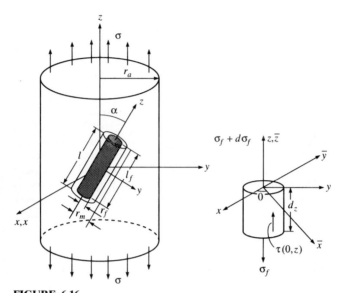

FIGURE 6.16
Representative volume element for an off-axis short fiber composite. (*From Chon and Sun* [6.17]. *Reprinted by permission of Chapman & Hall.*)

from the Cox model (recall Fig. 6.8) are recovered for the case of $\alpha = 0°$, and that the stress distribution curves are just shifted up or down as the angle α changes. Maximum values of shear stresses and fiber stresses normalized to the applied composite stress are shown for various angles α in Fig. 6.18. It is seen that the maximum interfacial shear stress, τ_{max}, occurs at some off-axis angle, that τ_{max} decreases with increasing E_f/G_m, and that the angle corresponding to τ_{max} increases with increasing E_f/G_m. Thus, the maximum interfacial shear stresses according to the Kelly-Tyson and Cox models are only maximum values for the case of $\alpha = 0°$. On the basis of these results, Chon and Sun suggest that if fiber failure is the composite failure mode, the matrix should be modified to reduce the ratio of E_f/G_m; but if failure is due to interfacial shear, E_f/G_m should be increased. In more recent work finite element analyses of off-axis short fiber composites, including the effects of fiber angle and fiber end geometry, were conducted by Sun and Wu [6.11].

Elastic constants for the off-axis aligned discontinuous fiber composite may be estimated by using equations developed earlier in this chapter and in Chaps. 2 and 3. Following the procedure outlined by Sun et al. [6.18] and Suarez et al. [6.14], the Cox model [Eq. (6.17)] is used to find the longitudinal modulus along the 1 direction. The transverse modulus, E_2, the in-plane shear modulus, G_{12}, and the major Poisson's ratio, ν_{12}, are assumed to be independent of fiber length [6.16] and are

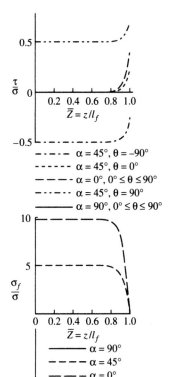

FIGURE 6.17
Variation of interfacial shear stress and fiber normal stress along fiber for Chon-Sun model at various off-axis angles. (*From Chon and Sun* [6.17]. *Reprinted by permission of Chapman & Hall.*)

calculated using the micromechanics equations developed in Chap. 3. The off-axis modulus of elasticity, E_x, is then found by substituting the Cox modulus, E_{C1}, for E_1 in the transformation equation [Eq. (2.39)], along with the calculated values of E_2, G_{12}, v_{12}, and θ. The other off-axis properties E_y, G_{xy}, and v_{xy} are found by using a similar approach. The resulting set of equations is of the form

$$E_x = f_1(E_{C1}, E_2, G_{12}, v_{12}, \theta)$$
$$E_y = f_2(E_{C1}, E_2, G_{12}, v_{12}, \theta)$$
$$G_{xy} = f_3(E_{C1}, E_2, G_{12}, v_{12}, \theta)$$
$$v_{xy} = f_4(E_{C1}, E_2, G_{12}, v_{12}, \theta)$$

(6.24)

A comparison of the predicted off-axis modulus, E_x, for graphite/epoxy with experimental values for various angles, θ, is shown in Fig. 6.19. It should be mentioned that the good agreement between theory and experiment seen in Fig. 6.19 was not possible as long as the fibers were assumed to be isotropic. Once the orthotropic nature of the graphite fibers was taken into account (i.e., $E_{f1} \gg E_{f2}$), the agreement between theory and experiment improved significantly. The same

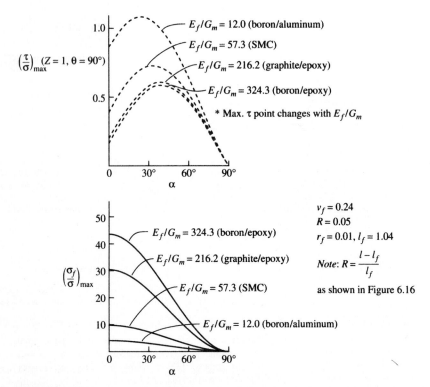

FIGURE 6.18
Variation of maximum interfacial shear stress and maximum fiber stress with off-axis angle from Chon-Sun model. (*From Chon and Sun [6.17]. Reprinted by permission of Chapman & Hall.*)

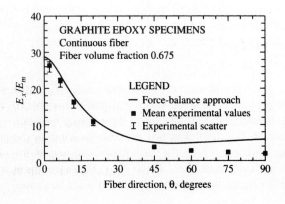

FIGURE 6.19
Comparison of predicted and measured off-axis modulus ratio, E_x/E_m, for graphite/epoxy. (*From Suarez et al. [6.14].*)

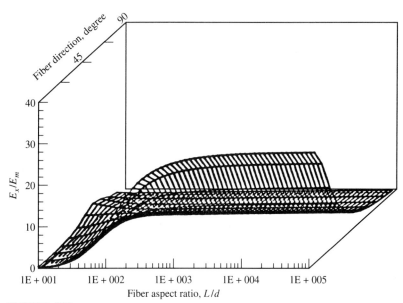

FIGURE 6.20
Tridimensional plot of E_x/E_m as a function of fiber aspect ratio and fiber orientation for graphite/epoxy. (*From Suarez et al.* [6.14].)

analysis was used to generate a tridimensional plot of the off-axis modulus, E_x, vs. the fiber aspect ratio and the fiber orientation, as shown in Fig. 6.20. Due to the assumption that E_2, G_{12}, and ν_{12} are independent of the fiber aspect ratio, L/d has little effect on the calculated E_x for fiber orientations other than those near $\theta = 0°$. As shown in the previous section, the fiber length required to attain the maximum composite stiffness at $\theta = 0°$ is quite small. Thus, the relatively low stiffness of practical short fiber composites is more likely to be caused by the off-axis orientation of the fibers than by the short length of the fibers. Another important factor that should not be overlooked is the fiber volume fraction. In most short fiber composites the maximum fiber volume fraction is quite low due to processing limitations. That is, the viscosity of the fiber/resin mixture must be kept below a certain limit for proper flow during the molding process. All these conclusions have important implications for the behavior of randomly oriented short fiber composites, which are discussed in the next section.

6.4 RANDOMLY ORIENTED DISCONTINUOUS FIBERS

If the fiber orientation in a composite is truly random in a three-dimensional sense, the composite exhibits three-dimensional isotropy.

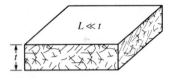 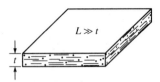

(a) Fiber length is less than thickness of part, so fibers are randomly oriented in three dimensions.

(b) Fiber length is greater than thickness of part, so fibers are randomly oriented in only two dimensions.

FIGURE 6.21
Three-dimensional and two-dimensional random orientations of fibers.

Such a situation is likely to exist when the fiber length, L, is much less than the thickness of the part, t, as shown in Fig. 6.21(a). Composites with low aspect ratio reinforcement such as whiskers or microfibers generally fall into this category. However, in many short fiber composite parts (e.g., panels made from sheet-molding compounds or resin transfer moldings) the fiber length is much greater than the thickness of the part, as shown in Fig. 6.21(b). In this case fiber orientation in the thickness direction is not possible and the material exhibits two-dimensional isotropy, or planar isotropy. The analysis of both types of materials will be discussed here, but the emphasis will be on the two-dimensional case.

One major conclusion from the previous section was that fiber orientation is more important than fiber length in the determination of off-axis elastic constants of unidirectional composites. Further support for this conclusion is provided by the observation that continuous fiber models give reasonably accurate predictions of elastic properties of randomly oriented fiber-reinforced composites. The concept of averaging the elastic constants over all possible orientations by integration was apparently introduced by Cox [6.2], who modeled paper as a planar mat of continuous fibers without matrix material. The Cox formulas for the averaged isotropic elastic constants of random arrays of fibers are given here for later reference, but they are not considered to be accurate enough for design use. For the two-dimensional case

$$\tilde{E} = \frac{E_f v_f}{3} \qquad \tilde{G} = \frac{E_f v_f}{8} \qquad \tilde{v} = \frac{1}{3} \tag{6.25}$$

and for the three-dimensional case

$$\tilde{E} = \frac{E_f v_f}{6} \qquad \tilde{G} = \frac{E_f v_f}{15} \qquad \tilde{v} = \frac{1}{4} \tag{6.26}$$

where $\tilde{E}$ = averaged Young's modulus for randomly oriented fiber composite

$\tilde{G}$ = averaged shear modulus for randomly oriented fiber composite

$\tilde{v}$ = averaged Poisson's ratio for randomly oriented fiber composite

Nielsen and Chen [6.19] used the averaging concept, along with micromechanics equations and transformation equations for a unidirectional continuous fiber-reinforced lamina, to analyze a planar isotropic composite. The geometrically averaged Young's modulus, which is assumed to be the same as the in-plane Young's modulus of the isotropic composite, is given by

$$\tilde{E} = \frac{\int_0^\pi E_x \, d\theta}{\int_0^\pi d\theta} \tag{6.27}$$

where the off-axis Young's modulus, E_x, is defined by Eq. (2.39), and the angle θ is defined in Fig. 2.6. Nielsen and Chen used a set of micromechanics equations for a unidirectional continuous fiber composite to calculate E_1, E_2, G_{12}, and v_{12} for use in Eq. (2.39). Figure 6.22 shows that the averaged modulus for the randomly oriented fiber composite is much lower than the corresponding longitudinal modulus, E_1, for most practical composites. Since the analysis is based on a continuous fiber model, the predicted reduction in modulus is due to fiber orientation, and not to fiber length. The equation that Nielsen and Chen used for E_2 was known to give values lower than measured values, so the predictions of Eq. (6.27) were also lower than the corresponding experimental values.

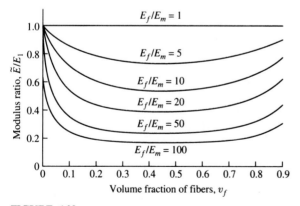

FIGURE 6.22
Dependence of modulus ratio, $\tilde{E}/E_1$, on fiber volume fraction for several values of E_f/E_m from Nielsen-Chen model. (*From Nielsen and Chen* [6.19]. *Copyright ASTM. Reprinted with permission.*)

The evaluation of Eq. (6.27) requires the integration of the expression for E_x given by Eq. (2.39), which is quite cumbersome. The integration is much simpler if the invariant forms of the transformed lamina stiffnesses are used. For example, the averaged value of the transformed lamina stiffness $\bar{Q}_{11}$ is given by

$$\tilde{Q}_{11} = \frac{\int_0^\pi \bar{Q}_{11}\, d\theta}{\int_0^\pi d\theta} = \frac{\int_0^\pi [U_1 + U_2 \cos 2\theta + U_3 \cos 4\theta]\, d\theta}{\pi} = U_1 \quad (6.28)$$

Similarly, $\tilde{Q}_{22} = U_1$, $\tilde{Q}_{12} = \tilde{Q}_{21} = U_4$, $\tilde{Q}_{66} = (U_1 - U_4)/2$, $\tilde{Q}_{16} = \tilde{Q}_{26} = 0$, and the stress-strain relations for any set of axes x, y in the plane are

$$\left\{ \begin{array}{c} \sigma_x \\ \sigma_y \\ \tau_{xy} \end{array} \right\} = \left[\begin{array}{ccc} U_1 & U_4 & 0 \\ U_4 & U_1 & 0 \\ 0 & 0 & (U_1 - U_4)/2 \end{array} \right] \left\{ \begin{array}{c} \epsilon_x \\ \epsilon_y \\ \gamma_{xy} \end{array} \right\} \quad (6.29)$$

Since this is an isotropic material, we can write

$$\tilde{Q}_{11} = U_1 = \frac{\tilde{E}}{1 - \tilde{v}^2} = \tilde{Q}_{22}$$

$$\tilde{Q}_{12} = U_4 = \frac{\tilde{v}\tilde{E}}{1 - \tilde{v}^2} \quad (6.30)$$

$$\tilde{Q}_{66} = \tilde{G} = \frac{\tilde{E}}{2(1 + \tilde{v})} = \frac{U_1 - U_4}{2}$$

Tsai and Pagano [6.20] and Halpin and Pagano [6.21] have obtained the same results by using invariant concepts along with quasi-isotropic laminate theory, which will be discussed later. Solving these equations for the isotropic engineering constants, we get

$$\tilde{E} = \frac{(U_1 - U_4)(U_1 + U_4)}{U_1}$$

$$\tilde{G} = \frac{U_1 - U_4}{2} \quad (6.31)$$

$$\tilde{v} = \frac{U_4}{U_1}$$

Using the equations relating the invariants in Eqs. (6.31) to the

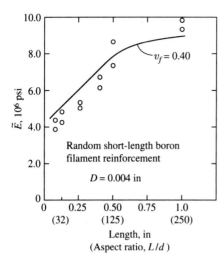

Random short-length boron
filament reinforcement

$D = 0.004$ in

Length, in
(Aspect ratio, L/d)

FIGURE 6.23
Dependence of Young's modulus of randomly oriented short fiber boron/epoxy composite on fiber aspect ratio. Comparison of predictions from Halpin-Tsai equations and invariant expressions with experimental data. (*From Halpin and Pagano* [6.21]. *Reprinted by permission of Technomic Publishing Co.*)

engineering constants E_1, E_2, G_{12}, and v_{12} for the orthotropic lamina [recall Eqs. (2.43) and (2.27)], Tsai and Pagano [6.20] also developed the following approximate expressions:

$$\tilde{E} = \tfrac{3}{8}E_1 + \tfrac{5}{8}E_2 \qquad \tilde{G} = \tfrac{1}{8}E_1 + \tfrac{1}{4}E_2 \qquad (6.32)$$

Rule of Thumb

These equations, along with the Halpin-Tsai equations for E_1 and E_2, were used to estimate the elastic moduli of randomly oriented boron fiber-reinforced epoxy, and the results compare favorably with experimental results (Fig. 6.23). Manera [6.22] also got good agreement with experimental results by using Eqs. (6.31) with a different set of micromechanics equations for E_1, E_2, G_{12}, and v_{12}.

Christensen and Waals [6.23] also used the averaging approach to find the isotropic elastic constants for continuous fiber composites with two-dimensional and three-dimensional random fiber orientation. This appears to be the first published report of the anaysis of a composite with three dimensionally oriented fibers although Cox [6.2] derived Eqs. (6.26) for the case of fibers without matrix material. Only the three-dimensional analysis of Christensen and Waals is summarized here since the two-dimensional analysis is quite similar to those which have already been discussed.

For the three-dimensional Christensen-Waals analysis the spherical coordinate system shown in Fig. 6.24 is used. An orthotropic, transversely isotropic composite with fibers oriented along the 1 direction is subjected to an arbitrary normal strain such as ϵ'_{33} along the 3' direction. For the purpose of the analysis the 3 axis is taken to be in the 1'2' plane. The basic premise of the solution is that the resulting ratio of stress to strain $\sigma'_{ij}/\epsilon'_{33}$ ($i, j = 1, 2, 3$) for a random orientation of fibers can be found by

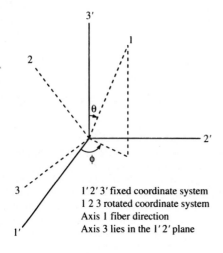

1′ 2′ 3′ fixed coordinate system
1 2 3 rotated coordinate system
Axis 1 fiber direction
Axis 3 lies in the 1′ 2′ plane

FIGURE 6.24
Spherical coordinates for three-dimensional Christensen-Waals analysis. (*From Christensen and Waals* [6.23]. *Reprinted by permission of Technomic Publishing Co.*)

calculating the average value of $\sigma'_{ij}/\epsilon'_{33}$ over all possible orientations of the fiber direction (1 axis) relative to the fixed x'_i axes. Using the three-dimensional stress-strain relationships for a generally orthotropic, transversely isotropic material [i.e., the stiffness matrix of Eq. (2.17) transformed to an arbitrary 1′2′3′ off-axis coordinate system], it can be shown that

$$\frac{\sigma'_{33}}{\epsilon'_{33}} = C_{11}\lambda_{31}^4 + (2C_{12} + 4C_{66})\lambda_{31}^2\lambda_{32}^2 + C_{22}\lambda_{32}^4 \tag{6.33}$$

and that

$$\frac{\sigma'_{22}}{\epsilon'_{33}} = C_{11}\lambda_{31}^2\lambda_{21}^2 + C_{12}(\lambda_{32}^2\lambda_{21}^2 + \lambda_{31}^2\lambda_{22}^2 + \lambda_{31}^2\lambda_{23}^2)$$

$$+ C_{22}\lambda_{32}^2\lambda_{22}^2 + 4C_{66}\lambda_{31}\lambda_{32}\lambda_{21}\lambda_{22} + C_{23}\lambda_{32}^2\lambda_{23}^2 \tag{6.34}$$

where the direction cosines λ_{ij} are given by

$$\lambda_{ij} = \begin{bmatrix} \sin\theta\cos\phi & -\cos\theta\cos\phi & \sin\phi \\ \sin\theta\sin\phi & -\cos\theta\sin\phi & -\cos\phi \\ \cos\theta & \sin\theta & 0 \end{bmatrix} \tag{6.35}$$

Averaging over all possible orientations of the fiber direction, we have

$$\left.\frac{\sigma'_{ij}}{\epsilon'_{33}}\right|_{\text{Random}} = \frac{\displaystyle\int_0^\pi \int_0^\pi \frac{\sigma'_{ij}}{\epsilon'_{33}} \sin\theta \, d\theta \, d\phi}{\displaystyle\int_0^\pi \int_0^\pi \sin\theta \, d\theta \, d\phi} \tag{6.36}$$

After substituting Eq. (6.33) in Eq. (6.36), we get

$$\frac{\sigma'_{33}}{\epsilon'_{33}}\Bigg|_{\text{Random}} = \frac{1}{15}(3C_{11} + 4C_{12} + 8C_{22} + 8C_{23}) \qquad (6.37)$$

For an equivalent homogeneous isotropic material the corresponding ratio of stress to strain is

$$\frac{\sigma'_{33}}{\epsilon'_{33}} = \frac{\tilde{E}(1 - \tilde{v})}{(1 + \tilde{v})(1 - 2\tilde{v})} \qquad (6.38)$$

Similarly, after substituting Eq. (6.34) in Eq. (6.36), we get

$$\frac{\sigma'_{22}}{\epsilon'_{33}}\Bigg|_{\text{Random}} = \frac{1}{15}(C_{11} + 8C_{12} + C_{22} - 4C_{66} + 5C_{23}) \qquad (6.39)$$

and the corresponding ratio of stress to strain for an equivalent homogeneous isotropic material is

$$\frac{\sigma'_{22}}{\epsilon'_{33}} = \frac{\tilde{v}\tilde{E}}{(1 + \tilde{v})(1 - 2\tilde{v})} \qquad (6.40)$$

Equating the ratio in Eq. (6.37) to that in Eq. (6.38), then equating the ratio in Eq. (6.39) to that in Eq. (6.40), and solving the two resulting equations simultaneously for the effective isotropic engineering constants, Christensen and Waals found that

$$\tilde{E} = \frac{[E_1 + (4v_{12}^2 + 8v_{12} + 4)K_{23}][E_1 + (4v_{12}^2 - 4v_{12} + 1)K_{23} + 6(G_{12} + G_{23})]}{3[2E_1 + (8v_{12}^2 + 12v_{12} + 7)K_{23} + 2(G_{12} + G_{23})]}$$

$$(6.41)$$

and

$$\tilde{v} = \frac{E_1 + (4v_{12}^2 + 16v_{12} + 6)K_{23} - 4(G_{12} + G_{23})}{4E_1 + (16v_{12}^2 + 24v_{12} + 14)K_{23} + 4(G_{12} + G_{23})} \qquad (6.42)$$

where K_{23} is the plane strain bulk modulus for dilatation in the 2–3 plane with $\epsilon_{11} = 0$, and the other properties are defined in Chap. 2. Christensen and Waals used previously developed micromechanics equations by Hashin [6.24, 6.25] and Hill [6.26] to calculate the five independent engineering constants E_1, v_{12}, G_{12}, G_{23}, and K_{23} which appear in Eqs. (6.41) and (6.42). Predictions from Eq. (6.41) for a glass/epoxy composite are shown in Fig. 6.25, along with the rule of mixtures prediction from Eq. (3.22) and the Cox prediction from Eq. (6.26). The prediction from the Cox model is well below that of the Christensen-Waals model, and the rule of mixtures prediction is much too high.

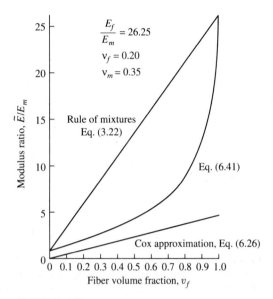

FIGURE 6.25
Comparison of Christensen-Waals three-dimensional analysis for Young's modulus of randomly oriented fiber composite with rule of mixtures and Cox approximation for a glass/epoxy composite. (*From Christensen and Waals* [6.23]. *Reprinted by permission of Technomic Publishing Co.*)

Using the same averaging technique, Christensen and Waals also developed a set of equations analogous to Eqs (6.41) and (6.42) for the two-dimensional case. The results are [6.23]

$$\tilde{E} = \frac{1}{u_1}(u_1^2 - u_2^2) \tag{6.43}$$

and

$$\tilde{v} = \frac{u_2}{u_1} \tag{6.44}$$

where

$$u_1 = \frac{3}{8}E_1 + \frac{G_{12}}{2} + \frac{(3 + 2v_{12} + 3v_{12}^2)G_{23}K_{23}}{2(G_{23} + K_{23})}$$

$$u_2 = \frac{1}{8}E_1 - \frac{G_{12}}{2} + \frac{(1 + 6v_{12} + v_{12}^2)G_{23}K_{23}}{2(G_{23} + K_{23})} \tag{6.45}$$

The results from Eqs. (6.43) to (6.45) for a glass/polystyrene composite are shown in Fig. 6.26. The Christensen-Waals model is seen to give much better agreement with the measurements than either the Cox

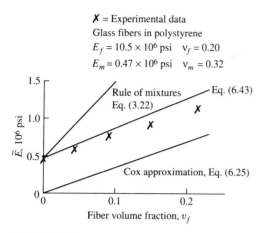

FIGURE 6.26
Comparison of Christensen-Waals two-dimensional analysis for Young's modulus of randomly oriented fiber composite with rule of mixtures and Cox approximation for a glass/polystyrene composite. (*From Christensen and Waals* [6.23]. *Reprinted by permission of Technomic Publishing Co.*)

model or the rule of mixtures, although none of the models takes into account the fiber length. Chang and Weng [6.27] also obtained good agreement with experimental results for glass/polyester sheet-molding compounds by using Eqs. (6.43) to (6.45). Christensen later presented simplified versions of these equations based on an asymptotic expansion [6.28, 6.29].

Up to this point, only continuous fiber models for randomly oriented fiber composites have been presented. Weng and Sun [6.30] used the Christensen-Waals equations along with micromechanics equations which were modified to account for the effect of fiber length. The effect of fiber length was modeled by using a so-called "fictitious fiber," which included the effect of matrix material at the ends of the fiber in the RVE shown in Fig. 6.2(*a*). The effects of varying stresses along the fiber were not accounted for, however, as it was assumed that the stresses were equal in the fiber and matrix portions of the fictitious fiber. The equation for the effective modulus of the fictitious fiber is analogous to Eq. (6.21) for the modified Cox model, except that the stress distribution along the fiber is assumed to be uniform. Figure 6.27 shows a comparison of the predictions of the modified Christensen-Waals theory with the original Christensen-Waals theory, the rule of mixtures, the Halpin-Tsai equations, and experimental data. For the glass/polyester sheet-molding compound material used, the effect of fiber length is apparently not very great, as the predictions of modified and original Christensen-Waals

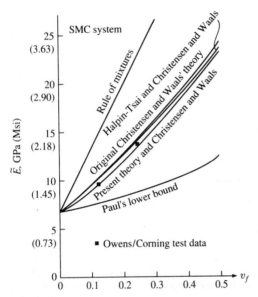

FIGURE 6.27

Comparison of various theories for prediction of Young's modulus of randomly oriented chopped glass/polyester sheet molding compound. (*From Weng and Sun* [6.30]. *Copyright ASTM. Reprinted with permission.*)

theories are almost the same. Both theories give predictions which are in good agreement with the experimental data.

The effects of fiber length and nonuniform stress distribution along the discontinuous fiber were accounted for by Sun et al. [6.31], who developed equations for the elastic moduli of two-dimensional randomly oriented short fiber composites as part of a study of vibration damping properties. A modified Cox model was used to determine E_1, while the other lamina properties were assumed to be independent of fiber length. The modified Cox model in this case is of the form

$$E_{MC1} = E_{f1}\left[1 - \frac{\tanh(\beta L/2)}{\beta L/2}\right]v_f\alpha + E_m v_m\gamma \qquad (6.46)$$

where α and γ are strain magnification factors which are determined from a finite element analysis. The modified Cox model for E_1, along with the rule of mixtures [Eq. (3.40)] for v_{12} and the Halpin-Tsai equations [Eqs. (3.57) and (3.58)] for E_2 and G_{12}, are used in transformation equations of the form described in Eqs. (6.24), which are then used in Eqs. (6.31) to determine the averaged isotropic engineering constants for the randomly oriented fiber composite. A tridimensional plot of the Young's modulus vs. the fiber aspect ratio, L/d, and the ratio E_f/E_m is

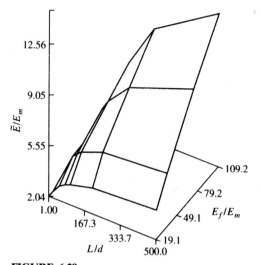

FIGURE 6.28
Tridimensional plot of $\tilde{E}/E_m$ as a function of L/d and E_f/E_m for randomly oriented short fiber composite. (*From Sun et al.* [6.31]. *Reprinted by permission of Technomic Publishing Co.*)

shown in Fig. 6.28. It is seen that high E_f/E_m and high L/d are required in order to have a high composite modulus. As with the aligned discontinuous case, the fiber aspect ratio required to attain maximum stiffness for given fiber and matrix materials is quite low.

The averaging technique has also been used to develop equations for predicting the strength of randomly oriented fiber composites. Lees [6.32] assumed that the angular dependence of the failure stress, σ_x, for such a material under uniaxial off-axis loading could be described by using the Maximum Stress Criterion. Lees also assumed that there are three failure mechanisms according to the Maximum Stress Criterion, each operating over a range of angles as follows [recalling Eqs. (4.3) for uniaxial off-axis loading]:

for $0 \leq \theta \leq \theta_1$, $\qquad \sigma_x = \dfrac{s_L^{(+)}}{\cos^2 \theta}$ $\qquad$ (longitudinal tensile failure)

for $\theta_1 \leq \theta \leq \theta_2$, $\qquad \sigma_x = \dfrac{s_{LT}}{\sin \theta \cos \theta}$ $\qquad$ (interfacial shear failure)

for $\theta_2 \leq \theta \leq \pi/2$, $\qquad \sigma_x = \dfrac{s_T^{(+)}}{\sin^2 \theta}$ $\qquad$ (transverse tensile failure)

where $\qquad \cot \theta_1 = \dfrac{s_L^{(+)}}{s_{LT}} \qquad$ and $\qquad \tan \theta_2 = \dfrac{s_T^{(+)}}{s_{LT}}$

For the case of the randomly oriented fiber composite Lees assumed that the average strength over all angles is given by

$$\tilde{\sigma}_x = \frac{2}{\pi} \left\{ \int_0^{\theta_1} \frac{s_L^{(+)}}{\cos^2 \theta} d\theta + \int_{\theta_1}^{\theta_2} \frac{s_{LT}}{\sin \theta \cos \theta} d\theta + \int_{\theta_2}^{\pi/2} \frac{s_T^{(+)}}{\sin^2 \theta} d\theta \right\} \quad (6.47)$$

After integrating and using Eq. (4.21) for $s_L^{(+)}$, then making some simplifying approximations, Lees found that

$$\tilde{\sigma}_x \simeq \frac{2s_{LT}}{\pi} \left[1 + \frac{s_T^{(+)}}{s_{mfl}} + \ln \frac{s_T^{(+)} s_{mfl}}{s_{LT}^2} \right] \quad (6.48)$$

Result is just a constant

where s_{mfl} is the matrix stress corresponding to the fiber failure strain. The same approach was later taken by Chen [6.33], who included a strength efficiency factor, ψ, to account for discontinuous fibers and obtained the equation

$$\tilde{\sigma}_x = \frac{2s_{LT}}{\pi} \left[2 + \ln \frac{\psi s_L^{(+)} s_T^{(+)}}{s_{LT}^2} \right] \quad (6.49)$$

Lees and Chen both reported reasonable agreement of their predictions with experimental data.

Another approach suggested by Halpin and Kardos [6.34] is based on the assumption that the strength of a randomly oriented fiber composite is the same as the strength of a quasi-isotropic laminate of the same material. Quasi-isotropic laminates, which are laminates of certain stacking sequences that behave in a planar isotropic manner, will be discussed later in Chap. 7 on laminates. Halpin and Kardos reported that the quasi-isotropic laminate model with the Maximum Strain Criterion for lamina failure gave good agreement with experimental data for a glass/epoxy composite [6.34].

Example 6.2. A graphite/epoxy composite with randomly oriented short fibers is made of the same constituent materials with the same fiber volume fraction as the material described in Examples 3.1, 3.3, and 4.3. Assuming that the in-plane shear strength $s_{LT} = 60$ MPa, and that the fiber length is much greater than the thickness of the material, estimate the Young's modulus, shear modulus, Poisson's ratio, and tensile strength of this composite.

Solution. From Eqs. (6.32), the Young's modulus is approximately

$$\tilde{E} = \tfrac{3}{8}E_1 + \tfrac{5}{8}E_2 = \tfrac{3}{8}(113) + \tfrac{5}{8}(5.65) = 45.9 \text{ GPa}$$

and the shear modulus is approximately

$$\tilde{G} = \tfrac{1}{8}E_1 + \tfrac{1}{4}E_2 = \tfrac{1}{8}(113) + \tfrac{1}{4}(5.65) = 15.54 \text{ GPa}$$

which means that the Poisson's ratio is

$$\tilde{\nu} = \frac{\tilde{E}}{2\tilde{G}} - 1 = \frac{45.9}{2(15.54)} - 1 = 0.47$$

From Eq. (6.48), the tensile strength is approximately

$$\tilde{\sigma}_x = \frac{2s_{LT}}{\pi}\left[1 + \frac{s_T^{(+)}}{s_{mfl}} + \ln\frac{s_T^{(+)}s_{mfl}}{s_{LT}^2}\right]$$

$$= \frac{2(60)}{\pi}\left[1 + \frac{66.9}{37.95} + \ln\frac{66.9(37.95)}{(60)^2}\right]$$

$$= 92.2 \text{ MPa}$$

Notice that the isotropic Young's modulus for the randomly oriented composite is much greater than the transverse modulus but less than half the longitudinal modulus of the corresponding orthotropic lamina. Likewise, the isotropic strength is greater than the orthotropic transverse strength but well below the orthotropic longitudinal strength. It is also important to remember that these predictions are based on randomly oriented continuous fibers, so that the differences between the isotropic properties and the orthotropic properties are due to fiber orientation, and not to fiber length.

Example 6.3. Determine the Young's modulus of a randomly oriented fiber composite if the unidirectional form of the composite has an off-axis Young's modulus which can be described by an equation of the form

$$E_x(\theta) = E_2 + (E_1 - E_2)[1 - (2\theta/\pi)^{1/3}]$$

where θ is the fiber angle in radians and E_1 and E_2 are the longitudinal and transverse Young's moduli, respectively, of the unidirectional composite.

Solution. The Young's modulus of the randomly oriented fiber composite, averaged over all angles is

$$\tilde{E} = \frac{2}{\pi}\int_0^{\pi/2} E_x(\theta)\, d\theta = \frac{2}{\pi}\int_0^{\pi/2}\{E_2 + (E_1 - E_2)[1 - (2\theta/\pi)^{1/3}]\}\, d\theta$$

$$= 0.25E_1 + 0.75E_2$$

If, say, $E_2 = 0.1E_1$ for a graphite/epoxy composite, then

$$\tilde{E} = 0.25E_1 + 0.75(0.1E_1) = 0.325E_1 \text{ or } 3.25E_2$$

These results again reflect the magnitude of the reduction in stiffness that can be expected because of fiber orientation effects alone since the fiber length has not been considered in this analysis.

PROBLEMS

6.1. A short fiber composite is to be modeled using the representative volume element (RVE) in Fig. 6.2(b). Assuming that the matrix is rigid-plastic in shear but that both the fiber and matrix are elastic in extension, develop an equation for the longitudinal modulus of the RVE. What values of the longitudinal modulus does the model give as the fiber length becomes very large? very small?

6.2. Using the result from Problem 6.1, develop an expression for the longitudinal modulus of the RVE shown in Fig. 6.2(a) which includes the effect of the matrix material at the fiber ends.

6.3. A graphite/epoxy single fiber test specimen is subjected to a uniaxial tensile stress which is increased until the fiber breaks up into pieces having a length of 0.625 mm. If the fiber has a diameter of 0.01 mm, a longitudinal modulus of 240 GPa, and an ultimate tensile strength of 2.5 GPa, what is the interfacial shear strength of the specimen? If the composite longitudinal modulus is 80 GPa, what applied composite stress is required to produce the condition above?

6.4. A linear elastic fiber of rectangular cross section is embedded in a linear elastic matrix material, and the composite is subjected to a uniaxial stress as shown in Fig. 6.29(a). The interfacial shear stress distribution along the

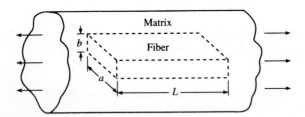

FIGURE 6.29(a)
Fiber with rectangular cross section embedded in matrix.

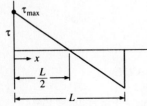

FIGURE 6.29(b)
Interfacial shear stress distribution along fiber shown in Fig. 6.29(a).

fiber is to be approximated by a linear function, as shown in Fig. 6.29(b). Determine the fiber length, L, that is required to develop the ultimate tensile stress, $s_{f1}^{(+)}$, at the midpoint of the fiber. Neglect the stress transmitted across the ends of the fiber.

6.5. A short fiber composite is made from boron fibers of length 0.125 in (3.175 mm) and diameter 0.0056 in. (0.142 mm) randomly oriented in a high modulus (HM) epoxy matrix with a fiber volume fraction of 0.4. Using the fiber and matrix properties in Tables 3.1 and 3.2, respectively, estimate the modulus of elasticity for the composite. Compare the modulus for the randomly oriented short fiber composite with the longitudinal and transverse moduli of an orthotropic aligned discontinuous fiber lamina of the same material.

6.6. Express the isotropic moduli $\tilde{E}$ and $\tilde{G}$ of a randomly oriented fiber composite in Eqs. (6.31) in terms of the orthotropic lamina stiffnesses Q_{ij}.

6.7. Determine the isotropic moduli $\tilde{E}$ and $\tilde{G}$ for a composite consisting of randomly oriented T300 graphite fibers in a 934 epoxy matrix if the fibers are long enough to be considered continuous. Use the properties in Table 2.2. Compare the values of $\tilde{E}$ and $\tilde{G}$ calculated from the invariant expressions [Eqs. (6.31)] with those calculated from the approximate expressions in Eqs. (6.32).

6.8. In order to reduce material costs, a composite panel is to be made by placing fibers in the matrix material in an X-pattern of $\pm\alpha$ as shown in Fig. 6.30 instead of randomly distributing the fibers over all angles. The X-pattern composite is to be designed so that it has at least 90% of the stiffness of the randomly oriented fiber composite along the longitudinal (L) axis. From tensile tests of a *unidirectional* composite consisting of the same fiber and matrix materials and the same fiber volume fraction, it is found that the off-axis Young's modulus of the composite can be described by the equation ~~composites~~

$$E_x(\theta) = 100 - 90 \sin\theta \quad \text{(GPa)} \qquad (0 \le \theta \le \pi/2)$$

whereas the Young's modulus of the matrix material is $E_m = 3.5$ GPa. Determine the angle α in Fig. 6.30 such that the longitudinal Young's

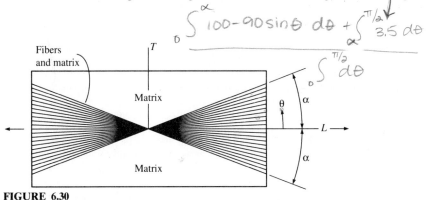

FIGURE 6.30
Composite panel with fibers arranged in X-pattern.

modulus of the X-pattern composite is equal to 90% of the Young's modulus of the randomly oriented fiber composite.

6.9. Determine the coefficient of thermal expansion for a randomly oriented fiber composite in terms of the longitudinal and transverse coefficients of thermal expansion α_1 and α_2 of the corresponding unidirectional composite lamina.

6.10. Using the Tsai-Hill Criterion, set up the equation for the averaged isotropic tensile strength for a randomly oriented fiber composite. The equation should be in terms of the angle θ and the unidirectional lamina strengths $s_L^{(+)}$, $s_T^{(+)}$, and s_{LT}.

REFERENCES

6.1. Kelly, A. and Tyson, W. R., "Tensile Properties of Fibre Reinforced Metals: Copper/Tungsten and Copper/Molybdenum," *Journal of the Mechanics and Physics of Solids*, **13**, 329–350 (1965).

6.2. Cox, H. L., "The Elasticity and Strength of Paper and Other Fibrous Materials," *British Journal of Applied Physics*, **3**, 72–79 (1952).

6.3. Rosen, B. W., "Composite Materials Analysis and Design," in T. J. Reinhart (ed.), *Engineered Materials Handbook, Volume 1 Composites*, Sec. 4, 173–281, ASM International, Materials Park, OH (1987).

6.4. Agarwal, B. D. and Broutman, L. J., *Analysis and Performance of Fiber Composites*, 2d ed., John Wiley & Sons, Inc., New York (1990).

6.5. Drzal, L. T., Rich, M. J., and Lloyd, P. F., "Adhesion of Graphite Fibers to Epoxy Matrices: I. The Role of Fiber Surface Treatment," *Journal of Adhesion*, **16**, 1–30 (1982).

6.6. Drzal, L. T., Rich, M. J., Koenig, M. F., and Lloyd, P. F., "Adhesion of Graphite Fibers to Epoxy Matrices: II. The Effect of Fiber Finish," *Journal of Adhesion*, **16**, 133–152 (1983).

6.7. Kelly, A., *Strong Solids*, 2d ed., Clarendon Press, Oxford, England (1973).

6.8. Rosen, B. W., "Mechanics of Composite Strengthening," in *Fiber Composite Materials*, Chap. 3, 37–75, American Society for Metals, Metals Park, OH (1965).

6.9. Gibson, R. F., Chaturvedi, S. K., and Sun, C. T., "Complex Moduli of Aligned Discontinuous Fibre Reinforced Polymer Composites," *Journal of Materials Science*, **17**, 3499–3509 (1982).

6.10. Hwang, S. J., *Finite Element Modeling of Damping in Discontinuous Fiber Composites*, M. S. Thesis, University of Idaho, Moscow, ID (1985).

6.11. Sun, C. T. and Wu, J. K., "Stress Distribution of Aligned Short Fiber Composites under Axial Load," *Journal of Reinforced Plastics and Composites*, **3**, 130–144 (1984).

6.12. Tyson, W. R. and Davies, G. J., "A Photoelastic Study of the Shear Stress Associated with the Transfer of Stress during Fiber Reinforcement," *British Journal of Applied Physics*, **16**, 199–205 (1965).

6.13. MacLaughlin, T. F., "A Photoelastic Analysis of Fiber Discontinuities in Composite Materials," *Journal of Composite Materials*, **2**(1), 44–45 (1968).

6.14. Suarez, S. A., Gibson, R. F., Sun, C. T., and Chaturvedi, S. K., "The Influence of Fiber Length and Fiber Orientation on Damping and Stiffness of Polymer Composite Materials," *Experimental Mechanics*, **26**(2), 175–184 (1986).

6.15. Hwang, S. J. and Gibson, R. F., "Micromechanical Modeling of Damping in Discontinuous Fiber Composites Using a Strain Energy/Finite Element Approach," *Journal of Engineering Materials and Technology*, **109**, 47–52 (1987).

6.16. Halpin, J. C., "Stiffness and Expansion Estimates for Oriented Short Fiber Composites, *Journal of Composite Materials*, **3**, 732–734 (1969).

6.17. Chon, C. T. and Sun, C. T., "Stress Distribution Along a Short Fiber in Fiber Reinforced Plastics," *Journal of Materials Science*, **15**, 931–938 (1980).

6.18. Sun, C. T., Gibson, R. F., and Chaturvedi, S. K., "Internal Damping of Polymer Matrix Composites under Off-Axis Loading," *Journal of Materials Science*, **20**, 2575–2585 (1985).

6.19. Nielsen, L. E. and Chen, P. E., "Young's Modulus of Composites Filled with Randomly Oriented Fibers," *Journal of Materials*, **3**(2), 352–358 (1968).

6.20. Tsai, S. W. and Pagano, N. J., "Invariant Properties of Composite Materials," in S. W. Tsai, J. C. Halpin, and N. J. Pagano (eds.), *Composite Materials Workshop*, 233–252, Technomic Publishing Co. Lancaster, PA (1968).

6.21. Halpin, J. C. and Pagano, N. J., "The Laminate Approximation for Randomly Oriented Fibrous Composites," *Journal of Composite Materials*, **3**, 720–724 (1969).

6.22. Manera, M., "Elastic Properties of Randomly Oriented Short Fiber-Glass Composites," *Journal of Composite Materials*, **11**, 235–247 (1977).

6.23. Christensen, R. M. and Waals, F. M., "Effective Stiffness of Randomly Oriented Fibre Composites," *Journal of Composite Materials*, **6**, 518–532 (1972).

6.24. Hashin, Z., "On Elastic Behavior of Fibre Reinforced Materials of Arbitrary Transverse Phase Geometry," *Journal of the Mechanics and Physics of Solids*, **13**, 119–134 (1965).

6.25. Hashin, Z., "Viscoelastic Fiber Reinforced Materials," *AIAA Journal*, **4**, 1411–1417 (1966).

6.26. Hill, R., "Theory of Mechanical Properties of Fiber-Strengthened Materials: I. Elastic Behavior," *Journal of the Mechanics and Physics of Solids*, **12**, 199–212 (1964).

6.27. Chang, D. C. and Weng, G. J., "Elastic Moduli of Randomly Oriented Chopped Fibre Composites with Filled Resin," *Journal of Materials Science*, **14**, 2183–2190 (1979).

6.28. Christensen, R. M., "Asymptotic Modulus Results for Composites Containing Randomly Oriented Fibers," *International Journal of Solids and Structures*, **12**, 537–544 (1976).

6.29. Christensen, R. M., *Mechanics of Composite Materials*, John Wiley & Sons, New York (1979).

6.30. Weng, G. J. and Sun, C. T., "Effects of Fiber Length on Elastic Moduli of Randomly Oriented Chopped Fiber Composites," in S. W. Tsai (ed.), *Composite Materials: Testing and Design (Fifth Conference)*, ASTM STP **674**, 149–162, American Society for Testing and Materials, Philadelphia, PA (1979).

6.31. Sun, C. T., Wu, J. K., and Gibson, R. F., "Prediction of Material Damping in Randomly Oriented Short Fiber Polymer Matrix Composites," *Journal of Reinforced Plastics and Composites*, **4**, 262–272, (1985).

6.32. Lees, J. K., "A Study of the Tensile Strength of Short Fiber Reinforced Plastics," *Polymer Engineering and Science*, **8**(3), 195–201 (1968).

6.33. Chen, P. E., "Strength Properties of Discontinuous Fiber Composites," *Polymer Engineering and Science*, **11**(1), 51–55 (1971).

6.34. Halpin, J. C. and Kardos, J. L., "Strength of Discontinuous Reinforced Composites: I. Fiber Reinforced Composites," *Polymer Engineering and Science*, **18**(6), 496–504 (1978).

CHAPTER

7

ANALYSIS OF LAMINATES (CLT)

Based on Kirchoff Hypothesis.

7.1 INTRODUCTION

While an understanding of lamina mechanical behavior is essential to the development of theories for the analysis of composite structures, the unidirectional lamina alone is generally not very useful as a structural element because of its poor transverse properties. Composite structures are more likely to be in the form of laminates consisting of multiple laminae, or plies, oriented in the desired directions and bonded together in a structural unit. The virtually limitless combinations of ply materials, ply orientations, and ply stacking sequences offered by laminated construction considerably enhance the design flexibility inherent in composite structures.

In this chapter the analysis of laminates will be introduced by considering a simplified theory of laminated beams in pure flexure. This will be followed by a discussion of the more general Classical Lamination Theory, which makes it possible to analyze the complex coupling effects that may occur in laminates. Other aspects of laminate analysis such as prediction of thermal and residual stresses, interlaminar stresses, and laminate strength are also discussed.

190

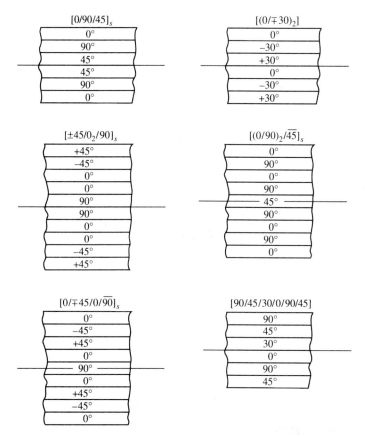

FIGURE 7.1
Examples of laminate stacking sequences and the corresponding laminate orientation codes.

Because of the need for adequate description of many possible combinations of ply orientations and stacking sequences in laminates, a laminate orientation code has evolved in the composites literature. The basis of the code is that ply angles, separated by slashes, are listed in order from the top surface to the bottom surface and enclosed in square brackets, as shown by the examples in Fig. 7.1. Note that symmetric laminates can be described by listing only the ply angles for the top half of the laminate and by using the subscript "s" outside the brackets, and that adjacent plies having the same orientations can be described by using a numerical subscript on the appropriate ply angle. In the case of symmetric laminates having an odd number of plies the center ply angle is denoted by an overbar. Sets of ply angles which are repeated in the laminate are identified by enclosing the set of angles in parentheses. The examples shown in Fig. 7.1 are for laminates consisting of plies of the same material. For hybrid laminates having plies of different materials additional subscripts on the ply angles may be used to identify the ply material.

7.2 THEORY OF LAMINATED BEAMS IN PURE FLEXURE

For the purpose of analysis, the simplest laminated structure is a laminated beam which is subjected to pure bending. A theory of laminated beams in pure flexure can be developed from the Bernoulli-Euler theory of elementary mechanics of materials. Although the application of this theory is quite restricted, it yields considerable insight into the analysis of laminated structures and provides a natural introduction to the more general Classical Lamination Theory, which is described in the next section. The theory described here is based on the analysis of Pagano [7.1].

A section of a rectangular laminated beam of depth h and width b is shown in Fig. 7.2 before and after the application of a bending moment M. The assumptions used in developing the analysis are as follows:

1. Plane sections which are initially normal to the longitudinal axis of the beam remain plane and normal during flexure.
2. The beam has both geometric and material property symmetry about the neutral surface (i.e., the plies are symmetrically arranged about the xy plane).
3. Each ply is linearly elastic with no shear coupling (i.e., ply orientations are either 0° or 90°).

Kirchoff
   hypoth.

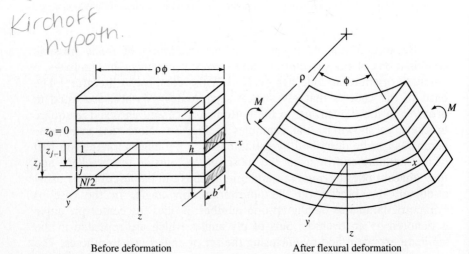

Before deformation   After flexural deformation

FIGURE 7.2
An element of a laminated beam before and after the application of a bending moment.

4. The plies are perfectly bonded together, so that no slip occurs at ply interfaces.

5. The only stress components present are σ_x and τ_{xz}.

As a result of assumption 1, the longitudinal normal strain at a distance z from the neutral surface is given by the familiar equation

$$\epsilon_x = \frac{(\rho + z)\phi - \rho\phi}{\rho\phi} = \frac{z}{\rho} \tag{7.1}$$

where ρ = radius of curvature of the neutral surface during flexure
ϕ = angle defined in Fig. 7.2
z = distance from neutral surface defined by the xy plane

From assumption 3 the longitudinal stress in the jth ply is given by

$$(\sigma_x)_j = (E_x)_j(\epsilon_x)_j \tag{7.2}$$

where $(E_x)_j$ is the Young's modulus of jth ply along the x directon and $(\epsilon_x)_j$ is the longitudinal strain in the jth ply along the x direction. From Eqs. (7.1) and (7.2), the longitudinal stress is seen to be

$$(\sigma_x)_j = (E_x)_j\frac{z}{\rho} \tag{7.3}$$

Static equilibrium requires that the applied bending moment M must be related to the longitudinal stresses by   *x-sec area*

$$M = 2\int_0^{h/2} \sigma_x zb\,dz \tag{7.4}$$

where the symmetry assumption 2 has been used. Substitution of Eq. (7.3) in Eq. (7.4) gives

$$M = \frac{2b}{3\rho}\sum_{j=1}^{N/2} (E_x)_j(z_j^3 - z_{j-1}^3) \longrightarrow \tag{7.5}$$

where N is the total number of plies and z_j is the distance from the neutral surface to the outside of the jth ply. For an even number of plies of uniform thickness $z_j = jh/N$ and Eq. (7.5) becomes

Even # plies
uniform
thickness.
$$M = \frac{2bh^3}{3\rho N^3}\sum_{j=1}^{N/2} (E_x)_j(3j^2 - 3j + 1) \tag{7.6}$$

Equation (7.6) can also be used for an odd number of plies if we simply divide each ply into two identical plies having half the thickness of the original ply, so that the total number of plies is now even.

Recall that for a homogeneous, isotropic beam the moment-curvature relation is given by

$\frac{1}{\rho} = \frac{M}{EI}$
$$M = \frac{E_f I_{yy}}{\rho} = \frac{E_f bh^3}{12\rho} \tag{7.7}$$

where $I_{yy} = \int z^2 \, dA = \dfrac{bh^3}{12}$ is the moment of inertia of cross section about the neutral axis (y axis)

A = cross-sectional area

E_f = effective flexural modulus of the beam (which is same as Young's modulus of beam material for a homogeneous, isotropic beam)

Combining Eqs. (7.5) and (7.7), we find that the effective flexural modulus of the laminated beam can be expressed as

Effective Flexural Modulus.

$$E_f = \frac{8}{h^3} \sum_{j=1}^{N/2} (E_x)_j (z_j^3 - z_{j-1}^3) \quad \text{Pure Bending} \tag{7.8}$$

or for an even number of plies we can combine Eqs. (7.6) and (7.7) to get

Even # of plies

$$E_f = \frac{8}{N^3} \sum_{j=1}^{N/2} (E_x)_j (3j^2 - 3j + 1) \tag{7.9}$$

Thus, the flexural modulus of the laminated beam, unlike the Young's modulus of the homogeneous isotropic beam, depends on the ply stacking sequence and the ply moduli. That is, if the properties do not change through the thickness of a beam, the flexural modulus is the same as the Young's modulus.

The deflections of laminated beams can now be calculated by using the flexural modulus in place of the Young's modulus in the beam deflection equations from elementary mechanics of materials. For example, the differential equation for the transverse deflection, w, of a laminated beam would be of the form

$$E_f I_{yy} \frac{d^2 w}{dx^2} = M \tag{7.10}$$

and the maximum deflection at the tip of the laminated cantilever beam in Fig. 7.3 would be given by the familiar equation

$$w_{\max} = \frac{PL^3}{3 E_f I_{yy}} \quad \text{at } x = L \tag{7.11}$$

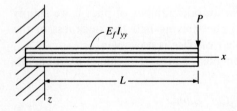

FIGURE 7.3
Cantilevered laminated beam under a concentrated tip load.

where P is the applied tip load and L is the beam length. The Euler buckling load, P_{cr}, for a laminated beam can be estimated by the formula

$$P_{cr} = \frac{\pi^2 E_f I_{yy}}{L_e^2} \qquad (7.12)$$

where L_e is the effective length which includes the effect of end conditions. Similarly, other beam equations involving the Young's modulus can now be modified for use with laminated beams.

An alternative expression for the stress in the jth ply can be obtained by combining Eqs. (7.3) and (7.7) and by eliminating the radius of curvature:

$$(\sigma_x)_j = \frac{M}{E_f I_{yy}}(E_x)_j z = \frac{Mz}{I_{yy}}\left[\frac{(E_x)_j}{E_f}\right] \qquad (7.13)$$

constant *correction term* *max stress*

Thus, the term in square brackets can be thought of as a correction term which when multiplied by the familiar homogeneous isotropic beam stress, Mz/I_{yy}, gives the stress in the jth ply of the laminated beam. Another important observation is that the maximum stress in the laminated beam does not always occur on the outer surface as it does in the homogeneous, isotropic beam. At each section in a laminated beam the ratio $M/E_f I_{yy}$ is constant, and the remaining term $(E_x)_j z$ determines the maximum stress. The maximum stress in the laminated beam therefore occurs in the ply having the greatest product of modulus $(E_x)_j$ and distance from the neutral axis, z. For the homogeneous isotropic beam the stress at a given point in the cross section depends only on the distance z, and the maximum stress occurs at the outer surface where z is the greatest. The stress distributions in homogeneous isotropic beams and laminated beams are compared schematically in Fig. 7.4.

Failure of laminated beams can be estimated by using the stress from Eq. (7.13) in one of the failure criteria that were discussed in Chap. 4. For example, if the jth ply is a longitudinal (0°) ply in compression,

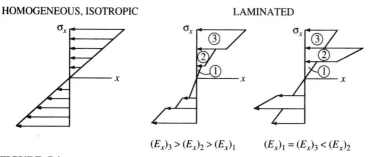

HOMOGENEOUS, ISOTROPIC   LAMINATED

$(E_x)_3 > (E_x)_2 > (E_x)_1 \qquad (E_x)_1 = (E_x)_3 < (E_x)_2$

FIGURE 7.4
Stress distributions in homogeneous, isotropic beams and in laminated beams.

failure in this ply according to the Maximum Stress Criterion will occur when $(\sigma_{xmax})_j = s_L^{(-)}$, where $(\sigma_{xmax})_j$ is the maximum stress in the jth ply at $z = z_j$. From Eq. (7.13), the applied bending moment which would cause this condition is

$$M_{max} = \frac{E_f I_{yy} s_L^{(-)}}{(E_1)_j z_j} \qquad (7.14)$$

where $(E_1)_j$ is the longitudinal modulus of the jth ply. Similarly, for a transverse (90°) ply in tension the maximum bending moment is

$$M_{max} = \frac{E_f I_{yy} s_T^{(+)}}{(E_2)_j z_j} \qquad (7.15)$$

Laminate failure would therefore occur when the bending moment reaches the value that would cause first ply failure. This value can be determined by applying the failure criterion to each ply until the lowest M_{max} is found. The internal bending moment can be related to the external applied loads by the equations of static equilibrium, so that the applied loads corresponding to first ply failure can also be determined. Since the maximum stress does not necessarily occur on the outer surface, first ply failure may occur in an interior ply. One of the difficulties encountered in inspection for ply failure in laminates is that only failures on the outer surfaces can be observed with the naked eye. Interior ply failures can only be detected by methods such as ultrasonic or X-ray inspection.

In most practical cases the applied loads on a beam would be such that not only bending moments, but also transverse shear forces would be developed. These transverse shear forces cause corresponding transverse shear stresses. In laminated beams the transverse shear stresses are often referred to as interlaminar shear stresses. Pagano [7.1] has also developed a mechanics of materials approach for estimating these interlaminar shear stresses, as summarized here.

Recall from mechanics of materials [7.2] that the bending moment, M, is related to the transverse shear force, V, by the equation

$$\frac{dM}{dx} = V \qquad (7.16)$$

Thus, the presence of the shear force implies that the bending moment must change along the length of the beam (the x direction). From Eq. (7.13), we see that if the bending moment changes with respect to x, so, too, must the normal stress, σ_x. This means that the normal stresses acting on the two faces of the jth ply in a differential element must be

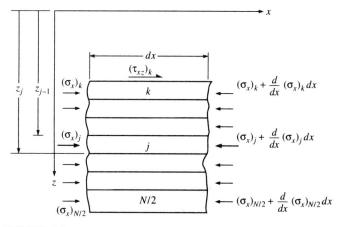

FIGURE 7.5
Differential element of a laminated beam showing interlaminar shear stress which is necessary for static equilibrium when the bending moment varies along the length.

different, as shown in Fig. 7.5. Since the element cannot be in static equilibrium under these normal stresses alone, the interlaminar shear stress, $(\tau_{xz})_k$, must act at the inner edge of the kth ply, as shown in Fig. 7.5. From static equilibrium of the element with respect to the forces along the x direction,

$$(\tau_{xz})_k \, dx + \sum_{j=k}^{N/2} \int_{z_{j-1}}^{z_j} (\sigma_x)_j \, dz - \sum_{j=k}^{N/2} \int_{z_{j-1}}^{z_j} \left[(\sigma_x)_j + \frac{d(\sigma_x)_j}{dx} dx \right] dz = 0 \quad (7.17)$$

or

$$(\tau_{xz})_k = \sum_{j=k}^{N/2} \int_{z_{j-1}}^{z_j} \frac{d(\sigma_x)_j}{dx} \, dz \quad (7.18)$$

Substituting Eqs. (7.13) and (7.16) in (7.18) and integrating, we find that the interlaminar stress at the inner edge of the kth ply is

$$(\tau_{xz})_k = \frac{V}{E_f I_{yy}} \sum_{j=k}^{N/2} \int_{z_{j-1}}^{z_j} (E_x)_j z \, dz \quad (7.19)$$

for a rectangular beam having an even number of plies of uniform thickness, $z_j = jh/N$, and Eq. (7.19) reduces to

$$(\tau_{xz})_k = \frac{3V}{2bh} \left[\frac{S}{E_f} \right] \quad (7.20)$$

Rectangular Beam w/ even # of plies

where

$$S = \frac{4}{N^2} \sum_{j=k}^{N/2} (E_x)_j (2j - 1) \quad (7.21)$$

Equation (7.20) is seen to be similar to the "mechanics of materials" equation for transverse shear stress in a homogeneous isotropic beam,

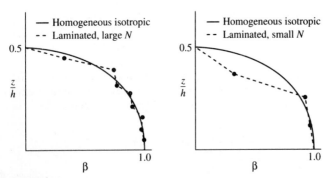

FIGURE 7.6
Variation of shear stress, as governed by the factor β, across half the beam thickness for homogeneous, isotropic beams and for laminated beams. Results are given for laminated beams with a small number of plies and a large number of plies.

which is

$$\tau_{xz} = \frac{3V}{2bh}\left[1 - 4\left(\frac{z}{h}\right)^2\right] \tag{7.22}$$

Thus, the transverse shear stress is given by

$$\tau_{xz} = \frac{3V}{2bh}\beta \tag{7.23}$$

where
$$\beta = \begin{cases} 1 - 4\left(\dfrac{z}{h}\right)^2 & \text{for a homogeneous isotropic beam} \\[2mm] \dfrac{S}{E_f} & \text{for a laminated beam} \end{cases}$$

The shear stress distribution, as governed by the variation of the factor β, is shown for both types of beams in Fig. 7.6. As the number of plies increases, the shear stress distribution for the laminated beam can be expected to approach the parabolic distribution described by Eq. (7.22). For a small number of plies, however, the laminated beam shear stress distribution departs significantly from the parabolic distribution.

Interlaminar stresses are responsible for an important failure mode in composites known as delamination. Recall that the failure criteria discussed in Chap. 4 were based only on in-plane stresses in the lamina. Both normal and shear components of the interlaminar stresses in laminated plates will be discussed later, along with failure criteria which include the interlaminar stresses.

Example 7.1. Determine the flexural and Young's moduli of E-glass/epoxy

N.A

laminated beams having stacking sequences of $[0/90/0]_s$ and $[90/0/90]_s$. The ply moduli are $E_1 = 5 \times 10^6$ psi (34.48 GPa) and $E_2 = 1.5 \times 10^6$ psi (10.34 GPa), and the plies all have the same thickness.

Solution. The total number of plies is $N = 6$ in each case, and only the stacking sequences are different. Since the ply thicknesses are all the same, we can use Eq. (7.9) for the flexural modulus in both cases. For the $[0/90/0]_s$ beam

$$E_f = \frac{8}{N^3} \sum_{j=1}^{N/2} (E_x)_j (3j^2 - 3j + 1)$$

or

$$E_f = \frac{8}{(6)^3} \{5[3(1)^2 - 3 + 1] + 1.5[3(2)^2 - 3(2) + 1] + 5[3(3)^2 - 3(3) + 1]\} \times 10^6 \text{ psi}$$

$$= 4.09 \times 10^6 \text{ psi} \quad (28.2 \text{ GPa})$$

The Young's modulus, or extensional modulus, can be estimated by using the rule of mixtures

$$E_x = E_1 v_1 + E_2 v_2$$

where v_1 = volume fraction of longitudinal (0°) plies
v_2 = volume fraction of transverse (90°) plies

Therefore,

$$E_x = [5(\tfrac{4}{6}) + 1.5(\tfrac{2}{6})] \times 10^6 \text{ psi} = 3.83 \times 10^6 \text{ psi} \quad (26.4 \text{ GPa})$$

For the $[90/0/90]_s$ beam

$$E_f = \frac{8}{(6)^3} \{1.5(1) + 5(7) + 1.5(19)\} \times 10^6 \text{ psi}$$

$$= 2.4 \times 10^6 \text{ psi} \quad (16.55 \text{ GPa})$$

and

$$E_x = [1.5(\tfrac{4}{6}) + 5(\tfrac{2}{6})] \times 10^6 \text{ psi} = 2.66 \times 10^6 \text{ psi} \quad (18.34 \text{ GPa})$$

Note that the flexural modulus depends on the stacking sequence and is not the same as the Young's modulus. The Young's modulus does not depend on the stacking sequence (i.e., the rule of mixtures gives the same result regardless of the ply stacking sequence, as long as the number of longitudinal and transverse plies remains unchanged).

Example 7.2. For the $[90/0/90]_s$ E-glass/epoxy beam described in Example 7.1, sketch the distribution of normal and shear stresses through the thickness of the beam. Assume a ply thickness of 0.01 in (0.254 mm).

Solution. The normal stress is given by Eq. (7.13), but the ratio $M/E_f I_{yy}$ is constant for a given cross section, and the stress distribution across the thickness is governed by the product $(E_x)_j z$. Thus, the stress distribution can be determined to within a constant $K_1 = M/E_f I_{yy}$ by finding the corresponding variation of $(E_x)_j z$.

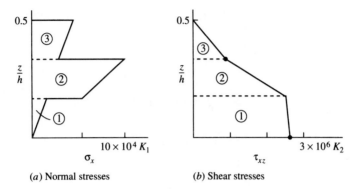

(a) Normal stresses   (b) Shear stresses

FIGURE 7.7
Stress distributions for the beam described in Example 7.2.

For the outer surface of ply number 3 (the outer transverse ply) the ply modulus is $(E_x)_3 = E_2 = 1.5 \times 10^6$ psi, $z = 0.03$ in and the stress is $\sigma_x = K_1(1.5 \times 10^6)(0.03) = 4.5 \times 10^4 K_1$.

For the inner surface of ply number 3, $z = 0.02$ in and the stress is $\sigma_x = K_1(1.5 \times 10^6)(0.02) = 3.0 \times 10^4 K_1$.

Similarly, for the outer surface of ply number 2 the stress is $\sigma_x = K_1(5 \times 10^6)(0.02) = 10 \times 10^4 K_1$.

For the inner surface of ply number 2, $\sigma_x = K_1(5 \times 10^6)(0.01) = 5 \times 10^4 K_1$.

For the outer surface of ply number 1, $\sigma_x = K_1(1.5 \times 10^6)(0.01) = 1.5 \times 10^4 K_1$.

For the inner surface of ply number 1 (on the neutral surface), $\sigma_x = 0$.

The predicted distribution of σ_x across the thickness is plotted in Fig. 7.7(a). It is seen that the maximum normal stress occurs not on the outer surface as in a homogeneous isotropic beam but, rather, at the outer edge of ply number 2.

The interlaminar shear stress at the inner surface of the kth ply for a beam with an even number of uniform thickness plies is given by Eqs. (7.20) and (7.21). For a given cross section, however, the ratio $3V/2bhE_f$ can be set equal to a constant, K_2, and the shear stress can be written as $(\tau_{xz})_k = K_2 S$, where S is defined by Eq. (7.21). The shear stress distribution can then be determined to within a constant K_2 by finding the variation of S across the thickness.

From Eqs. (7.20) and (7.21),

$$(\tau_{xz})_k = \frac{3V}{2bh}\left[\frac{S}{E_f}\right] = K_2 S$$

where

$$S = \frac{4}{N^2}\sum_{j=k}^{N/2}(E_x)_j(2j-1)$$

for $k = 1$

$$S = \frac{4}{(6)^2}\{1.5[2(1) - 1] + 5[2(2) - 1] + 1.5[2(3) - 1]\} \times 10^6$$

$$= 2.66 \times 10^6 \text{ psi} \quad (18.34 \text{ GPa})$$

for $k = 2$

$$S = \frac{4}{(6)^2}\{5[2(2) - 1] + 1.5[2(3) - 1]\} \times 10^6$$

$$= 2.5 \times 10^6 \text{ psi} \quad (17.24 \text{ GPa})$$

for $k = 3$

$$S = \frac{4}{(6)^2}\{1.5[2(3) - 1]\} \times 10^6$$

$$= 0.833 \times 10^6 \text{ psi} \quad (5.74 \text{ GPa})$$

Finally, for $k = 4$, Eq. (7.20) gives the shear stress at the inner surface of an "imaginary ply" whose inner surface is the same as the outer surface of ply number 3, or the outer surface of the laminate. Since there is no material in this "imaginary ply," $S = 0$ and the shear stress must be zero on the outer surface. This also satisfies the boundary condition that the outer surface must be stress-free. The predicted distribution of τ_{xz} across the thickness is plotted in Fig. 7.7(b). As with the shear stress in a homogeneous isotropic beam, the maximum shear stress occurs on the neutral surface and the shear stress at the outer surface is zero. The deviation from the parabolic distribution is substantial, however, because of the small number of plies.

7.3 THEORY OF LAMINATED PLATES WITH COUPLING

While the simplified theory of laminated beams in pure flexure is useful and instructive, it is restricted to symmetric laminates without coupling which are subjected to a single bending moment. In this section we will discuss the more general Classical Lamination Theory, which does not have these restrictions. Using this theory, we can analyze nonsymmetric laminates whose arbitrarily oriented plies may have various coupling effects which may lead to complex combinations of extensional, flexural, and torsional deformations. In addition, in-plane loading due to shear and axial forces and both bending and twisting moments are included. The most important limitation of the Classical Lamination Theory is that each ply is assumed to be in a state of plane stress and that interlaminar stresses are neglected.

What is now referred to as the Classical Lamination Theory has apparently evolved from work in the 1950s and 1960s by investigators such as Smith [7.3], Pister and Dong [7.4], Reissner and Stavsky [7.5], Stavsky [7.6], Lekhnitskii [7.7], and Stavsky and Hoff [7.8]. The major difference between this theory and the classical theory of homogeneous, isotropic plates [7.9] is in the form of the lamina stress-strain relationships. Other elements of the theory such as the deformation hypothesis,

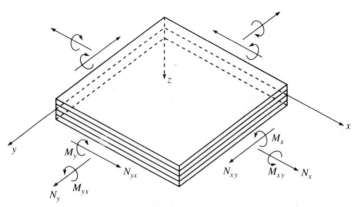

FIGURE 7.8
Coordinate system and stress resultants for laminated plate.

the equilibrium equations, and the strain-displacement relationships are the same as those used in the classical plate theory [7.9].

Although the laminate is made up of multiple laminae, it is assumed that the individual laminae are perfectly bonded together so as to behave as a unitary, nonhomogeneous, anisotropic plate. Interfacial slip is not allowed and the interfacial bonds are not allowed to deform in shear, which means that displacements across lamina interfaces are assumed to be continuous. These assumptions mean that the deformation hypothesis from the classical homogeneous plate theory can be used for the laminated plate. The laminate force-deformation equations resulting from this deformation hypothesis are now derived following the procedure outlined by Whitney [7.10]. Although Whitney has presented a general analysis including the equations of motion, only the static analysis will be considered here.

Figure 7.8 defines the coordinate system to be used in developing the laminated plate analysis. The xyz coordinate system is assumed to have its origin on the middle surface of the plate, so that the middle surface lies in the xy plane. The displacements at a point in the x, y, z directions are u, v, and w, respectively. The basic assumptions relevant to the present static analysis are [7.10]:

1. The plate consists of orthotropic laminae bonded together, with the principal material axes of the orthotropic laminae oriented along arbitrary directions with respect to the xy axes.

2. The thickness of the plate, t, is much smaller than the lengths along the plate edges, a and b.

3. The displacements u, v, and w are small compared with the plate thickness.

4. The in-plane strains ϵ_x, ϵ_y, and γ_{xy} are small compared with unity.
5. Transverse shear strains γ_{xz} and γ_{yz} are negligible.
6. Tangential displacements u and v are linear functions of the z coordinate.
7. The transverse normal strain ϵ_z is negligible.
8. Each ply obeys Hooke's law.
9. The plate thickness t is constant.
10. Transverse shear stresses τ_{xz} and τ_{yz} vanish on the plate surfaces defined by $z = \pm t/2$.

Assumption 5 is a result of the assumed state of plane stress in each ply, whereas assumptions 5 and 6 together define the Kirchhoff deformation hypothesis that normals to the middle surface remain straight and normal during deformation. According to assumptions 6 and 7, the displacements can be expressed as

$$u = u^0(x, y) + zF_1(x, y)$$

$$v = v^0(x, y) + zF_2(x, y) \qquad (7.24)$$

$$w = w^0(x, y) = w(x, y)$$

where u^0 and v^0 are the tangential displacements of the middle surface along the x and y directions, respectively. Due to assumption 7, the transverse displacement at the middle surface, $w^0(x, y)$, is the same as the transverse displacement of any point having the same x and y coordinates, so $w^0(x, y) = w(x, y)$. At this point it is appropriate to mention that in order to account for possible warping of the cross section of the laminate and resulting transverse shear deformations, it is necessary to use a so-called higher-order lamination theory. For example, Christensen [7.11] describes one such theory, which is based on the assumption that the displacements are *nonlinear* functions of the z coordinate as follows:

$$u = u^0(x, y) + z\psi_x(x, y) + z^2\zeta_x(x, y) + z^3\phi_x(x, y)$$

$$v = v^0(x, y) + z\psi_y(x, y) + z^2\zeta_y(x, y) + z^3\phi_y(x, y) \qquad (7.25)$$

$$w = w^0(x, y) + z\psi_z(x, y) + z^2\zeta_z(x, y)$$

Such a theory is beyond the scope of this book, however, and we will only develop the Classical Lamination Theory based on Eqs. (7.24). Substituting Eqs. (7.24) in the strain-displacement equations for the transverse shear strains and using assumption 5, we find that

$$\gamma_{xz} = \frac{\partial u}{\partial z} + \frac{\partial w}{\partial x} = F_1(x, y) + \frac{\partial w}{\partial x} = 0$$

$$\gamma_{yz} = \frac{\partial v}{\partial z} + \frac{\partial w}{\partial y} = F_2(x, y) + \frac{\partial w}{\partial y} = 0$$

(7.26)

and that

$$F_1(x, y) = -\frac{\partial w}{\partial x} \qquad F_2(x, y) = -\frac{\partial w}{\partial y}$$

(7.27)

Substituting Eqs. (7.24) and (7.27) in the strain-displacement relations for the in-plane strains, we find that

$$\epsilon_x = \frac{\partial u}{\partial x} = \epsilon_x^0 + z\kappa_x$$ *constant for laminate (considering in one dir)*

$$\epsilon_y = \frac{\partial v}{\partial y} = \epsilon_y^0 + z\kappa_y \qquad (7.28)$$

$$\gamma_{xy} = \frac{\partial u}{\partial y} + \frac{\partial v}{\partial x} = \gamma_{xy}^0 + z\kappa_{xy}$$

where the strains on the middle surface are

$$\epsilon_x^0 = \frac{\partial u^0}{\partial x} \qquad \epsilon_y^0 = \frac{\partial v^0}{\partial y} \qquad \gamma_{xy}^0 = \frac{\partial u^0}{\partial y} + \frac{\partial v^0}{\partial x} \qquad (7.29)$$

and the curvatures of the middle surface are

$$\kappa_x = -\frac{\partial^2 w}{\partial x^2} \qquad \kappa_y = -\frac{\partial^2 w}{\partial y^2} \qquad \kappa_{xy} = -2\frac{\partial^2 w}{\partial x \, \partial y} \qquad (7.30)$$

κ_x is a bending curvature associated with bending of the middle surface in the xz plane and κ_y is a bending curvature associated with bending of the middle surface in the yz plane. κ_{xy} is a twisting curvature associated with out-of-plane twisting of the middle surface, which lies in the xy plane before deformation.

Since Eqs. (7.28) give the strains at any distance z from the middle surface, the stresses along arbitrary xy axes in the kth lamina of a laminate may be found by substituting Eqs. (7.28) into the lamina stress-strain relationships from Eqs. (2.35) as follows:

$$\begin{Bmatrix} \sigma_x \\ \sigma_y \\ \tau_{xy} \end{Bmatrix}_k = \begin{bmatrix} \bar{Q}_{11} & \bar{Q}_{12} & \bar{Q}_{16} \\ \bar{Q}_{12} & \bar{Q}_{22} & \bar{Q}_{26} \\ \bar{Q}_{16} & \bar{Q}_{26} & \bar{Q}_{66} \end{bmatrix}_k \textit{ times } \begin{Bmatrix} \epsilon_x^0 + z\kappa_x \\ \epsilon_y^0 + z\kappa_y \\ \gamma_{xy}^0 + z\kappa_{xy} \end{Bmatrix} \qquad (7.31)$$

where the subscript k refers to the kth lamina. Comparing the laminated plate stresses in Eq. (7.31) with the laminated beam stress given by Eq.

(7.3), we notice several differences. The laminated beam analysis only gives the uniaxial stress, σ_x, due to the bending curvature, whereas the laminated plate analysis gives the two-dimensional lamina stresses σ_x, σ_y, and τ_{xy} due to bending and twisting curvatures and to the midplane biaxial extension and shear. In addition, the laminated plate analysis includes the stresses due to shear coupling, as discussed in Chap. 2.

In the laminated beam analysis Eq. (7.3) for lamina stress is seen to be of limited practical use because the curvature is not generally known and is difficult to measure. Thus, the lamina stress was related to the applied bending moment by using the static equilibrium relationship in Eq. (7.4). The result was that a more useful equation for stress, Eq. (7.13), was developed. The bending moment can be related to the loads on the structure by additional static equilibrium equations. Similarly, in the laminated plate analysis the midplane strains and curvatures in Eqs. (7.31) must be related to applied forces and moments by static equilibrium equations in order to make these equations more useful. In the laminated plate analysis, however, it is convenient to use forces and moments per unit length rather than forces and moments. The forces and moments per unit length shown in Fig. 7.8 are also referred to as stress resultants.

For example, the force per unit length, N_x, is given by

through thickness

$$N_x = \int_{-t/2}^{t/2} \sigma_x \, dz = \sum_{k=1}^{N} \left\{ \int_{z_{k-1}}^{z_k} (\sigma_x)_k \, dz \right\} \qquad (7.32)$$

and the moment per unit length, M_x, is given by

$$M_x = \int_{-t/2}^{t/2} \sigma_x z \, dz = \sum_{k=1}^{N} \left\{ \int_{z_{k-1}}^{z_k} (\sigma_x)_k z \, dz \right\} \qquad (7.33)$$

where t = laminate thickness
$\quad (\sigma_x)_k$ = stress in the kth lamina
$\quad z_{k-1}$ = distance from middle surface to inner surface of the kth lamina
$\quad z_k$ = corresponding distance from middle surface to outer surface of the kth lamina, as shown in Fig. 7.9

Substituting the lamina stress-strain relationships from Eqs. (7.31) in Eqs. (7.32) and (7.33), respectively, we find that

$$N_x = \sum_{k=1}^{N} \int_{z_{k-1}}^{z_k} \{ (\bar{Q}_{11})_k(\epsilon_x^0 + z\kappa_x) + (\bar{Q}_{12})_k(\epsilon_y^0 + z\kappa_y) + (\bar{Q}_{16})_k(\gamma_{xy}^0 + z\kappa_{xy}) \} \, dz$$

$$\qquad (7.34)$$

and

$$M_x = \sum_{k=1}^{N} \int_{z_{k-1}}^{z_k} \{ (\bar{Q}_{11})_k(\epsilon_x^0 + z\kappa_x) + (\bar{Q}_{12})_k(\epsilon_y^0 + z\kappa_y) + (\bar{Q}_{16})_k(\gamma_{xy}^0 + z\kappa_{xy}) \} z \, dz$$

$$\qquad (7.35)$$

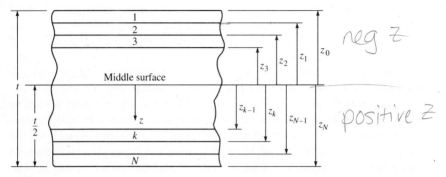

FIGURE 7.9
Laminated plate geometry and ply numbering system. *Caution:* The ply numbering system here is different from that used in Fig. 7.2 for the laminated beam analysis.

Combining terms and rearranging Eqs. (7.34) and (7.35), we find that

$$N_x = A_{11}\epsilon_x^0 + A_{12}\epsilon_y^0 + A_{16}\gamma_{xy}^0 + B_{11}\kappa_x + B_{12}\kappa_y + B_{16}\kappa_{xy} \qquad (7.36)$$

and

$$M_x = B_{11}\epsilon_x^0 + B_{12}\epsilon_y^0 + B_{16}\gamma_{xy}^0 + D_{11}\kappa_x + D_{12}\kappa_y + D_{16}\kappa_{xy} \qquad (7.37)$$

where the laminate extensional stiffnesses are given by

$$A_{ij} = \int_{-t/2}^{t/2} (\bar{Q}_{ij})_k \, dz = \sum_{k=1}^{N} (\bar{Q}_{ij})_k(z_k - z_{k-1}) \qquad (7.38)$$

the laminate coupling stiffnesses are given by

$$B_{ij} = \int_{-t/2}^{t/2} (\bar{Q}_{ij})_k z \, dz = \frac{1}{2} \sum_{k=1}^{N} (\bar{Q}_{ij})_k(z_k^2 - z_{k-1}^2) \qquad (7.39)$$

and the laminate bending stiffnesses are given by

$$D_{ij} = \int_{-t/2}^{t/2} (\bar{Q}_{ij})_k z^2 \, dz = \frac{1}{3} \sum_{k=1}^{N} (\bar{Q}_{ij})_k(z_k^3 - z_{k-1}^3) \qquad (7.40)$$

where the subscripts $i, j = 1$, 2, or 6. The other stress resultants can be written in similar form, and the complete set of equations can be expressed in matrix form as

$$
\begin{Bmatrix} N_x \\ N_y \\ N_{xy} \\ M_x \\ M_y \\ M_{xy} \end{Bmatrix} =
\begin{bmatrix}
A_{11} & A_{12} & A_{16} & B_{11} & B_{12} & B_{16} \\
A_{12} & A_{22} & A_{26} & B_{12} & B_{22} & B_{26} \\
A_{16} & A_{26} & A_{66} & B_{16} & B_{26} & B_{66} \\
B_{11} & B_{12} & B_{16} & D_{11} & D_{12} & D_{16} \\
B_{12} & B_{22} & B_{26} & D_{12} & D_{22} & D_{26} \\
B_{16} & B_{26} & B_{66} & D_{16} & D_{26} & D_{66}
\end{bmatrix}
\begin{Bmatrix} \epsilon_x^0 \\ \epsilon_y^0 \\ \gamma_{xy}^0 \\ \kappa_x \\ \kappa_y \\ \kappa_{xy} \end{Bmatrix} \qquad (7.41)
$$

or in partitioned form as

$$\left\{ \frac{N}{M} \right\} = \left[\begin{array}{c|c} A & B \\ \hline B & D \end{array} \right] \left\{ \frac{\epsilon^0}{\kappa} \right\} \tag{7.42}$$

From Eqs. (7.42), we can see that the extensional stiffness matrix [A] relates the in-plane forces {N} to the midplane strains {e^0} and the bending stiffness matrix [D] relates the moments {M} to the curvatures {k}. The coupling stiffness matrix [B] couples the in-plane forces {N} with the curvatures {κ} and the moments [M] with the midplane strains {ϵ^0}. A laminate having nonzero B_{ij} will bend or twist under in-plane loads. Such a laminate will also exhibit midplane stretching under bending and twisting moment loading. It can be easily shown that laminate geometric and material property symmetry with respect to the middle surface leads to the condition that all $B_{ij} = 0$ and that asymmetry about the middle surface leads to nonzero B_{ij}.

It is now clear that there may be coupling effects at both the lamina level and the laminate level, but the two types of coupling are not necessarily related. Lamina shear coupling is a result of anisotropic material behavior and the presence of 16 and 26 terms in the lamina stiffness or compliance matrices (recall Sec. 2.6). This type of coupling at the lamina level also leads to coupling at the laminate level due to terms like A_{16}, A_{26}, D_{16}, and D_{26}. On the other hand, the B_{ij}-type coupling at the laminate level is due to geometric and/or material property asymmetry with respect to the middle surface and is unrelated to material anisotropy. For example, it is possible for a laminate to have nonzero B_{ij} even with isotropic laminae if they are stacked in nonsymmetrical fashion, but the isotropic lamina properties lead to the condition $A_{16} = A_{26} = D_{16} = D_{26} = 0$. In the next section the nature of the stiffness matrices for several special types of laminates will be summarized.

7.4 STIFFNESS CHARACTERISTICS OF SELECTED LAMINATE CONFIGURATIONS

As shown in the previous section, the number of nonzero terms in the laminate stiffness matrices is reduced for certain laminate configurations. Symmetry or antisymmetry of geometric and material properties about the middle surface, ply orientations, and ply stacking sequences are all factors that govern the form of the laminate stiffness matrices. It is particularly important to be able to understand the effects of these factors on the type of coupling that may exist in the stiffness matrices of commonly used laminates.

Before beginning the discussion of special laminate configurations, it is useful to define several terms which are associated with special ply orientations. Although these ply orientations, by themselves, do not necessarily produce simplifications in the stiffness matrices, they are often used in combination with other terms to describe special laminates which do have simplified stiffness matrices. "Angle-ply" laminates have lamina orientations of either $+\theta$ or $-\theta$, where $0° \leq \theta \leq 90°$. Depending on ply stacking sequences, angle-ply laminates may be either symmetric, anti-symmetric, or asymmetric with respect to the middle surface. "Cross-ply" laminates consist of plies oriented at either $\theta = 0°$ or $\theta = 90°$. A balanced cross-ply laminate has equal numbers of $0°$ and $90°$ plies. Depending on the ply arrangement, cross-ply laminates may be either symmetric or asymmetric with respect to the middle surface, but not antisymmetric. Since all plies in a cross-ply laminate behave as specially orthotropic laminae, such a laminate will always have $A_{16} = A_{26} = D_{16} = D_{26} = 0$. However, since all plies in an angle-ply laminate behave as generally orthotropic laminae, the 16 and 26 terms may not vanish.

7.4.1 Symmetric Laminates

A symmetric laminate has both geometric and material property symmetry about the middle surface. That is, the ply material, ply orientation, and ply thickness at a positive distance z from the middle surface are identical to the corresponding values at an equal negative distance z from the middle surface. Examples of symmetric angle-ply and cross-ply laminates are shown in Figs. 7.10(a) and 7.10(b), respectively. Such a symmetry condition when substituted in Eqs. (7.39) leads to the major simplification that all $B_{ij} = 0$. This means that bending-stretching coupling will not be present in such laminates. Consequently, in-plane loads will not generate bending and twisting curvatures which cause out-of-plane warping, and bending or twisting moments will not produce an extension

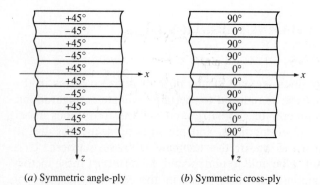

(a) Symmetric angle-ply   (b) Symmetric cross-ply

FIGURE 7.10
Examples of symmetric laminates. Ply orientations and material properties are symmetric about middle surface.

of the middle surface. This can be particularly important in structures which are subjected to changes in environmental conditions, where the resulting hygrothermal forces would lead to undesirable warping in nonsymmetric laminates.

Example 7.3. Determine the stiffness matrix for a $[+45/-45/-45/+45]$ symmetric angle-ply laminate consisting of 0.25-mm-thick unidirectional AS/3501 graphite/epoxy laminae. An exploded view of the laminate is shown in Fig. 7.11.

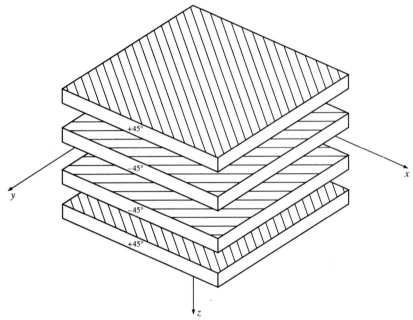

FIGURE 7.11
Exploded view of $[+45/-45/-45/+45]$ symmetric laminate.

Solution. From Table 2.2, the lamina engineering constants are

$$E_1 = 138 \text{ GPa} \qquad E_2 = 9 \text{ GPa} \qquad G_{12} = 6.9 \text{ GPa} \qquad v_{12} = 0.3$$

and
$$v_{21} = v_{12} \frac{E_2}{E_1} = 0.3 \frac{9.0}{138.0} = 0.0196$$

Substitution of the above engineering constants in Eqs. (2.27) yields the components of the lamina stiffness matrix associated with the principal material axes:

$$[Q] = \begin{bmatrix} 138.8 & 2.72 & 0 \\ 2.72 & 9.05 & 0 \\ 0 & 0 & 6.9 \end{bmatrix} \text{ GPa}$$

The transformed lamina stiffness matrices for the $+45°$ and $-45°$ plies are then found by substituting the above stiffnesses in Eqs. (2.36) or Eqs. (2.42) and (2.43). For the $+45°$ plies

$$[\bar{Q}]_{+45°} = \begin{bmatrix} 45.22 & 31.42 & 32.44 \\ 31.42 & 45.22 & 32.44 \\ 32.44 & 32.44 & 35.6 \end{bmatrix} \text{ GPa}$$

for the $-45°$ plies

$$[\bar{Q}]_{-45°} = \begin{bmatrix} 45.22 & 31.42 & -32.44 \\ 31.42 & 45.22 & -32.44 \\ -32.44 & -32.44 & 35.6 \end{bmatrix} \text{ GPa}$$

Note that the only difference between the stiffness matrices for the two plies is that the shear coupling terms (terms with subscripts 16 and 26) for the $-45°$ ply have the opposite sign from the corresponding terms for the $+45°$ ply. Before calculating the laminate stiffnesses, we must determine the distances from the middle surface on the various ply interfaces according to Fig. 7.9. The distances are $z_0 = -0.50$ mm, $z_1 = -0.25$ mm, $z_2 = 0$, $z_3 = 0.25$ mm, and $z_4 = 0.5$ mm. The laminate extensional stiffnesses are then found by substituting these distances, along with the lamina stiffnesses above, in Eqs. (7.38):

$$[A] = \begin{bmatrix} 45.22 & 31.42 & 0 \\ 31.42 & 45.22 & 0 \\ 0 & 0 & 35.6 \end{bmatrix} \text{ GPa-mm}$$

Similarly, the laminate coupling stiffnesses are found from Eqs. (7.39):

$$[B] = \begin{bmatrix} 0 & 0 & 0 \\ 0 & 0 & 0 \\ 0 & 0 & 0 \end{bmatrix} \text{ GPa-mm}^2$$

and the laminate bending stiffnesses are found from Eqs. (7.40):

$$[D] = \begin{bmatrix} 3.77 & 2.62 & 2.03 \\ 2.62 & 3.77 & 2.03 \\ 2.03 & 2.03 & 2.97 \end{bmatrix} \text{ GPa-mm}^3$$

7.4.2 Antisymmetric Laminates

An antisymmetric laminate has plies of identical material and thickness at equal positive and negative distances from the middle surface, but the ply orientations are antisymmetric with respect to the middle surface. That is, if the ply orientation at a positive distance z is $+\theta$, the ply orientation at an equal negative distance z is $-\theta$. Examples of antisymmetric angle-ply laminates are shown in Fig. 7.12. Note that the antisymmetric definition

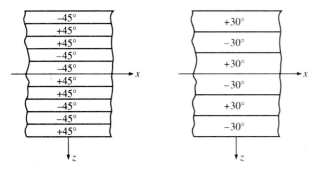

FIGURE 7.12
Examples of antisymmetric angle-ply laminates. Although ply orientations are antisymmetric about middle surface, the material distribution is symmetric.

has no meaning for a cross-ply laminate, which must be either symmetric or nonsymmetric. It can be shown that by substituting the antisymmetric condition into Eqs. (7.38) and (7.40), the coupling terms $A_{16} = A_{26} = D_{16} = D_{26} = 0$. From Eq. (7.39), it can also be shown that $B_{11} = B_{12} = B_{22} = B_{66} = 0$ for the antisymmetric angle-ply laminate.

Example 7.4. Determine the stiffness matrix for a $[-45/+45/-45/+45]$ antisymmetric angle-ply laminate consisting of the same 0.25-mm-thick unidirectional AS/3501 graphite/epoxy laminae that were used in Example 7.3. An exploded view of the laminate is shown in Fig. 7.13.

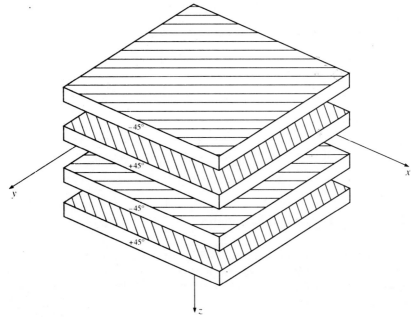

FIGURE 7.13
Exploded view of $[-45/+45/-45/+45]$ antisymmetric laminate.

Solution. Since the lamina orientations are still $+45°$ and $-45°$, the lamina stiffnesses are the same as those calculated in Example 7.3. The distances z_k are also the same as those shown in Example 7.3 since the ply thicknesses and the number of plies are the same. Laminate extensional stiffnesses are then found by substituting these values, along with the antisymmetric stacking sequence, in Eqs. (7.38):

$$[A] = \begin{bmatrix} 45.22 & 31.42 & 0 \\ 31.42 & 45.22 & 0 \\ 0 & 0 & 35.6 \end{bmatrix} \text{ GPa-mm}$$

Note that these results are identical to those in Example 7.3 because we still have two plies at $+45°$ and two plies at $-45°$. Thus, the stacking sequence has no effect on $[A]$ as long as the number of plies at a given orientation remains the same. The laminate coupling stiffnesses are found from Eqs. (7.39):

$$[B] = \begin{bmatrix} 0 & 0 & 4.055 \\ 0 & 0 & 4.055 \\ 4.055 & 4.055 & 0 \end{bmatrix} \text{ GPa-mm}^2$$

Note that due to the antisymmetry, $B_{11} = B_{22} = B_{12} = B_{66} = 0$ but B_{16} and B_{26} have nonzero values; this is true in general for antisymmetric angle-ply laminates. Thus, the antisymmetric laminate has extension-twisting and bending-shearing coupling, but the symmetric laminate does not. The laminate bending stiffnesses are found from Eqs. (7.40):

$$[D] = \begin{bmatrix} 3.77 & 2.62 & 0 \\ 2.62 & 3.77 & 0 \\ 0 & 0 & 2.97 \end{bmatrix} \text{ GPa-mm}^3$$

Note that D_{11}, D_{22}, D_{12}, and D_{66} are the same as the corresponding values in Example 7.3, but we now have $D_{16} = D_{26} = 0$. Thus, bending-twisting coupling is present in symmetric angle-ply laminates, but not in antisymmetric angle-ply laminates.

7.4.3 Quasi-Isotropic Laminates

Although it may seem unlikely, it is possible to use orthotropic laminae to construct a laminate which exhibits some elements of isotropic behavior. For example, if a laminate consists of three or more identical orthotropic laminae (i.e., all have the same material and geometric properties) which are oriented at the same angle relative to adjacent laminae, the extensional stiffness matrix $[A]$ will be isotropic, but the other stiffness matrices $[B]$ and $[D]$ will not necessarily have isotropic form. Such a laminate is called a quasi-isotropic, or planar isotropic laminate, and the angle between adjacent laminae must be π/N, where N is the total number of laminae. For example, $[60/0/-60]$ and $[90/45/0/-45]$ laminates are quasi-isotropic.

Recall that in Sec. 6.4 it was mentioned that randomly oriented fiber composites could be modeled as planar isotropic or quasi-isotropic laminates. Now it is clear that although a randomly oriented fiber composite must theoretically have an infinite number of fiber orientations to be isotropic, the behavior of such materials can be modeled by using a quasi-isotropic laminate having only three laminae, as in the $[60/0/-60]$ laminate.

Recall also that the stress-strain relationships for an isotropic lamina are given by Eqs. (2.26) with the additional requirements that $Q_{11} = Q_{22}$, $Q_{66} = (Q_{11} - Q_{12})/2$, and $Q_{16} = Q_{26} = 0$. Similarly, the extensional force-deformation relationships for the quasi-isotropic laminate are given by

$$\left\{ \begin{array}{c} N_x \\ N_y \\ N_{xy} \end{array} \right\} = \left[\begin{array}{ccc} A_{11} & A_{12} & 0 \\ A_{12} & A_{11} & 0 \\ 0 & 0 & (A_{11}-A_{12})/2 \end{array} \right] \left\{ \begin{array}{c} \epsilon_x^0 \\ e_y^0 \\ \gamma_{xy}^0 \end{array} \right\} \tag{7.43}$$

In general, such simplifications are not possible for the $[B]$ and $[D]$ matrices, as can be shown by calculating the stiffness matrices for quasi-isotropic laminates such as $[60/0/-60]$ or $[90/45/0/-45]$.

In Sec. 6.4 it was shown that the invariants could be useful in the development of the stress-strain relationships and equations for the engineering constants of a planar isotropic, randomly oriented fiber composite. Similarly, the invariants can be used in the study of quasi-isotropic laminates. For example, by substituting the lamina stiffnesses in terms of invariants from Eqs. (2.42) in Eqs. (7.38) for the laminate extensional stiffnesses, we find that

$$\begin{aligned} A_{11} &= A_{22} = U_1 t \\ A_{12} &= U_4 t \\ &\vdots \qquad\qquad \vdots \\ A_{66} &= \frac{(U_1 - U_4)t}{2} \end{aligned} \tag{7.44}$$

Using developments similar to those in Sec. 6.4, we can show that the effective extensional engineering constants for quasi-isotropic laminates are given by Eqs. (6.31).

Example 7.5. Determine the stiffness matrices and engineering constants for a quasi-isotropic $[60/0/-60]$ laminate consisting of the same laminae

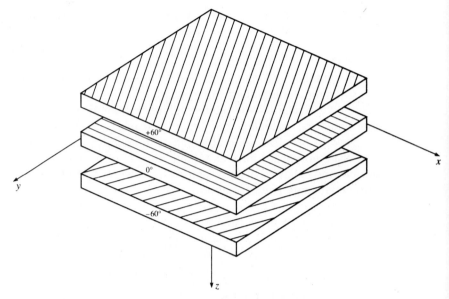

FIGURE 7.14
Exploded view of $[+60/0/-60]$ quasi-isotropic laminate,

that were described in Example 7.3. Figure 7.14 shows an exploded view of the laminate.

Solution. The required lamina stiffnesses are the Q_{ij} in Example 7.3 and the appropriate transformed stiffnesses for 60° and −60° from Eqs. (2.36) or Eqs. (2.42) and (2.43). Substituting these stiffnesses in Eqs. (7.38), we find that the laminate extensional stiffnesses are

$$[A] = \begin{bmatrix} 44.68 & 12.80 & 0 \\ 12.80 & 44.68 & 0 \\ 0 & 0 & 15.94 \end{bmatrix} \text{ GPa-mm}$$

It is easily shown that this matrix is of the isotropic form shown in Eqs (7.43). The laminate coupling stiffnesses from Eqs. (7.39) are

$$[B] = \begin{bmatrix} 0 & 0 & -1.96 \\ 0 & 0 & -5.06 \\ -1.96 & -5.06 & 0 \end{bmatrix} \text{ GPa-mm}^2$$

and the bending stiffnesses from Eqs. (7.40) are

$$[D] = \begin{bmatrix} 0.856 & 0.824 & 0 \\ 0.824 & 2.88 & 0 \\ 0 & 0 & 0.972 \end{bmatrix} \text{ GPa-mm}^3$$

It is seen that the matrices $[B]$ and $[D]$ do not have the isotropic form of Eqs. (7.43). It can also be shown that by changing the lamina orientations while maintaining equal angles between adjacent laminae (e.g., $[75/15/-45]$, $[30/-30/-90]$, or $[0/-60/-120]$), the A_{ij} remain unchanged but the B_{ij} and D_{ij} do not. Thus, the laminate is isotropic with respect to in-plane behavior only.

The engineering constants for the laminate may be found by using the invariants. By substituting the Q_{ij} from Example 7.3 in Eqs. (2.43), we find that

$$U_1 = 59.57\,\text{GPa} \qquad \text{and} \qquad U_4 = 17.07\,\text{GPa}$$

Substituting these results in Eqs. (6.31), we find that the engineering constants are

$$\tilde{E} = 54.68\,\text{GPa} \qquad \tilde{G} = 21.25\,\text{GPa} \qquad \text{and} \qquad \tilde{v} = 0.287$$

Note that the A_{ij} can also be found by using the invariants U_1 and U_4, along with the laminate thickness, t, in Eqs. (7.44).

7.5 DERIVATION AND USE OF LAMINATE COMPLIANCES

Since the applied loads are generally known rather than the deformations, it is often necessary to use the inverted form of the laminate force-deformation relationships shown in Eqs. (7.41) and (7.42). The use of the inverted equations means that we must deal with the laminate compliance matrix instead of the laminate stiffness matrix. In this section the inverted equations are derived and used to calculate the lamina stresses and strains due to known laminate loads. The inverted equations are also used in the derivation of the laminate engineering constants and in the comparison of predicted and measured laminate compliances.

7.5.1 Inversion of Laminate Force-Deformation Equations

The general laminate force-deformation equations shown in Eqs. (7.42) can be expressed as

$$\left\{ \begin{matrix} N \\ \hline M \end{matrix} \right\} = \left[\begin{matrix} A & \vdots & B \\ \hline B & \vdots & D \end{matrix} \right] \left\{ \begin{matrix} \epsilon^0 \\ \hline \kappa \end{matrix} \right\} = [E] \left\{ \begin{matrix} \epsilon^0 \\ \hline \kappa \end{matrix} \right\} \tag{7.45}$$

where the A_{ij}, B_{ij}, and D_{ij} make up the laminate stiffness matrix, $[E]$. The fully inverted form of this equation can be obtained directly by premultiplying both sides of the equation by the compliance matrix, which is the inverse of the stiffness matrix

$$\left\{ \begin{matrix} \epsilon^0 \\ \hline \kappa \end{matrix} \right\} = \left[\begin{matrix} A & \vdots & B \\ \hline B & \vdots & D \end{matrix} \right]^{-1} \left\{ \begin{matrix} N \\ \hline M \end{matrix} \right\} = [E]^{-1} \left\{ \begin{matrix} N \\ \hline M \end{matrix} \right\} \tag{7.46}$$

Alternatively, Eqs. (7.46) are derived below by the inversion of sub-divided smaller matrices, as shown by Halpin [7.12].

From Eqs. (7.45), the in-plane forces per unit length are

$$\{N\} = [A]\{\epsilon^0\} + [B]\{\kappa\} \qquad (7.47)$$

whereas the moments per unit length are

$$\{M\} = [B]\{\epsilon^0\} + [D]\{\kappa\} \qquad (7.48)$$

The midplane strains may be obtained from Eqs. (7.47) as

$$\{\epsilon^0\} = [A]^{-1}\{N\} - [A]^{-1}[B]\{\kappa\} \qquad (7.49)$$

Substitution of these strains in Eqs. (7.48) gives

$$\{M\} = [B][A]^{-1}\{N\} - [B][A]^{-1}[B]\{\kappa\} + [D]\{\kappa\} \qquad (7.50)$$

Equations (7.49) and (7.50) can be combined to give a partially inverted form of Eqs. (7.45) as follows:

$$\left\{ \begin{array}{c} \epsilon^0 \\ \hline M \end{array} \right\} = \left[\begin{array}{c|c} A^* & B^* \\ \hline C^* & D^* \end{array} \right] \left\{ \begin{array}{c} N \\ \hline \kappa \end{array} \right\} \qquad (7.51)$$

where

$$[A^*] = [A]^{-1}$$
$$[B^*] = -[A]^{-1}[B]$$
$$[C^*] = [B][A]^{-1}$$
$$[D^*] = [D] - [B][A]^{-1}[B]$$

Inverting the last set of partitioned Eqs. (7.51) to solve for the curvatures, we find that

$$\{\kappa\} = [D^*]^{-1}\{M\} - [D^*]^{-1}[C^*]\{N\} \qquad (7.52)$$

Now substituting Eqs. (7.52) in Eqs. (7.49), we have

$$\{\epsilon^0\} = ([A^*] - [B^*][D^*]^{-1}[C^*])\{N\} + [B^*][D^*]^{-1}\{M\} \qquad (7.53)$$

Equations (7.52) and (7.53) can now be combined in partitioned matrix form to give

$$\left\{ \begin{array}{c} \epsilon^0 \\ \hline \kappa \end{array} \right\} = \left[\begin{array}{c|c} A' & B' \\ \hline C' & D' \end{array} \right] \left\{ \begin{array}{c} N \\ \hline M \end{array} \right\} \qquad (7.54)$$

where

$$[A'] = [A^*] - [B^*][D^*]^{-1}[C^*]$$
$$[B'] = [B^*][D^*]^{-1}$$
$$[C'] = -[D^*]^{-1}[C^*] = [B']^T = [B']$$
$$[D'] = [D^*]^{-1}$$

and the compliance matrix is

$$\left[\begin{array}{c|c} A' & B' \\ \hline C' & D' \end{array} \right] = [E]^{-1} \qquad (7.55)$$

Since the stiffness matrix $[E]$ is symmetric, the compliance matrix must also be symmetric.

7.5.2 Determination of Lamina Stresses and Strains

Now that we have the inverted laminate force-deformation relationships in Eqs. (7.54) the calculation of lamina stresses and strains from known laminate forces and moments is a straightforward procedure. For a laminate at constant temperature and moisture content the stresses in the kth lamina are given by Eqs. (7.31), which can be written in abbreviated matrix notation as

$$\{\sigma\}_k = [\bar{Q}]_k(\{\epsilon^0\} + z\{\kappa\}) \tag{7.56}$$

where the midplane strains $\{\epsilon^0\}$ and curvatures $\{\kappa\}$ are given in terms of laminate forces and moments by Eqs. (7.54). The lamina stresses from Eqs. (7.56) can then be used in conjunction with a lamina strength criterion to check each lamina against failure. The analysis of hygrothermal stresses will be discussed later in Sec. 7.6, and laminate strength analysis will be presented in Sec. 7.8.

Example 7.6. The symmetric angle-ply laminate described in Example 7.3 is subjected to a single uniaxial force per unit length $N_x = 50$ MPa-mm. Determine the resulting stresses associated with the x and y axes in each lamina.

Solution. Due to symmetry, $[B] = 0$ and $[A'] = [A*] = [A]^{-1}$. Since $\{M\} = 0$ here,

$$\{\epsilon^0\} = [A']\{N\} = [A]^{-1}\{N\}$$

Using the inverse of the $[A]$ matrix from Example 7.3, we find that

$$\begin{Bmatrix} \epsilon_x^0 \\ \epsilon_y^0 \\ \gamma_{xy}^0 \end{Bmatrix} = \begin{bmatrix} 0.04276 & -0.0297 & 0 \\ -0.0297 & 0.04276 & 0 \\ 0 & 0 & 0.02809 \end{bmatrix} \begin{Bmatrix} 50 \\ 0 \\ 0 \end{Bmatrix} \times (10^{-3})$$

$$= \begin{Bmatrix} 0.002138 \\ -0.001485 \\ 0 \end{Bmatrix}$$

where $(\text{GPa-mm})^{-1} = 10^{-3} (\text{MPa-mm})^{-1}$. Substituting the above strains and the lamina stiffnesses from Example 7.3 in Eqs. (7.56), we find that the stresses in the $+45°$ plies are

$$\begin{Bmatrix} \sigma_x \\ \sigma_y \\ \tau_{xy} \end{Bmatrix} = \begin{bmatrix} 45.22 & 31.42 & 32.44 \\ 31.42 & 45.22 & 32.44 \\ 32.44 & 32.44 & 35.6 \end{bmatrix} \begin{Bmatrix} 0.002138 \\ -0.001485 \\ 0 \end{Bmatrix} \times 10^3 = \begin{Bmatrix} 50 \\ 0 \\ 21.2 \end{Bmatrix} \text{ MPa}$$

where 10^3 MPa = GPa. Similarly, the stresses in the $-45°$ plies are

$$\begin{Bmatrix} \sigma_x \\ \sigma_y \\ \tau_{xy} \end{Bmatrix} = \begin{bmatrix} 45.22 & 31.42 & -32.44 \\ 31.42 & 45.22 & -32.44 \\ -32.44 & -32.44 & 35.6 \end{bmatrix} \begin{Bmatrix} 0.002138 \\ -0.001485 \\ 0 \end{Bmatrix} \times 10^3 = \begin{Bmatrix} 50 \\ 0 \\ -21.2 \end{Bmatrix} \text{ MPa}$$

Note that since the curvatures vanish for this problem, the stresses do not depend on the distance z.

Example 7.7. The antisymmetric angle-ply laminate described in Example 7.4 is subjected to a single uniaxial force per unit length $N_x = 50$ MPa-mm. Determine the resulting stresses associated with the x and y axes in each lamina.

Solution. Since this laminate is not symmetric, we must invert the full stiffness matrix as in Eq. (7.46) or Eq. (7.54). Forming the full stiffness matrix from the $[A]$, $[B]$, and $[D]$ matrices in Example 7.4 and inverting, we find the resulting midplane strains and curvatures to be

$$
\begin{Bmatrix} \epsilon_x^0 \\ \epsilon_y^0 \\ \gamma_{xy}^0 \\ \kappa_x \\ \kappa_y \\ \kappa_{xy} \end{Bmatrix} ==
\begin{bmatrix}
0.04386 & -0.02861 & 0 & 0 & 0 & -0.02083 \\
-0.02861 & 0.04386 & 0 & 0 & 0 & -0.02083 \\
0 & 0 & 0.03284 & -0.02083 & -0.02083 & 0 \\
0 & 0 & -0.02083 & 0.52625 & -0.34331 & 0 \\
0 & 0 & -0.02083 & -0.34331 & 0.52625 & 0 \\
-0.02083 & -0.02083 & 0 & 0 & 0 & 0.39356
\end{bmatrix}
$$

$$
\times \begin{Bmatrix} 50 \\ 0 \\ 0 \\ 0 \\ 0 \\ 0 \end{Bmatrix} \times 10^{-3} = \begin{Bmatrix} 0.002193 & \text{mm/mm} \\ -0.001430 & \text{mm/mm} \\ 0 & \text{mm/mm} \\ 0 & \text{mm}^{-1} \\ 0 & \text{mm}^{-1} \\ -0.001042 & \text{mm}^{-1} \end{Bmatrix}
$$

where again the factor of 10^{-3} has been introduced for dimensional consistency. Due to the curvatures, the total strains and stresses now depend on the distance z (unlike Example 7.6). For example, at the top surface of the #1 ply ($-45°$), $z = -0.5$ mm and the resulting total strains are

$$\epsilon_x = \epsilon_x^0 + z\kappa_x = 0.002193 + (-0.5)(0) = 0.002193 \text{ mm/mm}$$

$$\epsilon_y = \epsilon_y^0 + z\kappa_y = -0.00143 + (-0.5)(0) = -0.00143 \text{ mm/mm}$$

$$\gamma_{xy} = \gamma_{xy}^0 + z\kappa_{xy} = 0 + (-0.5)(-0.001042) = 0.000521 \text{ mm/mm}$$

Similarly, at the bottom surface of the #1 ply ($-45°$), or at the top surface of the #2 ply ($+45°$), $z = -0.25$ mm and the strains are

$$\epsilon_x = 0.002193 + (-0.25)(0) = 0.002193 \text{ mm/mm}$$

$$\epsilon_y = -0.00143 + (-0.25)(0) = -0.00143 \text{ mm/mm}$$

$$\gamma_{xy} = 0 + (-0.25)(-0.001042) = 0.000261 \text{ mm/mm}$$

At the top surface of the #3 ply ($-45°$), or at the bottom surface of the #2 ply, $z = 0$ and the strains are

$$\epsilon_x = 0.002193 \text{ mm/mm}$$

$$\epsilon_y = -0.00143 \text{ mm/mm}$$

$$\gamma_{xy} = 0$$

At the top surface of the #4 ply ($+45°$), or at the bottom surface of the #3 ply ($-45°$), $z = 0.25$ mm and

$$\epsilon_x = 0.002193 \text{ mm/mm}$$

$$\epsilon_y = -0.00143 \text{ mm/mm}$$

$$\gamma_{xy} = -0.000261 \text{ mm/mm}$$

Finally, at the bottom of the #4 ply ($+45°$) $z = +0.5$ mm and

$$\epsilon_x = 0.002193 \text{ mm/mm}$$

$$\epsilon_y = -0.00143 \text{ mm/mm}$$

$$\gamma_{xy} = -0.000521 \text{ mm/mm}$$

The stresses at the top surface of the #1 ply ($-45°$) are then

$$\begin{Bmatrix} \sigma_x \\ \sigma_y \\ \tau_{xy} \end{Bmatrix} = \begin{bmatrix} 45.22 & 31.42 & -32.44 \\ 31.42 & 45.22 & -32.44 \\ -32.44 & -32.44 & 35.6 \end{bmatrix} \begin{Bmatrix} 0.002193 \\ -0.001430 \\ 0.000521 \end{Bmatrix} \times 10^3$$

$$= \begin{Bmatrix} 37.3 \\ -12.7 \\ -6.2 \end{Bmatrix} \text{ MPa}$$

where again 10^3 MPa = GPa. Similar calculations for the other plies yield the values shown in the following table:

Location	σ_x (MPa)	σ_y (MPa)	τ_{xy} (MPa)
#1 Top	37.3	−12.7	−6.2
#1 Bottom	45.8	−4.2	−15.5
#2 Top	62.7	12.7	34.0
#2 Bottom	54.2	4.2	24.7
#3 Top	54.2	4.2	−24.7
#3 Bottom	62.7	12.7	−34.0
#4 Top	45.8	−4.2	15.5
#4 Bottom	37.3	−12.7	6.2

Thus, the stress distribution across the thickness of the antisymmetric laminate is quite complex, even for simple uniaxial loading. This is typical for laminates which exhibit coupling.

7.5.3   Determination of Laminate Engineering Constants

It is sometimes more convenient to use effective laminate engineering constants rather than the laminate stiffnesses defined in Eqs. (7.38), (7.39), and (7.40). These effective laminate engineering constants may be derived by using laminate compliances. For example, the force-deformation relationships for a symmetric laminate under in-plane loads only are given by

$$
\begin{Bmatrix} N_x \\ N_y \\ N_{xy} \end{Bmatrix} = \begin{bmatrix} A_{11} & A_{12} & A_{16} \\ A_{12} & A_{22} & A_{26} \\ A_{16} & A_{26} & A_{66} \end{bmatrix} \begin{Bmatrix} \epsilon_x^0 \\ \epsilon_y^0 \\ \gamma_{xy}^0 \end{Bmatrix} \tag{7.57}
$$

and the corresponding inverted force-deformation relationships are

$$
\begin{Bmatrix} \epsilon_x^0 \\ \epsilon_y^0 \\ \gamma_{xy}^0 \end{Bmatrix} = \begin{bmatrix} A_{11}' & A_{12}' & A_{16}' \\ A_{12}' & A_{22}' & A_{26}' \\ A_{16}' & A_{26}' & A_{66}' \end{bmatrix} \begin{Bmatrix} N_x \\ N_y \\ N_{xy} \end{Bmatrix} \tag{7.58}
$$

The effective longitudinal Young's modulus of the laminate, E_x, governs the response of the laminate under the single axial load per unit length N_x with $N_y = N_{xy} = 0$ [Fig. 7.15(a)] and is defined as

$$
E_x = \frac{\sigma_x}{\epsilon_x^0} = \frac{N_x/t}{A_{11}'N_x} = \frac{1}{tA_{11}'} \tag{7.59}
$$

The effective transverse Young's modulus of the laminate, E_y, governs the response of the laminate under the single axial load per unit length N_y with $N_x = N_{xy} = 0$ [Fig. 7.15(b)] and is defined as

$$
E_y = \frac{\sigma_y}{\epsilon_y^0} = \frac{N_y/t}{A_{22}'N_y} = \frac{1}{tA_{22}'} \tag{7.60}
$$

The effective laminate in-plane shear modulus, G_{xy}, governs the laminate response under the pure shear load per unit length N_{xy} with $N_x = N_y = 0$

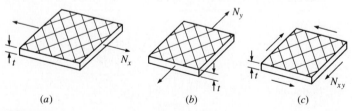

(a)   (b)   (c)

FIGURE 7.15
In-plane loading of symmetric laminate for defining the in-plane laminate engineering constants.

[Fig. 7.15(c)] and is defined as

$$G_{xy} = \frac{\tau_{xy}}{\gamma_{xy}^0} = \frac{N_{xy}/t}{A_{66}'N_{xy}} = \frac{1}{tA_{66}'} \tag{7.61}$$

Similarly, the effective laminate longitudinal Poisson's ratio is

$$\nu_{xy} = -\frac{A_{12}'}{A_{11}'} \tag{7.62}$$

and the effective laminate shear coupling ratios analogous to those given in Eqs. (2.40) and (2.41) for the orthotropic lamina are

$$\eta_{x,xy} = \frac{A_{16}'}{A_{11}'} \quad \text{and} \quad \eta_{xy,y} = \frac{A_{26}'}{A_{66}'} \tag{7.63}$$

Using similar derivations, the effective laminate flexural moduli may be expressed in terms of the flexural compliances. For the symmetric laminate subjected to bending only the laminate moment-curvature relationships are given by

$$\begin{Bmatrix} M_x \\ M_y \\ M_{xy} \end{Bmatrix} = \begin{bmatrix} D_{11} & D_{12} & D_{16} \\ D_{12} & D_{22} & D_{26} \\ D_{16} & D_{26} & D_{66} \end{bmatrix} \begin{Bmatrix} \kappa_x \\ \kappa_y \\ \kappa_{xy} \end{Bmatrix} \tag{7.64}$$

and the inverted forms are given by

$$\begin{Bmatrix} \kappa_x \\ \kappa_y \\ \kappa_{xy} \end{Bmatrix} = \begin{bmatrix} D_{11}' & D_{12}' & D_{16}' \\ D_{12}' & D_{22}' & D_{26}' \\ D_{16}' & D_{26}' & D_{66}' \end{bmatrix} \begin{Bmatrix} M_x \\ M_y \\ M_{xy} \end{Bmatrix} \tag{7.65}$$

Thus, when the laminate is subjected to a pure bending moment per unit length M_x with $M_y = M_{xy} = 0$ [Fig. 7.16(a)], the resulting curvature is

$$\kappa_x = D_{11}'M_x = D_{11}'\frac{M}{b} = \frac{1}{\rho_x} \tag{7.66}$$

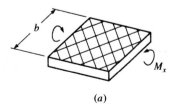
(a)

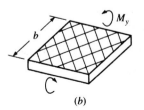
(b)

FIGURE 7.16
Bending moment loading of symmetric laminate for defining the laminate flexural moduli.

where M = total bending moment, which is $M_x b$
   b = laminate width
   ρ_x = radius of curvature = $1/\kappa_x$

For an equivalent homogeneous beam the moment-curvature relationship is

$$\frac{M}{E_{fx} I_{yy}} = \frac{1}{\rho_x} \qquad (7.67)$$

where I_{yy} = second moment of inertia of beam about neutral axis, which is $bt^3/12$
   E_{fx} = flexural modulus of beam along x direction

Recall that the flexural modulus was also defined by Eqs. (7.8) and (7.9) according to laminated beam theory. Combining Eqs. (7.66) and (7.67), we find that the flexural modulus is related to the laminate compliance D'_{11} by the equation

$$E_{fx} = \frac{12}{t^3 D'_{11}} \qquad (7.68)$$

Similarly, the flexural modulus along the y direction [Fig. 7.16(b)] is found to be

$$E_{fy} = \frac{12}{t^3 D'_{22}} \qquad (7.69)$$

Although the laminate stiffnesses A_{ij}, B_{ij}, and D_{ij} are meaningful parameters for all laminate configurations, the engineering constants may not be. Clearly, the use of effective engineering constants must be restricted to those cases where the deformations are similar to the deformations associated with the engineering constant being used. That is, in the above examples for symmetric laminates the $B_{ij} = 0$ and warping under in-plane loads or midplane extension under bending or twisting moments will not occur, so the deformations of the laminate under load would be similar to those for the equivalent homogeneous material. However, the use of engineering constants for the antisymmetric laminate may not be appropriate because of the complex deformations due to coupling effects.

Example 7.8. For the symmetric laminate described in Examples 7.3 and 7.6, determine the effective Young's moduli, in-plane shear modulus, longitudinal Poisson's ratio, and shear-coupling ratios associated with the x and y axes.

Solution. The effective longitudinal Young's modulus is given by Eq. (7.59):

$$E_x = \frac{1}{t A'_{11}} = \frac{1}{(1)(0.04276)} = 23.4 \text{ GPa}$$

Note that due to the $+45°$ ply orientations for this laminate, $E_x = E_y$. The effective in-plane shear modulus is given by Eq. (7.61):

$$G_{xy} = \frac{1}{tA'_{66}} = \frac{1}{(1)(0.02809)} = 35.6 \text{ GPa}$$

The effective longitudinal Poisson's ratio is given by Eq. (7.62):

$$\nu_{xy} = -\frac{A'_{12}}{A'_{11}} = -\frac{-0.0297}{0.04276} = 0.694$$

Since $A_{16} = A_{26} = 0$ for this laminate, it is seen from Eqs. (7.63) that the effective shear-coupling ratios $\eta_{x,xy} = \eta_{xy,y} = 0$. Due to the complex coupling effects acting in the antisymmetric laminate of Examples 7.4 and 7.7, the use of engineering constants for such a laminate would be questionable.

7.5.4 Comparison of Measured and Predicted Compliances

Experimental verification of the laminate theory can be done by applying known loads to laminate and by measuring resulting deformations, then comparing measured deformations with those predicted from the laminate theory. Alternatively, the compliances which are formed from ratios of strains to loads or ratios of curvature to moments for certain simple loading conditions can be experimentally determined and compared with predicted values. The latter approach has been used by Tsai [7.13], who reported results for cross-ply and angle-ply glass/epoxy laminates. Only the results for the angle-ply laminates will be discussed here.

In order to determine the compliances of the laminates under various loads, electrical resistance strain gage rosettes with gages oriented at 0° (x direction), 45°, and 90° (y direction) were attached on both sides of the test specimens (Fig. 7.17). From Eqs. (7.28), the measured normal

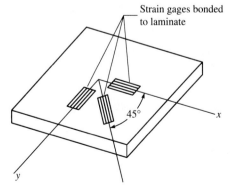

FIGURE 7.17
Strain gage rosette bonded to upper surface of laminate.

strains on the upper surface (where $z = -t/2$) are related to the corresponding midplane strains and curvatures by

$$\epsilon_x^U = \epsilon_x^0 - \frac{t}{2}\kappa_x \tag{7.70}$$

$$\epsilon_y^U = \epsilon_y^0 - \frac{t}{2}\kappa_y \tag{7.71}$$

where ϵ_x^U = measured normal strain along the x direction on upper surface

ϵ_y^U = measured normal strain along y direction on upper surface

Similarly, the normal strains on the lower surface (where $z = t/2$) are given by

$$\epsilon_x^L = \epsilon_x^0 + \frac{t}{2}\kappa_x \tag{7.72}$$

$$\epsilon_y^L = \epsilon_y^0 + \frac{t}{2}\kappa_y \tag{7.73}$$

where ϵ_x^L = measured normal strain along the x direction on lower surface

ϵ_y^L = measured normal strain along the y direction on lower surface

Equations (7.70) and (7.72) can be solved simultaneously for the midplane strain, ϵ_x^0, and curvature, κ_x, whereas ϵ_y^0 and κ_y can be determined from Eqs. (7.71) and (7.73). Although the surface shear strains γ_{xy}^U and γ_{xy}^L are not measured directly like the normal strains, they can be determined from the measured strains along 0°, 45°, and 90° and the strain transformation relationships similar to Eqs. (2.23). For example, from Eq. (2.33), the measured normal strain along the 45° direction on the upper surface, ϵ_{45}^U, is related to the corresponding strains along the x and y axes by

$$\epsilon_{45}^U = \epsilon_x^U \cos^2\theta + \epsilon_y^U \sin^2\theta + \gamma_{xy}^U \sin\theta\cos\theta \tag{7.74}$$

and substituting $\theta = 45°$ and solving for γ_{xy}^U, we find that

$$\gamma_{xy}^U = 2\epsilon_{45}^U - (\epsilon_x^U + \epsilon_y^U) \tag{7.75}$$

Thus, the shear strain is related to the measured strains on the right-hand side of Eq. (7.75). Similarly, for the lower surface

$$\gamma_{xy}^L = 2\epsilon_{45}^L - (\epsilon_x^L + \epsilon_y^L) \tag{7.76}$$

Now the last of Eqs. (7.28) is used to relate the surface shear strains to the midplane strains and curvatures:

$$\gamma_{xy}^{U} = \gamma_{xy}^{0} - \frac{t}{2}\kappa_{xy} \qquad (7.77)$$

$$\gamma_{xy}^{L} = \gamma_{xy}^{0} + \frac{t}{2}\kappa_{xy} \qquad (7.78)$$

These equations can be solved simultaneously for γ_{xy}^{0} and κ_{xy}, so that all midplane strains and curvatures can be determined from the six measured surface strains. For known loading conditions the compliances can then be found.

For a uniaxial loading test of such a strain-gaged specimen with $N_x \neq 0$ and $N_y = N_{xy} = M_x - M_y = M_{xy} = 0$ [Fig. 7.15(a)], Eqs. (7.54) can be used to determine six compliances from known loads, midplane strains, and curvatures as follows:

$$A'_{11} = \frac{\epsilon_x^0}{N_x} \qquad B'_{11} = \frac{\kappa_x}{N_x}$$

$$A'_{12} = \frac{\epsilon_y^0}{N_x} \qquad B'_{12} = \frac{\kappa_y}{N_x} \qquad (7.79)$$

$$A'_{16} = \frac{\gamma_{xy}^0}{N_x} \qquad B'_{16} = \frac{\kappa_{xy}}{N_x}$$

Similar data from a pure flexure test with $M_x \neq 0$ and $N_x = N_y = N_{xy} = M_y = M_{xy} = 0$ [Fig. 7.16(a)] can be used to find the six compliances:

$$B'_{11} = \frac{\epsilon_x^0}{M_x} \qquad D'_{11} = \frac{\kappa_x}{M_x}$$

$$B'_{12} = \frac{\epsilon_y^0}{M_x} \qquad D'_{12} = \frac{\kappa_y}{M_x} \qquad (7.80)$$

$$B'_{16} = \frac{\gamma_{xy}^0}{M_x} \qquad D'_{16} = \frac{\kappa_{xy}}{M_x}$$

All compliances can be determined from such tests. In addition, some compliances can be determined from more than one test (e.g., the B_{ij} in the above tests). A comparison of measured and predicted compliances of angle-ply glass/epoxy laminates having two or three plies of various

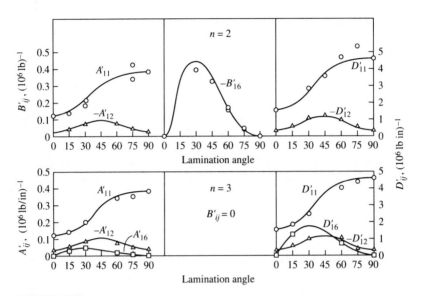

FIGURE 7.18
Measured and predicted compliances for glass/epoxy angle-ply laminates for various lamination angles. (*From Tsai* [7.13]).

lamination angles is shown in Fig. 7.18 from Tsai [7.13]. Predicted compliances were determined by using measured lamina properties as follows [7.13]:

$$E_1 = 7.8 \times 10^6 \, \text{psi} \quad (53.8 \, \text{MPa})$$

$$E_2 = 2.6 \times 10^6 \, \text{psi} \quad (17.9 \, \text{MPa})$$

$$G_{12} = 1.25 \times 10^6 \, \text{psi} \quad (8.6 \, \text{MPa})$$

$$\nu_{12} = 0.25$$

The agreement between measured and predicted values in Fig. 7.18 is quite good, which means that the laminate theory must be reasonably accurate. It is interesting to note that the predicted B_{ij}-type coupling occurs for the two ply antisymmetric laminate but not for the three ply symmetric laminate. Note also that the predicted A_{16}, A_{26}, D_{16}, D_{26}-type coupling occurs for the three ply laminate but not for the two ply laminate.

7.6 HYGROTHERMAL EFFECTS IN LAMINATES

The analysis of hygrothermal behavior of the lamina in Chap. 5 can now be extended to laminates. If we again restrict the discussion to polymer matrix composites, the two main effects of changes in the hygrothermal environment on laminate behavior are degradation of properties and changes in the stress and strain distributions. In this section the analysis of both these effects will be discussed, along with the prediction of laminate hygrothermal expansion coefficients. The basic assumption in all these discussions is that of linearity. That is, we assume that mechanical and hygrothermal effects can be treated separately and then combined using superposition. Coupling between the effects is ignored, as this would lead to nonlinear equations. Another key assumption used here is that temperature and moisture distributions in the laminate are uniform. That is, the temperature and moisture concentration are assumed to be the same for each ply in the laminate.

7.6.1 Hygrothermal Degradation of Laminates

The analysis of hygrothermal degradation in laminates involves the combination of the lamina degradation analysis in Sec. 5.2 with the laminate analysis described earlier in this chapter. For example, given a combination of temperature and moisture, Eq. (5.7) can be used to estimate the degraded matrix property, which is then substituted in the appropriate micromechanics equations to estimate the degraded lamina properties such as E_1, E_2, G_{12}, and v_{12}. The degraded lamina properties are then used in Eqs. (7.38) to (7.40) to find the corresponding degraded laminate stiffnesses. Hygrothermal properties such as the coefficients of thermal expansion and hygroscopic expansion can also be degraded using empirical equations such as Eq. (5.34). These properties would then be used to estimate hygrothermal stresses, as shown in the next section.

7.6.2 Hygrothermal Stresses in Laminates

In Sec. 5.3 the analysis of hygrothermal stresses in an isolated lamina due to temperature and moisture was developed. We now consider the corresponding lamina stresses due to interaction with other laminae in the laminate. Hygrothermal stresses are not only generated during use of composite materials in various environmental conditions, but are generated during fabrication as well. The hygrothermal stresses induced by fabrication are usually referred to as residual stresses. Composites are processed, or cured, at elevated temperatures and then cooled to room temperature. Due to differences between fiber and matrix CTEs (coefficient of thermal expansion) in the lamina and differences between lamina CTEs in the laminate, residual stresses of fabrication may occur at both the micromechanical and macromechanical levels. One particularly important result of residual stresses is that in nonsymmetric laminates the B_{ij}-type coupling can cause residual warping of the cured laminate.

In Sec. 7.5.2 the lamina stresses without hygrothermal effects were found by using Eqs. (7.56). As shown in Sec. 5.3, however, when changes in temperature and moisture concentration occur, the total strains in the kth lamina are given by

$$\{\epsilon\}_k = [\bar{S}]_k\{\sigma\}_k + \{\alpha\}_k \Delta T + \{\beta\}_k c \tag{7.81}$$

and the resulting stresses are given by

$$\{\sigma\}_k = [\bar{Q}]_k(\{\epsilon\}_k - \{\alpha\}_k \Delta T - \{\beta\}_k c) \tag{7.82}$$

where the subscript k refers to the kth lamina. As shown in Sec. 5.3, if the lamina is completely restrained by adjacent laminae, the total strains $\{\epsilon\}_k = 0$ and the resulting hygrothermal stresses are given by

$$\{\sigma\}_k = [\bar{Q}]_k(-\{\alpha\}_k \Delta T - \{\beta\}_k c) \tag{7.83}$$

In a laminate, however, the total lamina strains generally do not vanish, but are instead given by

$$\{\epsilon\}_k = \{\epsilon^0\} + z\{\kappa\} \tag{7.84}$$

and the resulting stresses, including hygrothermal effects, are given by

$$\{\sigma\}_k = [\bar{Q}]_k(\{\epsilon^0\} + z\{\kappa\} - \{\alpha\}_k \Delta T - \{\beta\}_k c) \tag{7.85}$$

Following the procedure outlined in Eqs. (7.32) to (7.42), the resultant laminate forces per unit length are found by integrating Eqs. (7.85) through the thickness of the laminate:

$$\{N\} = \int \{\sigma\}_k \, dz = \int [\bar{Q}]_k(\{\epsilon^0\} + z\{\kappa\} - \{\alpha\}_k \Delta T - \{\beta\}_k c) \, dz$$

$$= [A]\{\epsilon^0\} + [B]\{\kappa\} - \{N^T\} - \{N^M\} \tag{7.86}$$

where the thermal forces due to temperature change are given by

$$\{N^T\} = \int [\bar{Q}]_k\{\alpha\}_k \Delta T \, dz = (\Delta T) \sum_{k=1}^{N} [\bar{Q}]_k\{\alpha\}_k(z_k - z_{k-1}) \tag{7.87}$$

and the hygroscopic forces due to moisture are given by

$$\{N^M\} = \int [\bar{Q}]_k\{\beta\}_k c \, dz = (c) \sum_{k=1}^{N} [\bar{Q}]_k\{\beta\}_k(z_k - z_{k-1}) \tag{7.88}$$

Similarly, the resultant moments per unit length are

$$\{M\} = \int \{\sigma\}_k z \, dz = \int [\bar{Q}]_k(\{\epsilon^0\} + z\{\kappa\} - \{\alpha\}_k \Delta T - \{\beta\}_k c) z \, dz$$

$$= [B]\{\epsilon^0\} + [D]\{\kappa\} - \{M^T\} - \{M^M\} \tag{7.89}$$

where the thermal moments due to temperature changes are

$$\{M^T\} = \int [\bar{Q}]_k\{\alpha\}_k \Delta T z \, dz = \frac{\Delta T}{2} \sum_{k=1}^{N} [\bar{Q}]_k\{\alpha\}_k(z_k^2 - z_{k-1}^2) \tag{7.90}$$

and the hygroscopic moments due to moisture are given by

$$\{M^M\} = \int [\bar{Q}]_k \{\beta\}_k cz \, dz = \frac{c}{2} \sum_{k=1}^{N} [\bar{Q}]_k \{\beta\}_k (z_k^2 - z_{k-1}^2) \qquad (7.91)$$

Rearranging Eqs. (7.86) and (7.89), we find that

$$\{N\} + \{N^T\} + \{M^M\} = [A]\{\epsilon^0\} + [B]\{\kappa\} \qquad (7.92)$$

and

$$\{M\} + \{M^T\} + \{M^M\} = [B]\{\epsilon^0\} + [D]\{\kappa\} \qquad (7.93)$$

or

$$\left\{ \begin{array}{c} N^E \\ \hline M^E \end{array} \right\} = \left[\begin{array}{c|c} A & B \\ \hline B & D \end{array} \right] \left\{ \begin{array}{c} \epsilon^0 \\ \kappa \end{array} \right\} \qquad (7.94)$$

where the total effective forces (mechanical plus hygrothermal) are

$$\{N^E\} = \{N\} + \{N^T\} + \{N^M\} \qquad (7.95)$$

and the total effective moments (mechanical plus hygrothermal) are

$$\{M^E\} = \{M\} + \{M^T\} + \{M^M\} \qquad (7.96)$$

Alternatively, the inverted forms of Eqs. (7.94) are given by

$$\left\{ \begin{array}{c} \epsilon^0 \\ \hline \kappa \end{array} \right\} = \left[\begin{array}{c|c} A' & B' \\ \hline B' & D' \end{array} \right] \left\{ \begin{array}{c} N^E \\ \hline M^E \end{array} \right\} \qquad (7.97)$$

Thus, the lamina stresses for combined mechanical and hygrothermal loading are determined by using a procedure similar to that outlined in Sec. 7.5.2. That is, the midplane strains and curvatures are determined from the total effective forces and moments according to Eqs. (7.97); then the lamina stresses are determined from Eqs. (7.85).

Example 7.9. The antisymmetric angle-ply laminate described in Example 7.4 is heated from 20°C (68°F) to 100°C (212°F). Assuming that the lamina properties do not change over this temperature range, determine the hygrothermal stresses.

Solution. From Table 5.2, the lamina CTEs associated with the principal material axes are

$$\alpha_1 = 0.88 \times 10^{-6}/°C \qquad \alpha_2 = 31.0 \times 10^{-6}/°C$$

The CTEs associated with the +45° and −45° lamina orientations are found by using the transformations in Eqs. (5.22):

$$\left\{ \begin{array}{c} \alpha_x \\ \alpha_y \\ \alpha_{xy}/2 \end{array} \right\}_{+45°} = \left[\begin{array}{ccc} 0.5 & 0.5 & -1.0 \\ 0.5 & 0.5 & 1.0 \\ 0.5 & -0.5 & 0 \end{array} \right] \left\{ \begin{array}{c} 0.88 \\ 31.0 \\ 0 \end{array} \right\} \times 10^{-6} = \left\{ \begin{array}{c} 15.94 \\ 15.94 \\ -15.06 \end{array} \right\} \times 10^{-6}/°C$$

$$\left\{ \begin{array}{c} \alpha_x \\ \alpha_y \\ \alpha_{xy}/2 \end{array} \right\}_{-45°} = \left[\begin{array}{ccc} 0.5 & 0.5 & 1.0 \\ 0.5 & 0.5 & -1.0 \\ -0.5 & 0.5 & 0 \end{array} \right] \left\{ \begin{array}{c} 0.88 \\ 31.0 \\ 0 \end{array} \right\} \times 10^{-6} = \left\{ \begin{array}{c} 15.94 \\ 15.94 \\ 15.06 \end{array} \right\} \times 10^{-6}/°C$$

Next, the thermal forces due to temperature change are found by substituting the above values and the lamina stiffnesses from Example 7.3 in Eqs. (7.87). Note also that the third element in column vector $\{\alpha\}_k$ in Eq. (7.87) is α_{xy}, not $\alpha_{xy}/2$ as in the above transformations. Since $z_k - z_{k-1} = t/4$ for all laminae,

$$\{N^T\} = ([\bar{Q}]_{+45°}\{\alpha\}_{+45°} + [\bar{Q}]_{-45°}\{\alpha\}_{-45°})2(\Delta T)(t/4)$$

or

$$\begin{Bmatrix} N_x^T \\ N_y^T \\ N_{xy}^T \end{Bmatrix} = \begin{bmatrix} 45.22 & 31.42 & 32.44 \\ 31.42 & 45.22 & 32.44 \\ 32.44 & 32.44 & 35.6 \end{bmatrix} \begin{Bmatrix} 15.94 \\ 15.94 \\ -30.12 \end{Bmatrix} (10^{-6})(2)(80)(0.25)$$

$$+ \begin{bmatrix} 45.22 & 31.42 & -32.44 \\ 31.42 & 45.22 & -32.44 \\ -32.44 & -32.44 & 35.6 \end{bmatrix} \begin{Bmatrix} 15.94 \\ 15.94 \\ 30.12 \end{Bmatrix} (10^{-6})(2)(80)(0.25)$$

$$= \begin{Bmatrix} 1.956 \\ 1.956 \\ 0 \end{Bmatrix} \times 10^{-2} \text{ GPa-mm}$$

Similarly, the thermal moments are found from Eqs. (7.90) as

$$\{M^T\} = ([\bar{Q}]_{-45°}\{\alpha\}_{-45°}(z_1^2 - z_0^2) + [\bar{Q}]_{+45°}\{\alpha\}_{+45°}(z_2^2 - z_1^2)$$

$$+ [\bar{Q}]_{-45°}\{\alpha\}_{-45°}(z_3^2 - z_2^2) + [\bar{Q}]_{+45°}\{\alpha\}_{+45°}(z_4^2 - z_3^2)\frac{\Delta T}{2}$$

or

$$\begin{Bmatrix} M_x^T \\ M_y^T \\ M_{xy}^T \end{Bmatrix} = \begin{Bmatrix} 0 \\ 0 \\ -3.81 \end{Bmatrix} \times 10^{-4} \text{ GPa-mm}^2$$

From Eqs. (7.95) and (7.96), we have $\{N^E\} = \{N^T\}$ and $\{M^E\} = \{M^T\}$. Using these results along with the compliances from Example 7.7 in Eqs. (7.97), we find that the midplane strains and curvatures are

$$\begin{Bmatrix} \epsilon_x^0 \\ \epsilon_y^0 \\ \gamma_{xy}^0 \\ \kappa_x \\ \kappa_y \\ \kappa_{xy} \end{Bmatrix} = \begin{Bmatrix} 3.06 & \text{mm/mm} \\ 3.06 & \text{mm/mm} \\ 0 & \text{mm/mm} \\ 0 & \text{mm}^{-1} \\ 0 & \text{mm}^{-1} \\ -9.65 & \text{mm}^{-1} \end{Bmatrix} \times 10^{-4}$$

Note that the thermal twisting moment, M_{xy}^T, causes a corresponding twisting curvature, κ_{xy}, which means that the laminate will warp under the temperature change. Stresses along the x and y axes are now found by substituting the above midplane strains and curvatures, along with the lamina stiffnesses from Example 7.3, in Eqs. (7.85). Stresses at the top and bottom of each ply are given in the following table:

Location	α_x (MPa)	σ_y (MPa)	τ_{xy} (MPa)
#1 Top	−11.8	−11.8	−5.7
#1 Bottom	−3.9	−3.9	−14.3
#2 Top	11.8	11.8	31.6
#2 Bottom	3.9	3.9	23.0
#3 Top	3.9	3.9	−23.0
#3 Bottom	11.8	11.8	−31.6
#4 Top	−3.9	−3.9	14.3
#4 Bottom	−11.8	−11.8	5.7

As with Example 7.7, the stress distribution is quite complex because of the coupling effect.

7.6.3 Laminate Hygrothermal Expansion Coefficients

The effective hygrothermal expansion coefficients for the laminate can be calculated directly by combining the definitions of the coefficients with the appropriate laminate equations. For example, the effective coefficient of thermal expansion of a laminate along the x direction is

$$\alpha_x = \frac{\epsilon_x^0}{\Delta T} \tag{7.98}$$

For a symmetric laminate with $B_{ij} = 0$ the midplane strain along the x direction due to a temperature change ΔT only is given by the first of Eqs. (7.97):

$$\epsilon_x^0 = A_{11}' N_x^T + A_{12}' N_y^T + A_{16}' N_{xy}^T \tag{7.99}$$

The desired thermal expansion coefficient, α_x, is then found by substituting the thermal forces from Eqs. (7.87) in Eq. (7.99) and then by substituting the result in Eq. (7.98). It is important to note that this procedure effectively relates the laminate CTE to lamina CTEs, lamina stiffnesses, laminate compliances, and laminate geometry. The temperature change, ΔT, will cancel out since it appears in both the numerator and denominator. Similar results can be obtained for other thermal and hygroscopic expansion coefficients. As with the effective laminate engineering constants, it is appropriate to restrict the use of the effective hygrothermal expansion coefficients to those cases where the deformations are similar to the deformations associated with the particular coefficient being used. For example, it is probably not a good practice to use such coefficients to describe the hygrothermal behavior of a laminate which exhibits significant warping due to coupling effects.

7.7 INTERLAMINAR STRESSES

One of the key limitations of the Classical Lamination Theory is that each ply is assumed to be in-plane stress in the xy plane (Fig. 7.8), and that interlaminar stresses associated with the z axis are neglected. Such interlaminar stresses can cause delamination, or separation of the laminae, which is a failure mode that we have not previously considered. In this section three-dimensional stress analyses which yield the interlaminar stresses will be discussed, and the resulting interlaminar stresses will be used later in a laminate strength analysis.

A state of plane stress actually does exist in the laminae of a laminate in regions sufficiently far away from geometric discontinuities such as free edges. A three-dimensional elasticity solution by Pipes and Pagano [7.14] has shown, however, that even in a laminate under simple uniaxial loading (Fig. 7.19), there is a "boundary layer" region along the free edges where a three-dimensional state of stress exists, and that the boundary layer thickness is roughly equal to the laminate thickness.

The behavior of interlaminar stresses near a free edge in a laminate will be demonstrated here by using the three stress equilibrium equations

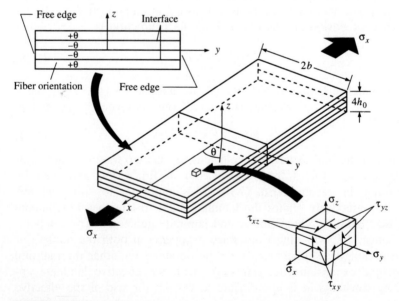

FIGURE 7.19
Pipes and Pagano model for analysis of interlaminar stresses in a laminate under uniaxial extension. (*From Pipes and Pagano* [7.14].)

from the theory of elasticity:

$$\frac{\partial \sigma_x}{\partial x} + \frac{\partial \tau_{xy}}{\partial y} + \frac{\partial \tau_{xz}}{\partial z} = 0 \qquad (7.100)$$

$$\frac{\partial \tau_{yx}}{\partial x} + \frac{\partial \sigma_y}{\partial y} + \frac{\partial \tau_{yz}}{\partial z} = 0 \qquad (7.101)$$

$$\frac{\partial \tau_{zx}}{\partial x} + \frac{\partial \tau_{zy}}{\partial y} + \frac{\partial \sigma_z}{\partial z} = 0 \qquad (7.102)$$

For the uniaxially loaded laminate in Fig. 7.19 we now consider a region near the free edges, where $y = \pm b$, and assume that the stresses do not vary along the loading direction (the x axis). It follows that $\partial \sigma_x / \partial x = 0$ and from Eq. (7.100), the interlaminar shear stress, $\tau_{xz}(z)$, is given by

$$\tau_{xz}(z) = -\int_{-t/2}^{z} \frac{\partial \tau_{xy}}{\partial y} dz \qquad (7.103)$$

We now assume that the in-plane shear stress, τ_{xy}, has a constant value given by the Classical Lamination Theory in the interior regions of the laminae. As we move along the y direction toward a free edge, however, τ_{xy} must decrease to zero at the stress-free surfaces where $y = \pm b$. Thus, as $y \to \pm b$, $\partial \tau_{xy} / \partial y$ must increase. It follows from Eq. (7.103) that τ_{xz} must increase from zero in the interior region to a very large value as $y \to \pm b$, as shown in Fig. 7.20. The region where these rapid changes take place is referred to as the interlaminar stress boundary layer region, as shown in Fig. 7.20. From Eqs. (7.101) and (7.102), respectively, the other interlaminar stresses as

$$\tau_{yz}(z) = -\int_{-t/2}^{z} \frac{\partial \sigma_y}{\partial y} dz \qquad (7.104)$$

and

$$\sigma_z(z) = -\int_{-t/2}^{z} \frac{\partial \tau_{yz}}{\partial y} dz \qquad (7.105)$$

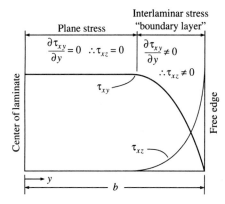

FIGURE 7.20
Schematic representation of in-plane shear stress and interlaminar shear stress distributions at ply interface.

Pipes and Pagano [7.14] used a finite difference numerical scheme to solve the three governing field equations which are generated by combining the three-dimensional versions of the stress equilibrium equations, the lamina stress-strain relationships, and the strain-displacement relations. The equations were solved subject to stress-free boundary conditions along the free edges of a four layer $\pm 45°$ graphite/epoxy laminate under uniform axial strain, ϵ_x. Figure 7.21 shows the complete stress results obtained by Pipes and Pagano [7.14]. It is important to note that the in-plane stresses σ_x and τ_{xy} from the three-dimensional analysis agree with those predicted by the Classical Lamination Theory in the central portion of the laminate, but both stresses drop in the boundary layer region near the free edge. On the other hand, the interlaminar stresses σ_z, τ_{xz}, and τ_{yz} are all equal to zero in the central portion of the laminate but change rapidly near the free

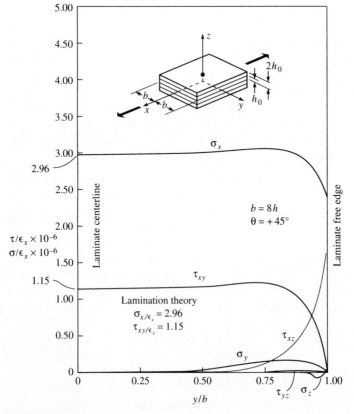

FIGURE 7.21
Distributions of all stresses from Pipes and Pagano analysis. (*From Pipes and Pagano* [7.14].)

edge. The shear stress τ_{xz} is the largest of the interlaminar stresses, as it appeared to grow without bound at $y/b = 1.0$. Pipes and Pagano suspected that a singularity for this stress component exists at the free edge, but it was not possible to prove the existence of such a singularity with the approximate finite difference solution. Analytical proof of the existence of these singularities was published later by Wang and Choi [7.15, 7.16].

The numerical results of Pipes and Pagano [7.14] for a variety of laminate cross-sectional aspect ratios led to the conclusion that the boundary layer region of three-dimensional stresses extends inward approximately one laminate thickness from the free edge. This conclusion was later verified experimentally by Pipes and Daniel [7.17], who used a Moiré technique to measure displacements along the x direction on the surface of the laminate. The measured surface displacement profiles, which also clearly indicated the presence of the boundary layer, agreed closely with those predicted by the Pipes and Pagano analysis.

It has been shown both analytically and experimentally that the lamina stacking sequence influences interlaminar stresses and, consequently, delamination in laminates. Pipes and Pagano [7.18] used an approximate elasticity solution to study the effect of the stacking sequence on the interlaminar shear stress in ±45° laminates, as shown in Fig. 7.22. It is clear from Fig. 7.22 that when layers having the same orientation are stacked together (which increases the apparent layer thickness), the interlaminar shear stress, τ_{xz}, is higher than for the case where layers of opposite orientation are stacked together. In a separate paper Pagano and Pipes [7.19] showed that a change in the stacking sequence can actually cause the interlaminar normal stress, σ_z, to change from tensile to compressive. Since tensile interlaminar normal stresses would tend to cause separation of the plies, while compressive interlaminar normal stresses would tend to keep the plies together, stacking sequences which produce the former stress state should have lower strengths than those producing the latter stress state. Experimental results such as those by Whitney and Browning [7.20] and Whitney and Kim [7.21] seem to support this conclusion. Ply orientation also has a strong effect on interlaminar stresses, as shown by Pipes and Pagano [7.14].

Since the publication of the Pipes and Pagano solution, a number of investigators have used other methods to study the "free-edge" interlaminar stress phenomenon. Rybicki [7.22], Wang and Crossman [7.23], Herakovich [7.24], and Hwang and Gibson [7.25] all used three-dimensional finite element analyses to investigate interlaminar stresses. The quarter domain finite element model used by Hwang and Gibson [7.25] for the analysis of the original Pipes and Pagano [7.14] laminate is shown in Fig. 7.23. Finite element stress distributions near the free edge

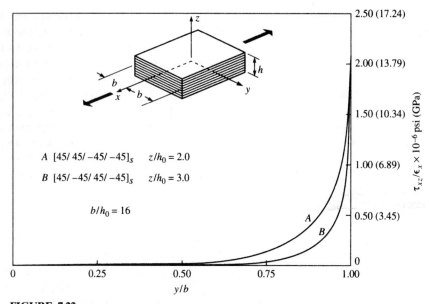

FIGURE 7.22
Effect of stacking sequence on interlaminar shear stress. (*From Pipes and Pagano* [7.18].
Reprinted by permission of The American Society of Mechanical Engineers.)

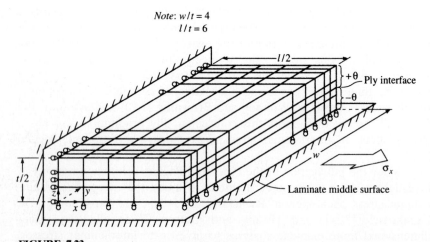

FIGURE 7.23
Quarter domain finite element model of laminate used by Hwang and Gibson to analyze the
Pipes-Pagano problem. (*From Hwang and Gibson* [7.25]. *Reprinted by permission of
Elsevier Publishers, Ltd.*)

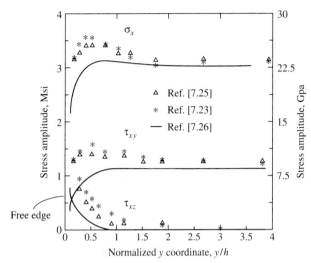

FIGURE 7.24
Comparison of stress distributions near the free edge. (*From Hwang and Gibson* [7.25]. *Reprinted by permission of Elsevier Science Publishers, Ltd.*)

from Wang and Crossman [7.23] and Hwang and Gibson [7.25] are compared with those from an empirical solution derived from the theory of elasticity by Hwang [7.26] in Fig. 7.24. The empirical elasticity solution by Hwang [7.26] is based on a similar solution by Whitney [7.27], which, in turn, is an attempt to fit the finite difference results of Pipes and Pagano [7.14] with relatively simple empirical equations which satisfy the stress equilibrium equations [Eqs. (7.100) to (7.102)] and the free edge boundary conditions. Thus, the empirical results shown by the solid curve in Fig. 7.24 should be very close to the original Pipes and Pagano results. Although the stress distributions from the two finite element models show good agreement with each other, both sets of stresses are seen to be greater than those from the empirical solution near the free edge. Improved approximate polynomial solutions have been proposed by Conti and De Paulis [7.28].

7.8 LAMINATE STRENGTH ANALYSIS

Recall that in Sec. 4.2 we discussed several multiaxial strength criteria for estimating the strength of individual laminae under in-plane stresses. Such strength criteria can also be used on a ply-by-ply basis for a laminate to determine which ply fails first under in-plane loads. In Sec. 7.7, however, we have seen that interlaminar stresses in laminates also have to be taken into account because they may lead to a different mode of failure known as delamination. This section deals with the analysis of both first ply failure due to in-plane stresses and delamination due to interlaminar stresses. The mechanical behavior of the laminate after first ply failure and subsequent ply failures is also discussed.

7.8.1 First Ply Failure Due to In-Plane Stresses

The prediction of first ply failure due to in-plane stresses is a straightforward application of the appropriate multiaxial lamina strength criterion in combination with the lamina stress analysis from the Classical Lamination Theory. The loads corresponding to first ply failure are not necessarily the laminate failure loads, however, since a laminate generally has plies at several orientations. That is, there will usually be a sequence of ply failures at different loads culminating in ultimate laminate failure when all plies have failed. Thus, the ultimate load-carrying capacity of the laminate may be significantly higher than the first ply failure load, and prediction of laminate failure based on first ply failure may be too conservative.

In the analysis of first ply failure and subsequent ply failures the stiffness matrices for the failed plies and the corresponding laminate stiffness matrix must be modified after each ply failure to reflect the effects of those failures. Figure 7.25 shows a piecewise linear laminate

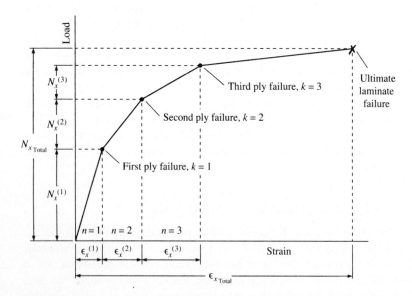

FIGURE 7.25
Load-strain curve for uniaxially loaded laminate showing multiple ply failures leading up to ultimate laminate failure.

load-deformation curve with several "knees" due to ply failures. The total forces and moments at the kth knee in the curve are related to the corresponding forces and moments for the nth section of such a curve (where $n \leq k$) by the summation

$$\left\{ \begin{array}{c} N \\ M \end{array} \right\}_{\text{Total}} = \sum_{n=1}^{k} \left\{ \begin{array}{c} N^{(n)} \\ M^{(n)} \end{array} \right\} \tag{7.106}$$

where the superscript (n) on a parameter denotes the particular value of that parameter associated with the nth section. The corresponding midplane strains and curvatures are given by

$$\left\{ \begin{array}{c} \epsilon^{0} \\ \kappa \end{array} \right\}_{\text{Total}} = \sum_{n=1}^{k} \left\{ \begin{array}{c} \epsilon^{0(n)} \\ \kappa^{(n)} \end{array} \right\} \tag{7.107}$$

Using the piecewise linear assumption, the load-deformation relationship for the nth section can be approximated by modifying Eqs. (7.42) as

$$\left\{ \begin{array}{c} N^{(n)} \\ \hline M^{(n)} \end{array} \right\} = \left[\begin{array}{c|c} A^{(n)} & B^{(n)} \\ \hline B^{(n)} & D^{(n)} \end{array} \right] \left\{ \begin{array}{c} \epsilon^{0(n)} \\ \hline \kappa^{(n)} \end{array} \right\} \tag{7.108}$$

where the $[A^{(n)}]$, $[B^{(n)}]$, and $[D^{(n)}]$ are the modified stiffness matrices after the $(n-1)$th ply failure. But the calculation of these modified laminate stiffnesses requires that we know the modified ply stiffnesses, $[Q^{(n)}]$, and before we can modify the ply stiffness matrices, we must know the type of failure. That is, if the ply failure is caused by the in-plane shear stress exceeding the shear strength, the shear modulus and the transverse modulus of that ply may be severely degraded by longitudinal cracks; but the longitudinal modulus may not be affected significantly by these cracks. Alternatively, all the ply stiffnesses for the failed ply could be equated to zero in the calculation of the degraded laminate stiffnesses.

Halpin [7.12] has used a procedure similar to the one outlined above to analyze the uniaxial stress-strain response of a $[0/\pm 45/90]_s$ glass/epoxy laminate. The Maximum Strain Criterion was used to predict ply failure, and the ply stiffnesses of the failed plies were set equal to zero. The predicted stress-strain curve shows good agreement with the corresponding experimental data, as shown in Fig. 7.26. Notice that the curve has two "knees"—the first one at the strain corresponding to failure of the $90°$ plies and the second one at the strain corresponding to failure of the $\pm 45°$ plies. The knee for the $\pm 45°$ ply failure is more distinct than the one for the $90°$ ply failure because the laminate has twice as many $\pm 45°$ plies as it does $90°$ plies. Ultimate laminate failure occurs at the longitudinal failure strain for the $0°$ plies. It is also interesting to note that the experimental data do not show as much of a change in slope at

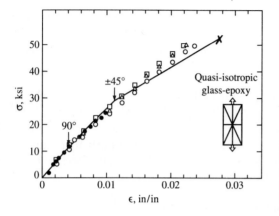

FIGURE 7.26
Comparison of predicted and measured stress-strain response of $[0/\pm45/90]_s$ glass/epoxy laminate. (*From Halpin* [7.12].)

the knees as the theoretical curve does. This may be due to the fact that actual ply failure occurs gradually over a finite strain range, whereas instantaneous ply failure at a single strain level is assumed in the analysis. The same reasoning may explain the absence of jumps in the stress-strain curve after ply failure [7.29]. A horizontal jump would be predicted if the test data were taken under load-control, whereas a vertical jump would be predicted for a displacement-control test. Hahn and Tsai [7.29] have observed that the knee in the stress-strain curve for cross-ply laminates is quite obvious if the 90° plies are all stacked adjacent to each other, but

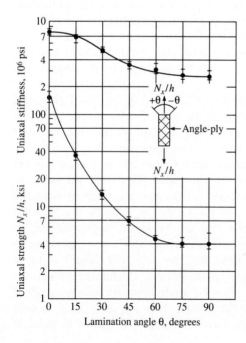

FIGURE 7.27
Comparison of predicted and measured uniaxial strength and stiffness of glass/epoxy angle-ply laiminate. (*From Tsai* [7.30].)

the knee is not so obvious if the 0° and 90° plies are arranged in an alternating 0°/90° sequence. Restraint of the failed 90° plies by the remaining 0° plies was thought to be more effective in the alternating 0°/90° sequence, making the failure of the 90° plies more gradual.

The in-plane strength of ±θ angle-ply laminates may also be analyzed using a multiaxial lamina strength criterion and the Classical Lamination Theory, but the piecewise linear approximation of the stress-strain curve is not needed. This is because all plies fail simultaneously in the angle-ply laminate, and the stress-strain curve does not have the characteristic knees shown in Figs. 7.25 and 7.26. Tsai [7.30] has used the Tsai-Hill Criterion to predict the strength of glass/epoxy angle-ply laminates as a function of the lamination angle θ, and the predictions are seen to agree well with experimental data in Fig. 7.27. The predicted laminate stiffness A_{11} also shows good agreement with the prediction from the Classical Lamination Theory in Fig. 7.27.

Example 7.10. A $[90/0/90]_s$ laminate consisting of the AS/3501 laminae described in Example 7.3 is subjected to tensile uniaxial loading along the x direction. Using the Maximum Strain Criterion, find the loads corresponding to first ply failure and subsequent ply failures; then plot the load-strain curve up to failure.

Solution. The failure strains are found by substituting the data from Tables 2.2 and 4.1 in Eqs. (4.1):

$$e_L^{(+)} = \frac{s_L^{(+)}}{E_1} = \frac{1448}{138 \times 10^3} = 0.0105$$

$$e_T^{(+)} = \frac{s_T^{(+)}}{E_2} = \frac{48.3}{9 \times 10^3} = 0.0054$$

Using these results in the Maximum Strain Criterion, we see that first ply failure occurs at a strain $\epsilon_x = e_T^{(+)} = 0.0054$. To find the corresponding load N_x at first ply failure, it is necessary to find the initial laminate stiffness matrix $[A^{(1)}]$. From Example 7.3, the lamina stiffness matrix for the 0° plies is given by

$$[\bar{Q}]_{0°} = [Q] = \begin{bmatrix} 138.8 & 2.72 & 0 \\ 2.72 & 9.05 & 0 \\ 0 & 0 & 6.9 \end{bmatrix} \text{ GPa}$$

The stiffness matrix for the 90° plies is formed by simply interchanging the 11 and 22 terms in the stiffness matrix for the 0° plies:

$$[\bar{Q}]_{90°} = \begin{bmatrix} 9.05 & 2.72 & 0 \\ 2.72 & 138.8 & 0 \\ 0 & 0 & 6.9 \end{bmatrix} \text{ GPa}$$

For the first section of the load-strain curve the laminate stiffness matrix is therefore

$$[A^{(1)}] = [\bar{Q}]_{0°}(2)(0.25) + [\bar{Q}]_{90°}(4)(0.25) = 0.5[\bar{Q}]_{0°} + [\bar{Q}]_{90°}$$

or

$$[A^{(1)}] = \begin{bmatrix} 78.45 & 4.08 & 0 \\ 4.08 & 143.3 & 0 \\ 0 & 0 & 10.35 \end{bmatrix} \text{ GPa-mm}$$

At first ply failure the laminate load-deformation equations can be written as

$$\begin{Bmatrix} N_x^{(1)} \\ 0 \\ 0 \end{Bmatrix} = \begin{bmatrix} 78.45 & 4.08 & 0 \\ 4.08 & 143.3 & 0 \\ 0 & 0 & 10.35 \end{bmatrix} \begin{Bmatrix} 0.0054 \\ \epsilon_y^{(1)} \\ \gamma_{xy}^{(1)} \end{Bmatrix}$$

These equations can be solved simultaneously to get the following values of loads and strains at first ply failure:

$$N_x^{(1)} = 0.423 \text{ GPa-mm}; \qquad \epsilon_y^{(1)} = -0.000154; \qquad \gamma_{xy}^{(1)} = 0$$

We will now demonstrate two different approaches for modifying the laminate stiffness matrix after first ply failure.

(a) In the first approach we simply set all ply stiffnesses equal to zero for the failed 90° plies. The adjusted laminate stiffness matrix is then

$$[A^{(2)}] = 0.5[\bar{Q}]_{0°} = \begin{bmatrix} 69.4 & 1.36 & 0 \\ 1.36 & 4.52 & 0 \\ 0 & 0 & 3.45 \end{bmatrix} \text{ GPa-mm}$$

Now the 0° ply failure and the ultimate laminate failure occurs at a strain level $\epsilon_x = e_L^{(+)} = 0.0105$, which means that the strain increment for the second section of the load-strain curve is

$$\epsilon_x^{(2)} = e_L^{(+)} - \epsilon_x^{(1)} = 0.0105 - 0.0054 = 0.0051$$

The load-deformation equations describing the second section of the load-strain curve are

$$\begin{Bmatrix} N_x^{(2)} \\ 0 \\ 0 \end{Bmatrix} = \begin{bmatrix} 69.4 & 1.36 & 0 \\ 1.36 & 4.52 & 0 \\ 0 & 0 & 3.45 \end{bmatrix} \begin{Bmatrix} 0.0051 \\ \epsilon_y^{(2)} \\ \gamma_{xy}^{(2)} \end{Bmatrix}$$

and the simultaneous solution of these equations yields the results

$$N_x^{(2)} = 0.352 \text{ GPa-mm}; \qquad \epsilon_y^{(2)} = -0.00153; \qquad \gamma_{xy}^{(2)} = 0$$

The total laminate failure load is then

$$N_{x\text{Total}} = N_x^{(1)} + N_x^{(2)} = 0.423 + 0.352 = 0.775 \text{ GPa-mm}$$

and the load-strain curve is shown as curve (a) in Fig. 7.28.

(b) In the second approach we set only $E_2 = G_{12} = \nu_{21} = 0$ for the failed 90° plies, but we assume that E_1 for the 90° plies is not affected by the transverse failure. According to these assumptions,

$$[\bar{Q}_{22}]_{90°} = E_1 = 138 \text{ GPa}; \qquad [\bar{Q}_{11}]_{90°} = [\bar{Q}_{12}]_{90°} = [\bar{Q}_{66}]_{90°} = 0$$

and the adjusted laminate stiffness matrix is

$$[A^{(2)}] = \begin{bmatrix} 69.4 & 1.36 & 0 \\ 1.36 & 143.3 & 0 \\ 0 & 0 & 3.45 \end{bmatrix} \text{ GPa-mm}$$

The laminate load-deformation equations for the second section are

$$\left\{ \begin{array}{c} N_x^{(2)} \\ 0 \\ 0 \end{array} \right\} = \begin{bmatrix} 69.4 & 1.36 & 0 \\ 1.36 & 143.3 & 0 \\ 0 & 0 & 3.45 \end{bmatrix} \left\{ \begin{array}{c} 0.0051 \\ \epsilon_y^{(2)} \\ \gamma_{xy}^{(2)} \end{array} \right\}$$

and the resulting loads and strains for the second section are

$$N_x^{(2)} = 0.354 \text{ GPa-mm}; \qquad \epsilon_y^{(2)} = -0.000048; \qquad \gamma_{xy}^{(2)} = 0$$

The total load at laminate failure is

$$N_{x\text{Total}} = 0.423 + 0.354 = 0.777 \text{ GPa-mm}$$

and the load-strain curve is shown as curve (b) in Fig. 7.28. It is interesting to note that although the assumptions regarding degradation of the failed plies are quite different for curves (a) and (b), the predicted load-strain curves for the two approaches are virtually the same. In general, differences in predictions from the two approaches would depend on ply properties and stacking sequences. It is also interesting to note that we might intuitively

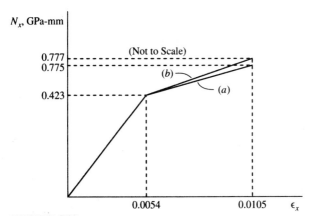

FIGURE 7.28
Predicted load-strain curves for Example 7.10.

expect approach (*a*) to be more conservative than approach (*b*), and this turns out to be the case.

7.8.2 Delamination Due to Interlaminar Stresses

Delamination due to interlaminar stresses can reduce the failure stress of the laminate below that predicted by the in-plane failure criteria discussed in the previous section. Failure by delamination is not necessarily the same as the initiation of delamination, however. The initiation of delamination is generally followed by stable delamination growth, which eventually leads to unstable growth and ultimate failure. The onset of delamination can be predicted by using either mechanics of materials approaches or fracture mechanics approaches. Fracture mechanics is also the preferred analytical treatment for delamination growth and failure. In this section we will discuss mechanics of materials approaches to the prediction of delamination initiation, and fracture mechanics will be covered later in Chap. 9.

The average stress criterion of Kim and Soni [7.31] was one of the first mechanics of materials approaches to the prediction of the onset of delamination. This criterion is based on the premise that delamination will begin once the average value of the interlaminar tensile normal stress, $\bar{\sigma}_z$, near the free edge reaches the interlaminar tensile strength, $s_Z^{(+)}$. A similar criterion for failure of notched laminates had been proposed previously by Whitney and Nuismer [7.32]. In the Kim-Soni Criterion the averaging is done over a critical length, b_0, as shown in Eq. (7.109) and Fig. 7.29:

$$\bar{\sigma}_z = \frac{1}{b_0} \int_{b-b_0}^{b} \sigma_z(y, 0)\, dy = s_Z^{(+)} \tag{7.109}$$

The distance *b* was the half-width of the laminate, as shown in Fig. 7.19,

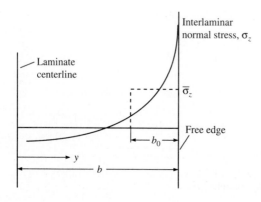

FIGURE 7.29

Graphical interpretation of average interlaminar normal stress near free edge according to the Kim-Soni Criterion.

and the critical length b_0 is assumed to be equal to one ply thickness. Due to the difficulty of measuring $s_Z^{(+)}$, Kim and Soni assumed that $s_Z^{(+)} = s_T^{(+)}$. Although this criterion provided reasonably accurate predictions of the onset of delamination in composites where the tensile normal stress, σ_z, was the dominant interlaminate stress, a more general criterion was needed for cases where delamination may be affected by interlaminar shear stresses as well.

The need for a more general criterion for predicting the onset of delamination was recognized by Brewer and Lagace [7.33], who proposed the Quadratic Delamination Criterion

$$\left(\frac{\bar{\sigma}_{xz}}{s_{XZ}}\right)^2 + \left(\frac{\bar{\sigma}_{yz}}{s_{YZ}}\right)^2 + \left(\frac{\bar{\sigma}_z^t}{s_Z^{(+)}}\right)^2 + \left(\frac{\bar{\sigma}_z^c}{s_Z^{(-)}}\right)^2 = 1 \qquad (7.110)$$

where $\bar{\sigma}_{xz}, \bar{\sigma}_{yz}$ = average interlaminar shear stresses

$\bar{\sigma}_z^t, \bar{\sigma}_z^c$ = average interlaminar tensile and compressive normal stresses, respectively

s_{XZ}, s_{YZ} = interlaminar shear strengths

$s_Z^{(+)}, s_Z^{(-)}$ = interlaminar tensile and compressive strengths, respectively

Each of the average stress components in this case is defined as

$$\bar{\sigma}_{ij} = \frac{1}{\lambda_{\text{avg}}} \int_0^{\lambda_{\text{avg}}} \sigma_{ij} \, d\lambda \qquad (7.111)$$

where λ = distance from some reference point (in this case the free edge)

λ_{avg} = averaging dimension

σ_{ij} = stress component σ_{xz}, σ_{yz}, σ_z^t, or σ_z^c and the overbar denotes its average value.

Brewer and Lagace found that for the $[\pm 15_n]_s$, $[\pm 15_n/0_n]_s$, and $[0_n/\pm 15_n]_s$ AS1/3501-6 graphite/epoxy laminates tested the second and fourth terms in Eq. (7.110) were negligible, so that the Quadratic Delamination Criterion took on the simplified form

$$\left(\frac{\bar{\sigma}_{xz}}{s_{XZ}}\right)^2 + \left(\frac{\bar{\sigma}_z^t}{s_Z^{(+)}}\right)^2 = 1 \qquad (7.112)$$

Transverse isotropy was assumed, so that $s_Z^{(+)} = s_T^{(+)} = 53.9$ MPa. The parameters λ_{avg} and s_{XZ} were used as curve-fitting parameters to obtain the best agreement with experimental data. In the corresponding experiments laminate specimens were tested under displacement-control and the instantaneous drop in the tensile load at delamination onset was observed. The "best-fit" parameters for all laminate configurations tested were $\lambda_{\text{avg}} = 0.178$ mm and $s_{XZ} = 105$ MPa. Further support for the validity of the Quadratic Delamination Criterion and the assumption of transverse isotropy was discovered with the observation that the best-fit value of s_{XZ} was the same as s_{LT}, the in-plane shear strength of this material.

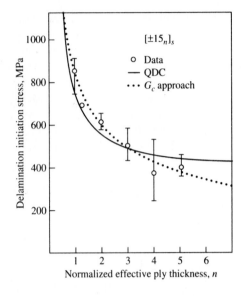

FIGURE 7.30
Predicted and measured delamination initiation stresses for $[+15_n]_s$ laminates. (*From Brewer and Lagace* [7.33].)

Although the value of λ_{avg} was not assumed to be equal to the ply thickness as in the Kim-Soni analysis, the best-fit value of 0.178 mm was of the same order as the ply thickness. A comparison of the measured and predicted delamination onset stresses for various normalized effective ply thicknesses, n, are shown for the $[\pm15_n]_s$ laminate in Fig. 7.30. Specimens were made by stacking single plies of the same orientation together to form a ply with greater effective thickness, and the value of n is this effective ply thickness divided by the single ply thickness. Also shown in Fig. 7.30 are the predictions from a fracture mechanics approach, which will be discussed later.

Catastrophic failure of laminated structures is not the only undesirable result of delamination. The reduction in stiffness of a laminate during delamination growth may make the structure unsafe even if fracture does not occur. Conversely, stiffness loss can be used to characterize the growth of delamination. Thus, analytical models are needed for estimating this stiffness loss during delamination.

O'Brien [7.34] has developed an analysis of stiffness reduction in symmetric laminates during delamination based on a simple "rule of mixtures" and the Classical Lamination Theory. Recall from Eq. (7.59) that the effective longitudinal Young's modulus of a symmetric laminate is given by

$$E_x = \frac{1}{tA'_{11}} \tag{7.113}$$

This equation was used by O'Brien to model the stiffness of the laminate

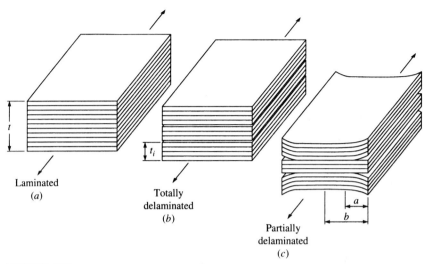

FIGURE 7.31
Rule of mixtures analysis of stiffness loss due to delamination. (*From O'Brien* [7.34].
Copyright ASTM. Reprinted with permission.)

without delaminations, as shown in Fig. 7.31(*a*). The corresponding
stiffness of a laminate which has been totally delaminated along one or
more interfaces [Fig. 7.31(*b*)], but whose sublaminates must still have the
same longitudinal strain, is given by the rule of mixtures formula

$$E_{td} = \frac{\sum_{i=1}^{m} E_{xi} t_i}{t} \tag{7.114}$$

where E_{td} = longitudinal Young's modulus of a laminate totally delaminated along one or more interfaces
E_{xi} = longitudinal Young's modulus of ith sublaminate formed by
the delamination
t_i = thickness of the ith sublaminate
m = number of sublaminates formed by the delamination

The longitudinal Young's modulus of a laminate which has been partially
delaminated along the same interfaces [Fig. 7.31(*c*)] is given by the rule
of mixtures formula

$$E = (E_{td} - E_x)\frac{a}{b} + E_x \tag{7.115}$$

where E = longitudinal Young's modulus of a laminate partially delaminated along one or more interfaces
a = distance that delamination extends in from free edge
b = half-width of laminate

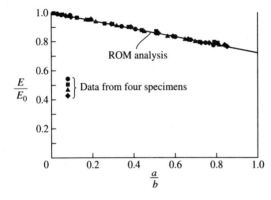

FIGURE 7.32
Predicted and measured laminate stiffness as a function of delamination size. (*From O'Brien* [7.34]. *Copyright ASTM. Reprinted with permission.*)

A more general form of Eq. (7.115) is given by Eq. (7.116):

$$E = (E_{td} - E_x)\frac{A_d}{A_t} + E_x \tag{7.116}$$

where A_d = delaminated area
A_t = total interfacial area

The predicted values of E normalized to the initial modulus, E_0, are compared with measured values of E/E_0 for various delamination sizes in $[\pm30/\pm30/90/\overline{90}]_s$ graphite/epoxy laminates in Fig. 7.32, and the agreement is seen to be excellent. As shown in Fig. 7.32, complete delamination of this laminate would result in a 25.8 percent reduction in the laminate stiffness. Such a loss of stiffness would lead to an undesirable increase in the deflection of the structure under load.

In this section we have only been concerned with delamination near free edges in laminates, but interlaminar stresses and delamination may occur at other discontinuities such as holes, ply drops, and joints (see Fig. 7.33 from Ref. [7.35]). Low velocity impact (e.g., dropping a wrench) on

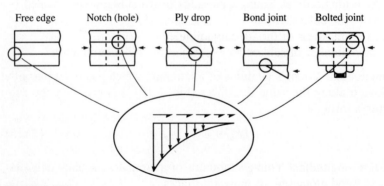

FIGURE 7.33
Interlaminar stresses occur at a variety of discontinuities in composite structures. (*From Newaz* [7.35].)

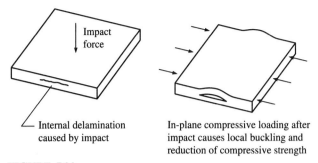

Internal delamination
caused by impact

In-plane compressive loading after
impact causes local buckling and
reduction of compressive strength

FIGURE 7.34
Reduction of in-plane compressive strength of laminate after transverse impact.

a composite structure may cause internal delaminations, which may reduce the in-plane compressive strength (Fig. 7.34). The analysis of delamination under such complex states of stress generally requires the use of fracture mechanics and finite element numerical approaches, some of which are discussed in Chap. 9. The reader is also encouraged to refer to numerous analytical and experimental studies in several recent books [7.35–7.37].

7.9 DEFLECTION AND BUCKLING OF LAMINATES

This section is concerned with the analysis of transverse deflections of laminates under transverse loading and the prediction of laminate buckling forces. Transverse deflections of laminates due to bending are generally much larger than in-plane deflections because flexural stiffnesses are lower than extensional stiffnesses. Thus, transverse deflections are an important design consideration, and the development of analytical models for predicting such deflections are of interest. Buckling of laminates is an instability which is characterized by excessive transverse deflections under in-plane compressive or shear forces. The general equilibrium equations governing transverse deflections involve both in-plane and out-of-plane forces, but the coupling between in-plane forces and transverse deflections is usually taken into account only for the buckling analysis or for large deflection analysis. In the analysis of small transverse deflections alone the out-of-plane forces are the most important because the laminate is normally designed in such a way that the in-plane forces are less than the corresponding buckling loads. Only a brief introduction to deflection and buckling is given here. For more detailed coverage of these subjects the reader is referred to the works of Whitney [7.10], Lekhnitskii [7.7], Vinson and Sierakowski [7.38], and Liessa [7.39].

7.9.1   Analysis of Small Transverse Deflections

The analysis of transverse deflections of laminated plates has its basis in the Classical Laminatinon Theory, which was outlined in Sec. 7.3 and in the differential equations of equilibrium. In order to develop the differential equations governing plate deflections, it is convenient to use an infinitesimal element, as shown in Figs. 7.35(*a*), (*b*), and (*c*) from Halpin [7.12]. In-plane stress resultants are shown in Fig. 7.35(*a*), moment resultants are shown in Fig. 7.35(*b*), and transverse shear stress resultants are shown in Fig. 7.35(*c*). Transverse shear stress resultants were not considered in Sec. 7.3, but they must be considered here in the transverse deflection analysis. In these diagrams it is assumed that the transverse deflections are small, so that the out-of-plane components of the in-plane resultants N_x, N_y, and N_{xy} are negligible. However, these out-of-plane components will be considered in the next section on buckling analysis. Along with the stress and moment resultants such as those defined previously in Fig. 7.8 and in Eqs. (7.32 to 7.33), the transverse shear stress resultants Q_x and Q_y are similarly defined as

$$Q_x = \int_{-t/2}^{t/2} \tau_{xz}\, dz \tag{7.117}$$

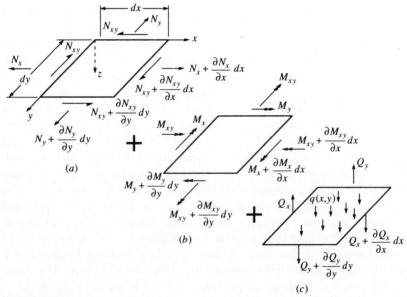

(a)

(b)

(c)

FIGURE 7.35
Stress resultants and external loads acting on laminate. (*From Halpin* [7.12].)

and
$$Q_y = \int_{-t/2}^{t/2} \tau_{yz}\, dz \qquad (7.118)$$

Also included in Fig. 7.35 is a distributed transverse load $q(x, y)$. Following the derivation by Halpin [7.12], for static equilibrium according to Newton's second law, the summation of forces along the x direction must be

$$N_x\, dy + \frac{\partial N_x}{\partial x} dx\, dy + N_{xy}\, dx + \frac{\partial N_{xy}}{\partial y} dx\, dy - N_x\, dy - N_{xy}\, dx = 0 \quad (7.119)$$

Equation (7.119) may be simplified as

$$\frac{\partial N_x}{\partial x} + \frac{\partial N_{xy}}{\partial y} = 0 \qquad (7.120)$$

The summation of forces along the y direction yields

$$N_y\, dx + \frac{\partial N_x}{\partial y} dx\, dy + N_{xy}\, dy + \frac{\partial N_{xy}}{\partial x} dx\, dy - N_y\, dx - N_{xy}\, dy = 0 \quad (7.121)$$

or
$$\frac{\partial N_y}{\partial y} + \frac{\partial N_{xy}}{\partial x} = 0 \qquad (7.122)$$

The summation of forces along the z direction gives

$$Q_x\, dy + \frac{\partial Q_x}{\partial x} dx\, dy + Q_y\, dx + \frac{\partial Q_y}{\partial y} - Q_x\, dy - Q_y\, dx + q(x, y) = 0 \quad (7.123)$$

or
$$\frac{\partial Q_x}{\partial x} + \frac{\partial Q_y}{\partial y} + q(x, y) = 0 \qquad (7.124)$$

The summation of moments about the x axis yields

$$-M_y\, dx - \frac{\partial M_y}{\partial y} dy\, dx - M_{xy}\, dy - \frac{\partial M_{xy}}{\partial x} dx\, dy + Q_y\, dx\, dy$$

$$+ \frac{\partial Q_y}{\partial y} dy\, dx\, dy + q(x, y)\, dx\, dy\, dy/2 + Q_x\, dy\, dy/2$$

$$+ \frac{\partial Q_x}{\partial x} dx\, dy\, dy/2 + M_y\, dx + M_{xy}\, dy - Q_x\, dy\, dy/2 = 0 \quad (7.125)$$

Simplifying and neglecting products of differentials, we get

$$\frac{\partial M_y}{\partial y} + \frac{\partial M_{xy}}{\partial x} = Q_y \qquad (7.126)$$

A similar summation of moments about the y axis gives

$$\frac{\partial M_x}{\partial x} + \frac{\partial M_{xy}}{\partial y} = Q_x \tag{7.127}$$

Substitution of Eqs. (7.126) and (7.127) in Eq. (7.124) yields

$$\frac{\partial^2 M_x}{\partial x^2} + 2\frac{\partial^2 M_{xy}}{\partial x\, \partial y} + \frac{\partial^2 M_y}{\partial y^2} + (q(x, y) = 0 \tag{7.128}$$

Equations (7.120), (7.122), and (7.128) are the differential equations of equilibrium of the plate in terms of stress and moment resultants. The corresponding equilibrium equations in terms of displacements can be derived by substituting the laminate force-deformation equations (7.41), the strain-displacement relations (7.29), and the curvature-displacement equations (7.30) in Eqs. (7.120), (7.122), and (7.128). The resulting equations are

$$A_{11}\frac{\partial^2 u^0}{\partial x^2} + 2A_{16}\frac{\partial^2 u^0}{\partial x\, \partial y} + A_{66}\frac{\partial^2 u^0}{\partial y^2} + A_{16}\frac{\partial^2 v^0}{\partial x^2} + (A_{12} + A_{66})\frac{\partial^2 v^0}{\partial x\, \partial y}$$

$$+ A_{26}\frac{\partial^2 v^0}{\partial y^2} - B_{11}\frac{\partial^3 w}{\partial x^3} - 3B_{16}\frac{\partial^3 w}{\partial x^2\, \partial y}$$

$$- (B_{12} + 2B_{66})\frac{\partial^3 w}{\partial x\, \partial y^2} - B_{26}\frac{\partial^3 w}{\partial y^3} = 0 \tag{7.129}$$

$$A_{16}\frac{\partial^2 u^0}{\partial x^2} + (A_{12} + A_{66})\frac{\partial^2 u^0}{\partial x\, \partial y} + A_{26}\frac{\partial^2 u^0}{\partial y^2} + A_{66}\frac{\partial^2 v^0}{\partial x^2}$$

$$+ 2A_{26}\frac{\partial^2 v^0}{\partial x\, \partial y} + A_{22}\frac{\partial^2 v^0}{\partial y^2} - B_{16}\frac{\partial^3 w}{\partial x^3} - (B_{12} + 2B_{66})\frac{\partial^3 w}{\partial x^2\, \partial y}$$

$$- 3B_{26}\frac{\partial^3 w}{\partial x\, \partial y^2} - B_{22}\frac{\partial^3 w}{\partial y^3} = 0 \tag{7.130}$$

$$D_{11}\frac{\partial^4 w}{\partial x^4} + 4D_{16}\frac{\partial^4 w}{\partial x^3\, \partial y} + 2(D_{12} + 2D_{66})\frac{\partial^4 w}{\partial x^2\, \partial y^2} + 4D_{26}\frac{\partial^4 w}{\partial x\, \partial y^3}$$

$$+ D_{22}\frac{\partial^4 w}{\partial y^4} - B_{11}\frac{\partial^3 u^0}{\partial x^3} - 3B_{16}\frac{\partial^3 u^0}{\partial x^2\, \partial y} - (B_{12} + 2B_{66})\frac{\partial^3 u^0}{\partial x\, \partial y^2}$$

$$- B_{26}\frac{\partial^3 u^0}{\partial y^3} - B_{16}\frac{\partial^3 v^0}{\partial x^3} - (B_{12} + 2B_{66})\frac{\partial^3 v^0}{\partial x^2\, \partial y} - 3B_{26}\frac{\partial^3 v^0}{\partial x\, \partial y^2}$$

$$- B_{22}\frac{\partial^3 v^0}{\partial y^3} = q(x, y) \tag{7.131}$$

Note that the in-plane displacements u^0 and v^0 are coupled with the

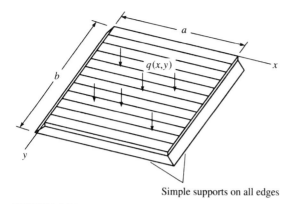

Simple supports on all edges

FIGURE 7.36
Simply supported, specially orthotropic plate with distributed loading.

transverse displacements, w, when the coupling stiffnesses, B_{ij}, are present. For symmetric laminates with $B_{ij} = 0$ Eq. (7.131) alone becomes the governing equation for transverse displacements. These governing partial differential equations must be solved subject to the appropriate boundary conditions. In the general case, when the in-plane displacements are coupled with the transverse displacements, the boundary conditions must be a combination of boundary conditions for a planar theory of elasticity problem and boundary conditions for a plate bending problem [7.10]. In this section, however, we will restrict the discussion to bending of symmetric laminated plates. That is, we will only consider transverse displacements according to Eq. (7.131) with all $B_{ij} = 0$.

Let us now consider the case of transverse deflection of the rectangular, specially orthotropic plate which is simply supported on all edges and loaded with a distributed load, $q(x, y)$, as shown in Fig. 7.36. For a specially orthotropic plate all $B_{ij} = 0$, $A_{16} = A_{26} = D_{16} = D_{26} = 0$ and Eq. (7.131) becomes

$$D_{11} \frac{\partial^4 w}{\partial x^4} + 2(D_{12} + 2D_{66}) \frac{\partial^4 w}{\partial x^2 \, \partial y^2} + D_{22} \frac{\partial^4 w}{\partial y^4} = q(x, y) \quad (7.132)$$

For the simply supported boundary condition the transverse displacements and bending moments must vanish at the edges. In order to use the bending moment boundary conditions to solve the differential equation for displacements, however, the bending moments must be expressed in terms of displacements. Such expressions can be obtained from Eqs. (7.30) and (7.41) for the specially orthotropic plate as follows:

$$M_x = D_{11} \kappa_x + D_{12} \kappa_y = -D_{11} \frac{\partial^2 w}{\partial x^2} - D_{12} \frac{\partial^2 w}{\partial y^2} \quad (7.133)$$

and
$$M_y = D_{12}\kappa_x + D_{22}\kappa_y = -D_{12}\frac{\partial^2 w}{\partial x^2} - D_{22}\frac{\partial^2 w}{\partial y^2}$$

(7.134)

Thus, along $x = 0$ and $x = a$,

$$w = 0$$

(7.135)

and
$$M_x = -D_{11}\frac{\partial^2 w}{\partial x^2} - D_{12}\frac{\partial^2 w}{\partial y^2} = 0$$

and along $y = 0$ and $y = b$,

$$w = 0$$

(7.136)

and
$$M_y = -D_{12}\frac{\partial^2 w}{\partial x^2} - D_{22}\frac{\partial^2 w}{\partial y^2} = 0$$

Several approaches to the solution of such problems have been proposed [7.10, 7.38]. This simplest method involves the use of double Fourier sine series to represent both the load $q(x, y)$ and the displacements $w(x, y)$. If the load can be represented as

$$q(x, y) = \sum_{m=1}^{\infty} \sum_{n=1}^{\infty} q_{mn} \sin\frac{m\pi x}{a} \sin\frac{n\pi y}{b}$$

(7.137)

then it can be shown that the differential equation and the boundary conditions are satisfied by solutions of the form

$$w(x, y) = \sum_{m=1}^{\infty} \sum_{n=1}^{\infty} w_{mn} \sin\frac{m\pi x}{a} \sin\frac{n\pi y}{b}$$

(7.138)

Substitution of Eqs. (7.138) and (7.137) in Eq. (7.132) yields the displacement coefficients

$$w_{mn} = \frac{a^4 q_{mn}}{\pi^4[D_{11}m^4 + 2(D_{12} + 2D_{66})(mnR)^2 + D_{22}(nR)^4]}$$

(7.139)

where the plate aspect ratio $R = a/b$ [7.10]. The Fourier coefficients q_{mn} can be found for the particular assumed load distribution [7.9, 7.10]. For the uniform load $q(x, y) = q_0$, a constant, it can be shown that the Fourier coefficients are

$$q_{mn} = \frac{16q_0}{\pi^2 mn} \qquad \text{for } m, n = 1, 3, 5, \ldots$$

(7.140)

and
$$q_{mn} = 0 \qquad \text{for } m, n = 2, 4, 6, \ldots$$

Displacements $w(x, y)$ for the uniformly loaded, simply supported plate may now be found by substituting Eqs. (7.139) and (7.140) in Eq.

(7.138). Moment resultants may be found by substsituting these same equations in Eqs. (7.30) and then substituting the result in Eqs. (7.41). Finally, lamina stresses may be found by combining Eqs. (7.30), (7.31), (7.138), (7.139), and (7.140).

For boundary conditions such as clamped edges or free edges, exact series solutions similar to Eq. (7.138) are generally not possible. For such cases approximate solutions must be derived using approaches such as the Rayleigh-Ritz method or the Galerkin method. For a detailed discussion of these methods and other boundary conditions the reader is referred to the book by Whitney [7.10].

7.9.2 Buckling Analysis

In the derivations of Eqs. (7.120), (7.122), and (7.124) the coupling between the in-plane forces N_x, N_y, and N_{xy} and the out-of-plane deflections, w, was ignored because of the assumption of small displacements. In order to develop the equations to predict buckling under in-plane loads, however, this coupling must be considered. Such equations can be derived by assuming the differential element of Fig. 7.35(a) to be oriented in a general out-of-plane position, as shown in Fig. 7.37. Using Fig. 7.37, and taking into account the vertical components of the

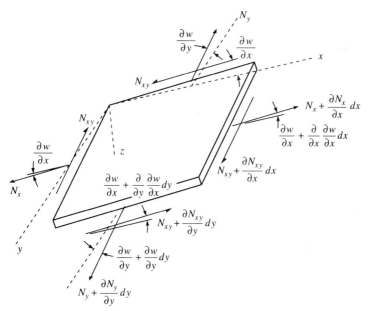

FIGURE 7.37
Differential element of laminate in out-of-plane position for buckling analysis.

in-plane forces, the summation of forces in the z direction now becomes

$$\frac{\partial Q_x}{\partial x} + \frac{\partial Q_y}{\partial y} + q(x, y) + N_x \frac{\partial^2 w}{\partial x^2} + 2N_{xy} \frac{\partial^2 w}{\partial x\, \partial y} + N_y \frac{\partial^2 w}{\partial y^2} = 0 \quad (7.141)$$

Note that Eq. (7.141) consists of the terms from Eq. (7.124) plus the terms involving the in-plane forces. Combining Eqs. (7.141), (7.126), and (7.127), we find that

$$\frac{\partial^2 M_x}{\partial x^2} + 2\frac{\partial^2 M_{xy}}{\partial x\, \partial y} + \frac{\partial^2 M_y}{\partial y^2} + N_x \frac{\partial^2 w}{\partial x^2}$$

$$+ 2N_{xy} \frac{\partial^2 w}{\partial x\, \partial y} + N_y \frac{\partial^2 w}{\partial y^2} + q(x, y) = 0 \quad (7.142)$$

Equation (7.142) consists of all the terms in Eq. (7.128) plus the terms involving the in-plane forces. Substitution of Eqs. (7.141), (7.29), and (7.30) in Eq. (7.142) yields the eqution

$$D_{11} \frac{\partial^4 w}{\partial x^4} + 4D_{16} \frac{\partial^4 w}{\partial x^3\, \partial y} + 2(D_{12} + 2D_{66}) \frac{\partial^4 w}{\partial x^2\, \partial y^2} + 4D_{26} \frac{\partial^4 w}{\partial x\, \partial y^3}$$

$$+ D_{22} \frac{\partial^4 w}{\partial y^4} - B_{11} \frac{\partial^3 u^0}{\partial x^3} - 3B_{16} \frac{\partial^3 u^0}{\partial x^2\, \partial y} - (B_{12} + 2B_{66}) \frac{\partial^3 u^0}{\partial x\, \partial y^2}$$

$$- B_{26} \frac{\partial^3 u^0}{\partial y^3} - B_{16} \frac{\partial^3 v^0}{\partial x^3} - (B_{12} + 2B_{66}) \frac{\partial^3 v^0}{\partial x^2\, \partial y} - 3B_{26} \frac{\partial^3 v^0}{\partial x\, \partial y^2}$$

$$- B_{22} \frac{\partial^3 v^0}{\partial y^3} = q(x, y) + N_x \frac{\partial^2 w}{\partial x^2} + 2N_{xy} \frac{\partial^2 w}{\partial x\, \partial y} + N_y \frac{\partial^2 w}{\partial y^2} \quad (7.143)$$

Note that Eq. (7.143) consists of the terms in Eq. (7.131) and the additional terms due to the in-plane forces.

We now consider the case of buckling of a rectangular, simply supported, specially orthotropic plate under a single compressive axial load, $N_x = -N$, as shown in Fig. 7.38. In this case the loads

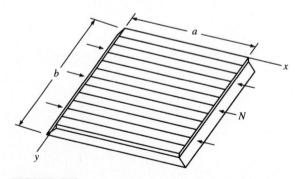

FIGURE 7.38
Simply supported, specially orthotropic plate under compressive uniaxial in-plane loading.

$N_y = N_{xy} = q(x, y) = 0$, all $B_{ij} = 0$, the stiffnesses $A_{16} = A_{26} = D_{16} = D_{26} = 0$ and Eq. (7.143) becomes

$$D_{11}\frac{\partial^4 w}{\partial x^4} + 2(D_{12} + 2D_{66})\frac{\partial^4 w}{\partial x^2 \partial y^2} + D_{22}\frac{\partial^4 w}{\partial y^4} = -N\frac{\partial^2 w}{\partial x^2} \qquad (7.144)$$

For the simply supported boundary condition described previously by Eqs. (7.135) and (7.136) we may assume a solution of the form

$$w(x, y) = w_{mn} \sin\frac{m\pi x}{a} \sin\frac{n\pi y}{b} \qquad (7.145)$$

The mode shape for a particular buckling mode is described by the subscripts m and n since m is the number of half-sine waves along the x direction and n is the number of half-sine waves along the y direction. Substitution of this solution in the governing differential equation (7.144) leads to the equation

$$w_{mn}\pi^2[(D_{11}m^4 + 2(D_{12} + 2D_{66})(mnR)^2 + D_{22}(bR)^4)] = w_{mn}Na^2m^2 \quad (7.146)$$

where again $R = a/b$. This equation has the trivial solution $w_{mn} = 0$, which is of no interest. For nontrivial solutions the critical buckling load must be

$$N_{cr} = \frac{\pi^2}{a^2 m^2}[(D_{11}m^4 + 2(D_{12} + 2D_{66})(mnR)^2 + D_{22}(nR)^4)] \qquad (7.147)$$

where the smallest buckling load occurs for $n = 1$, and the lowest value of the load corresponding to a particular value of m can only be determined if the D_{ij} and the plate dimensions a and b are known. As shown in Fig.

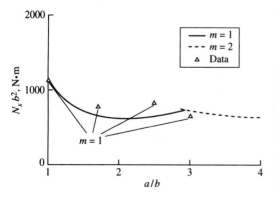

FIGURE 7.39
Comparison of predicted and measured normalized buckling load, $N_x b^2$, vs. plate aspect ratio, a/b, for $[0_{12}]$ graphite/epoxy laminates. (*From Hatcher and Tuttle* [7.40]. *Reprinted by permission of The American Society of Mechanical Engineers.*)

7.39, Hatcher and Tuttle [7.40] have compared experimentally determined buckling loads for simply supported, specially orthotropic graphite/epoxy panels with predicted buckling loads from Eq. (7.147). The value $n = 1$ is used for all predicted curves, and the curves for $m = 1$ and $m = 2$ are shown. Although the predictions are reasonably accurate for this case, it was found that the agreement for some other laminate configurations was not quite as good. Difficulties in simulating the simply supported boundary conditions and in measuring the critical buckling loads, along with other problems such as the existence of imperfections in the test panels, were cited as possible reasons for the disagreement.

For laminates other than specially orthotropic and boundary conditions other than simply supported, closed-form solutions similar to Eq. (7.147) are generally not possible, and approximate methods such as Rayleigh-Ritz or Galerkin must be used. Exceptions include the antisymmetric cross-ply and antisymmetric angle-ply laminates, which do admit closed-form solutions [7.41].

7.10   SELECTION OF LAMINATE DESIGNS

When designing with conventional isotropic materials, the problem of material selection is usually solved by simply looking up the appropriate properties of candidate materials in a handbook. The selection of a composite laminate design can be a formidable task, however, due to the large number of available fiber and matrix materials and the endless variety of laminate configurations. Fortunately, the laminate selection process can be accelerated by the use of computer software for laminate analysis and by the use of so-called carpet plots.

In order to use the laminate analysis equations which were derived and discussed earlier in this chapter, extensive matrix algebra is obviously required. In addition, proper evaluation of laminate designs requires numerous repetitive calculations resulting from changes in loading conditions, material properties, and/or laminate geometry. These computational requirements are ideally suited for solutions by digital computers, and a variety of software packages for laminate analysis now exist. A list of some of the available software packages is given in Table 7.1. Most of these programs have been developed for use on microcomputers, and many of them will do both micromechanical analysis of laminae and laminate analysis according to the Classical Lamination Theory. The two basic approaches used in many of these programs are (1) stress and strain analysis for prescribed loads and (2) first ply failure analysis and/or ultimate laminate failure analysis according to one of the multiaxial lamina strength criteria that were discussed in Chap. 4. In addition, some of the programs will do specialized tasks such as deflection and buckling

TABLE 7.1
Laminate analysis software

Name	Description	Computer	Vendor
ASCA	Micromechanical analysis of laminae and macromechanical analysis of laminate, including interlaminar stresses, effective properties, and stresses around transverse ply cracks. Plots stresses and strains.	IBM PC	(1)
COMPCAL	Micromechanical and macromechanical analysis, including elastic constants, thermal and transport properties, stresses and strains, strength analysis using quadratic interaction criterion.	IBM PC	(1)
ICAN	Integrated composite analyzer for analysis of laminae and laminate, including strength analysis, hygrothermal effects, interlaminar stresses.	IBM 370 Mainframe	(2)
LAMINATE ONE	Laminae stress and strain analysis and evaluation of laminate properties.	IBM PC	(3)
LAMPCAL	Bending, buckling, and free vibration analysis of laminated plates according to the Classical Lamination Theory and the shear deformation plate theory.	Apple, IBM PC	(1)
MIC-MAC	Both micromechanical and macromechanical analysis of laminate. Loading and material properties easily varied.	Apple, IBM PC	(4)
PC LAMINATE	Interactive code for stress and strain analysis of laminates subjected to in-plane and bending loads with strength analysis using several popular criteria.	IBM PC	(1)
PROMAL	Teaching aid for students in introductory mechanics of composites course, consisting of four programs for (1) matrix algebra, (2) materials database, (3) lamina stiffnesses for various ply angles, and (4) standard laminate analysis.	IBM PC	(5)

Vendors: (1) Technomic Publishing Co., Lancaster, PA 17604, (2) COSMIC, University of Georgia, Athens, GA 30602, (3) Alef Systems, Redondo Beach, CA 90277, (4) Think Composites, Dayton, OH 45419, (5) Kern International, Duxbury, MA 02332.

analyses, interlaminar stress calculations, effective property calculations, and hygrothermal and transport property calculations. Such programs are indispensable in design and analysis because hand calculations are not only too time consuming, but the possibilities for errors in such hand calculations are endless.

For composite structures having complex geometries the preferred analytical tool is the finite element method. The use of the finite element

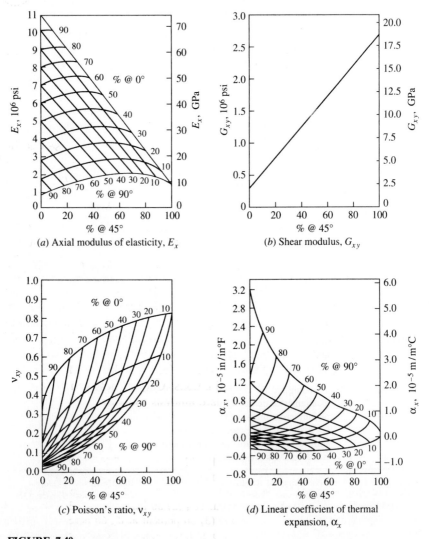

(a) Axial modulus of elasticity, E_x

(b) Shear modulus, G_{xy}

(c) Poisson's ratio, ν_{xy}

(d) Linear coefficient of thermal expansion, α_x

FIGURE 7.40
Carpet plots for $[0_i/\pm45_j/90_k]$ Kevlar®/epoxy laminates. (*From Peters et al.* [7.43]. *Reprinted by permission of The Society for the Advancement of Material and Process Engineering.*)

method in micromechanical analysis has already been discussed in Chap. 3. Macromechanical finite element analysis of laminated structures is also widely used, and most of the popular finite element codes have special elements such as orthotropic three-dimensional solid elements, orthotropic two- and three-dimensional shell elements, and orthotropic axisymmetric solid elements [7.42].

Although computer software gives the designer great flexibility in the selection of materials and laminate geometries, graphical representations which show the range of properties which can be attained with different laminate configurations are also helpful. One type of graphical representation, known as a carpet plot, is particularly useful. For example, if the ply orientations in a laminate are restricted to certain angles such as $0°$, $\pm 45°$, and $90°$, then a carpet plot can be generated which shows how a given laminate property depends on the percentages of the plies at the various orientations. The carpet plots in Fig. 7.40 from Ref. [7.43] show how E_x, G_{xy}, v_{xy}, and α_x for $[0_i/\pm 45_j/90_k]$ Kevlar®/epoxy laminates vary with the percentages of the plies at the three angles. In this case i is the number of $0°$ plies, j is the number of $\pm 45°$ plies, and k is the number of $90°$ plies. Therefore, the percentage of $0°$ plies is $i/(i+j+k)$, the percentage of $\pm 45°$ plies is $j/(i+j+k)$, and the percentage of $90°$ plies is $k/(i+j+k)$.

For example, the various ply combinations that will give a certain value of longitudinal modulus, E_x, can be determined by drawing a horizontal line in Fig. 7.40(a) at the value of E_x and then reading off the percentage of the plies at the three angles corresponding to a particular point on the line. Obviously, there are many possible combinations that will give the same value of E_x, and the design flexibility inherent in composite construction is again demonstrated. Carpet plots for laminate strength are also widely used. Since there would normally be more than one design constraint, an iterative approach involving the repeated use of carpet plots for several different properties may be needed for the selection of the required ply combinations. Carpet plots can be quickly generated using the output of laminate analysis software.

Example 7.11. The reader should be able to use Fig. 7.40(a) to verify that a Kevlar®/epoxy laminate with a longitudinal modulus of $E_x = 30$ GPa can be obtained with the following ply combinations; (1) 35 percent at $0°$, 0 percent at $\pm 45°$, and 65 percent at $90°$; (2) 30 percent at $0°$, 30 percent at $\pm 45°$, and 40 percent at $90°$; and (3) 30 percent at $0°$, 60 percent at $\pm 45°$, and 10 percent at $90°$. These are only three of many possible combinations which will give the same result. Additional design constraints may also be taken into account with other carpet plots. For example, if a shear modulus, G_{xy}, of at least 5.0 GPa is needed, Fig. 7.40(b) indicates that the laminate should have at least 20 percent of its plies at ±45°. Thus, laminates (2) and (3) above both satisfy the constraints on E_x and G_{xy}, but laminate (1) does not.

PROBLEMS

7.1. A laminated $[0/90/0/90]_s$ graphite/epoxy beam is 1 mm thick, is 20 mm wide, and has 0.125-mm thick plies. The lamina properties are

$$E_1 = 180 \text{ GPa} \qquad s_L^{(+)} = 1700 \text{ MPa}$$

$$E_2 = 10 \text{ GPa} \qquad s_L^{(-)} = 1400 \text{ MPa}$$

$$G_{12} = 7 \text{ GPa} \qquad s_T^{(+)} = 40 \text{ MPa}$$

$$v_{12} = 0.28 \qquad s_T^{(-)} = 230 \text{ MPa}$$

(*a*) Determine the flexural modulus for the beam.

(*b*) How could the flexural modulus be improved without changing the ply materials, the number of plies, or the ply orientations?

(*c*) Using the Maximum Stress Criterion for each ply, determine the magnitude of the maximum allowable bending moment that the beam can withstand. Which ply fails first?

(*d*) What type of analysis would be required if the ply orientations are $[+45/90/-45/0]_s$?

7.2. The laminated beam shown in Fig. 7.41 is made up of two outer plies of material "*A*" having Young's modulus E_A, two inner plies of material "*B*" having Young's modulus E_B, and a honeycomb core of negligible stiffness. Materials A and B are isotropic, but they have different thicknesses. The laminate is symmetric about the middle surface. Find the expression for the flexural modulus in terms of the given properties and the dimensions shown in Fig. 7.41.

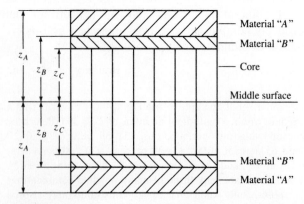

FIGURE 7.41
Laminated beam for Problem 7.2.

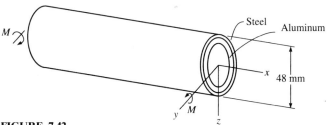

FIGURE 7.42
Composite tube for Problem 7.3.

7.3. A thin-walled composite tube having an outside diameter of 48 mm is made by securely bonding an aluminum tube inside a steel tube, as shown in Fig. 7.42. Determine the maximum allowable bending moment, M, that the composite tube can withstand without exceeding the yield stress of either the steel or the aluminum. The tube properties are:

	Steel	**Aluminum**
Young's modulus, GPa	200	78
Yield stress, MPa	259	98
Wall thickness, mm	3	6

7.4. Determine the stiffness matrix for a $[+45/-45]$ antisymmetric laminate consisting of 0.25-mm thick unidirectional AS/3501 graphite/epoxy plies.

7.5. Show that for symmetric laminates the coupling stiffnesses, B_{ij}, must all be equal to zero.

7.6. By expanding the $[A]$ matrix in terms of ply stiffnesses, show that a "balanced" cross-ply laminate having equal numbers of $0°$ and $90°$ plies is not necessarily quasi-isotropic.

7.7. A $[-60/0/60]$ laminate and a $[0/45/90]$ laminate both consist of 1.0-mm thick plies having the following properties: $E_1 = 181$ GPa, $E_2 = 10.3$ GPa, $G_{12} = 7.17$ GPa, $v_{12} = 0.28$. Plot the A_{ij} for both laminates as a function of the orientation in order to determine which, if any, of the laminates is quasi-isotropic.

7.8. The $[+45/-45]$ laminate described in Problem 7.4 is subjected to a uniaxial force per unit length $N_x = 30$ MPa-mm. Find the resulting stresses and strains in each ply along the x and y directions.

7.9. A $[0/90]_s$ laminate is subjected to a single bending moment per unit length, M_x. If the laminate is unconstrained, so that bending along both the x and y directions occurs freely, determine the ply stresses, $(\sigma_x)_k$, in terms of the moment, M_x, the bending stiffnesses, D_{ij}, the ply stiffnesses, Q_{ij}, and the distance from the middle surface, z. Determine the ply stresses $(\sigma_x)_k$ in terms of M_x, z, and a numerical coefficient if the properties are $E_1 = 129$ GPa, $E_2 = 12.8$ GPa, $G_{12} = 4.6$ GPa, $v_{12} = 0.313$, and $t = 1$ mm.

7.10. The laminate described in Problem 7.9 is subjected to a single bending moment per unit length, M_x, and the two edges on which M_x acts are fixed so that bending along the x direction occurs freely but bending along the y direction is prevented. That is, the longitudinal curvature is unconstrained (i.e., $\kappa_x \neq 0$), but the transverse curvature is constrained (i.e., $\kappa_y = 0$). Determine the ply stresses $(\sigma_x)_k$ as in Problem 7.9 (give equations and numerical results) and compare with the results of Problem 7.9.

7.11. A $[90/0/90]_s$ laminate is fabricated from laminae consisting of isotropic fibers $(E_f = 220\text{ GPa}, \ v_f = 0.25)$ embedded in an isotropic matrix $(E_m = 3.6\text{ GPa}, \ v_m = 0.4)$. Each lamina is 0.25 mm thick, and the 0.01-mm diameter fibers have been precoated with a 0.00125-mm thick sizing, which is the same as the matrix material. The precoated fibers are arranged in the closest possible packing array in the matrix. Using both micromechanics and laminate analysis, find the laminate engineering constants E_x, E_y, G_{xy}, and v_{xy}. The laminate x axis is parallel to the 0° lamina orientation.

optional.

7.12. An antisymmetric angle-ply $[+\theta/-\theta]$ laminate is to be made of graphite/epoxy and designed to have a laminate coefficient of thermal expansion, α_x, as close to zero as possible. Determine the ply orientation θ needed to meet this requirement. The lamina properties are as follows:

Compute N_x^T

$\theta = 8.3$
8.4 *+5 pts*
8.5 *E.C.*
8.525
8.55
8.575
8.6
8.7

$$E_1 = 138\text{ GPa} \qquad \text{lamina thickness} = 0.125\text{ mm}$$
$$E_2 = 8.96\text{ GPa} \qquad \alpha_1 = -0.3 \times 10^{-6}\text{ m/m/K}$$
$$G_{12} = 7.1\text{ GPa} \qquad \alpha_2 = 28.1 \times 10^{-6}\text{ m/m/K}$$
$$v_{12} = 0.3$$

This problem requires extensive calculations, and the use of a computer is recommended.

7.13. Repeat Problem 7.12 for a Kevlar®/epoxy composite having lamina properties as follows:

$$E_1 = 76\text{ GPa} \qquad \text{lamina thickness} = 0.125\text{ mm}$$
$$E_2 = 5.5\text{ GPa} \qquad \alpha_1 = -4.0 \times 10^{-6}\text{ m/m/K}$$
$$G_{12} = 2.3\text{ GPa} \qquad \alpha_2 = 79.0 \times 10^{-6}\text{ m/m/K}$$
$$v_{12} = 0.34$$

7.14. The distribution of the in-plane shear stress, τ_{xy}, along the y direction at a particular distance z from the middle surface of a uniaxially loaded laminate is idealized, as shown in Fig. 7.43. The interlaminar stress boundary layer

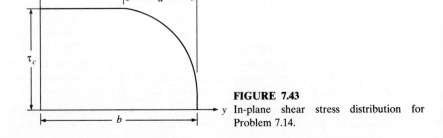

FIGURE 7.43 In-plane shear stress distribution for Problem 7.14.

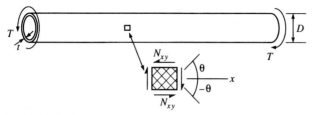

FIGURE 7.44
Filament wound shaft for Problem 7.15.

region is assumed to extend inward from the free edge at $y = b$ by a distance "a," the in-plane shear stress $\tau_{xy} = \tau_c$ in the region $0 \le y \le (b - a)$ is assumed to be the shear stress from the Classical Lamination Theory, and the in-plane shear stress in the boundary layer region $(b - a) \le y \le b$ is assumed to be of the form

$$\tau_{xy} = \frac{\tau_c}{a^2}(y - b)(b - 2a - y)$$

For the same location, determine the distribution of the interlaminar shear stress, τ_{xz}, along the y direction.

7.15. A filament-wound composite drive shaft for a helicopter transmits a torque T which generates shear loading of the shaft material, as shown in Fig. 7.44. The shaft is to be designed as a hollow tube with a two ply $[+\theta/-\theta]$ laminated wall. If the outside diameter, the length, and the material density are fixed, use invariants to determine the angle θ which should be used to maximize the shear stiffness-to-weight ratio, A_{66}/W, where A_{66} is the laminate shear stiffness and W is the shaft weight. It may be assumed that the shaft diameter, D, is much greater than the wall thickness, t.

7.16. Develop a "parallel axis theorem" for the effective laminate stiffnesses A''_{ij}, B''_{ij}, and D''_{ij} associated with the (x'', z'') axes, which are parallel to the original (x, z) axes, as shown in Fig. 7.45. Express the new A''_{ij}, B''_{ij}, and D''_{ij} in terms of the original A_{ij}, B_{ij}, and D_{ij} for the (x, z) axes and the distance d between the parallel axes, where $z'' = z + d$.

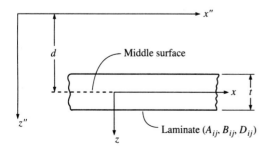

FIGURE 7.45
Laminate with parallel axes for Problem 7.16.

7.17. A $[0/90/0]_s$ laminate consisting of AS/3501 graphite/epoxy laminae is subjected to uniaxial loading along the x direction. Use the Maximum Strain Criterion to find the loads corresponding to first ply failure and ultimate laminate failure; then plot the load-strain curve up to failure. Compare these results with those of Example 7.10 and discuss any differences.

7.18. Prove that for the specially orthotropic plate shown in Fig. 7.36 under the loading described by Eq. (7.137) the solution given by Eq. (7.138) satisfies the differential equation (7.132) and the boundary conditions in Eqs. (7.135) and (7.136).

7.19. Find expressions for the moments M_x, M_y, and M_{xy} and the stresses $(\sigma_x)_k$, $(\sigma_y)_k$, and $(\tau_{xy})_k$ in the kth ply of the uniformly loaded, specially orthotropic laminate with simply supported edges shown in Fig. 7.36.

7.20. Derive the differential equation and the boundary conditions governing the small transverse deflections of a simply supported, rectangular, symmetric angle-ply laminate which is subjected to distributed loading. If the loading is described by Eq. (7.137), does a solution of the form given in Eq. (7.138) satisfy this differential equation and boundary conditions? Why?

7.21. Derive the coupled differential equations and the boundary conditions governing the small transverse deflections of a simply supported, rectangular, antisymmetric angle-ply laminate which is subjected to distributed loading. Propose solutions for the displacements u, v, and w which satisfy the differential equations and boundary conditions.

7.22. A simply supported, specially orthotropic plate is subjected to an in-plane compressive load per unit length N_x and an in-plane tensile load per unit length $N_x = -0.5N_x$, as shown in Fig. 7.46. Derive the expression for the critical buckling load.

7.23. Derive the differential equation and the boundary conditions governing the buckling of a simply supported, rectangular, symmetric angle-ply laminate which is subjected to a uniaxial in-plane load, N_x. Does a solution of the form given in Eq. (7.145) satisfy this differential equation and boundary conditions? Why?

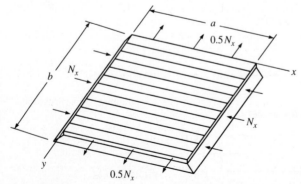

FIGURE 7.46
Simply supported, specially orthotropic plate under in-plane loads for Problem 7.22.

7.24. Using the carpet plots of Fig. 7.40, select the percentages of $0°$, $\pm 45°$, and $90°$ plies that are needed in a $[0_i/\pm 45_j/90_k]$ laminate if the longitudinal modulus, E_x, is to be at least 20 GPa, the in-plane shear modulus, G_{xy}, is to be at least 10 GPa, and the longitudinal coefficient of thermal expansion, α_x, is equal to zero.

REFERENCES

7.1. Pagano, N. J., "Analysis of the Flexure Test of Bidirectional Composites," *Journal of Composite Materials,* **1,** 336–342 (1967).

7.2. Higdon, A., Ohlsen, E. H., Stiles, W. B., Weese, J. A., and Riley, W. F., *Mechanics of Materials,* 3d ed. John Wiley & Sons, New York (1976).

7.3. Smith, C. B., "Some New Types of Orthotropic Plates Laminated of Orthotropic Material," *Journal of Applied Mechanics,* **20,** 286–288 (1953).

7.4. Pister, K. S. and Dong, S. B., "Elastic Bending of Layered Plates," in *Proceedings of the American Society of Civil Engineers* (*Journal of Engineering Mechanics Division*), **85,** EM4, 1–10 (1959).

7.5. Reissner, E. and Stavsky, Y., "Bending and Stretching of Certain Types of Heterogeneous Aeolotropic Elastic Plates," *Journal of Applied Mechanics,* **28,** 402–408 (1961).

7.6. Stavsky, Y., "On the General Theory of Heterogeneous Aeolotropic Plates," *Aeronautical Quarterly,* **15,** 29–38 (1964).

7.7. Lekhnitskii, S. G., *Anisotropic Plates,* Transl. from the 2d Russian ed. by S. W. Tsai and T. Cheron, Gordon and Breach Science Publishers, New York (1968).

7.8. Stavsky, Y. and Hoff, N. J., "Mechanics of Composite Structures," in A. G. H. Dietz (ed.), *Composite Engineering Laminates,* MIT Press, Cambridge, MA (1969).

7.9. Timoshenko, S. A. and Woinowsky-Krieger, S., *Theory of Plates and Shells,* 2d ed., McGraw-Hill, Inc., New York (1959).

7.10. Whitney, J. M., *Structural Analysis of Laminated Plates,* Technomic Publishing Co., Inc., Lancaster, PA (1987).

7.11. Christensen, R. M., *Mechanics of Composite Materials,* John Wiley & Sons, Inc., New York (1979).

7.12. Halpin, J. C., *Primer on Composite Materials: Analysis,* Technomic Publishing Co. Lancaster, PA (1984).

7.13. Tsai, S. W., *Structural Behavior of Composite Materials,* NASA CR-71 (1964).

7.14. Pipes, R. B. and Pagano, N. J., "Interlaminar Stresses in Composite Laminates under Uniform Axial Extension," *Journal of Composite Materials,* **4,** 538–548 (1970).

7.15. Wang, S. S. and Choi, I., "Boundary Layer Effects in Composite Laminates, Part 1: Free Edge Singularities," *Journal of Applied Mechanics,* **49,** 541–548 (1982).

7.16. Wang, S. S. and Choi, I., "Boundary Layer Effects in Composite Laminates, Part 2: Free Edge Solutions and Basic Characteristics," *Journal of Applied Mechanics,* **49,** 549–560 (1982).

7.17. Pipes, R. B. and Daniel, I. M., "Moiré Analysis of the Interlaminar Shear Edge Effect in Laminated Composites," *Journal of Composite Materials,* **5,** 255–259 (1971).

7.18. Pipes, R. B. and Pagano, N. J., "Interlaminar Stresses in Composite Laminates—An Approximate Elasticity Solution," *Journal of Applied Mechanics,* **41,** Series E (3), 668–672 (1974).

7.19. Pagano, N. J. and Pipes, R. B., "The Influence of Stacking Sequence on Laminate Strength," *Journal of Composite Materials,* **5**(1), 50–57 (1971).

7.20. Whitney, J. M. and Browning, C. E., "Free-Edge Delamination of Tensile Coupons," *Journal of Composite Materials,* **6**(2), 300–303 (1972).

7.21. Whitney, J. M. and Kim, R. Y., "Effect of Stacking Sequence on the Notched Strength of Laminated Composites," *Composite Materials: Testing and Design (Fourth Conference)*, ASTM STP **617**, 229–242, American Society for Testing and Materials, Philadelphia, PA, (1977).

7.22. Rybicki, E. F., "Approximate Three Dimensional Solutions for Symmetric Laminates under In-plane Loading," *Journal of Composite Materials*, **5**(3), 354–360 (1971).

7.23. Wang, A. S. D. and Crossman, F. W., "Some New Results on Edge Effect in Symmetric Composite Laminates," *Journal of Composite Materials*, **11**, 92–106 (1977).

7.24. Herakovich, C. T., "On the Relationship between Engineering Properties and Delamination of Composite Materials," *Journal of Composite Materials*, **15**, 336–348 (1981).

7.25. Hwang, S. J. and Gibson, R. F., "Contribution of Interlaminar Stresses to Damping in Thick Composites under Uniaxial Extension," *Composite Structures*, **20**, 29–35 (1992).

7.26. Hwang, S. J., *Characterization of the Effects of Three Dimensional States of Stress on Damping of Laminated Composites*, Ph.D. Dissertation, Mechanical Engineering Department, University of Idaho (1988).

7.27. Whitney, J. M., "Free Edge Effects in the Characterization of Composite Materials," *Analysis of the Test Methods for High Modulus Fibers and Composites*, ASTM STP 521, American Society for Testing and Materials, Philadelphia, PA, 167–180 (1973).

7.28. Conti, P. and De Paulis, A., "A Simple Model to Simulate the Interlaminar Stresses Generated Near the Free Edge of a Composite Laminate," in W. S. Johnson (ed.), *Delamination and Debonding of Materials*, ASTM STP 876, 35–51, American Society for Testing and Materials, Philadelphia, PA (1985).

7.29. Hahn, H. T. and Tsai, S. W., "On the Behavior of Composite Laminates after Initial Failures," *Journal of Composite Materials*, **8**(3), 288–305 (1974).

7.30. Tsai, S. W., *Strength Characteristics of Composite Materials*, NASA CR-224 (1965).

7.31. Kim, R. Y. and Soni, S. R., "Experimental and Analytical Studies on the Onset of Delamination in Laminated Composites," *Journal of Composite Materials*, **18**, 70–80 (1984).

7.32. Whitney, J. M. and Nuismer, R. J., "Stress Fracture Criteria for Laminated Composites Containing Stress Concentrations," *Journal of Composite Materials*, **8**, 253–265 (1974).

7.33. Brewer, J. C. and Lagace, P. A., "Quadratic Stress Criterion for Initiation of Delamination," *Journal of Composite Materials*, **22**, 1141–1155 (1988).

7.34. O'Brien, T. K., "Characterization of Delamination Onset and Growth in a Composite Laminate," in K. L. Reifsnider (ed.), *Damage in Composite Materials*, ASTM STP **775**, 140–167, American Society for Testing and Materials, Philadelphia, PA (1982).

7.35. Newaz, G. M. (ed), *Delamination in Advanced Composites*, Technomic Publishing. Co., Lancaster, PA (1991).

7.36. Johnson, W. S. (ed.), *Delamination and Debonding of Materials*, ASTM STP **876**, American Society for Testing and Materials, Philadelphia, PA (1985).

7.37. Pagano, N. J. (ed.), *Interlaminar Response of Composite Laminates*, Vol. 5, Composite Laminates Series, R. B. Pipes (Series ed.), Elsevier Science Publishers, Amsterdam (1989).

7.38. Vinson, J. R. and Sierakowski, R. L., *The Behavior of Structures Composed of Composite Materials*, Martinus Mijhoff Publishers, Dordrecht, The Netherlands (1986).

7.39. Leissa, A. W., *Buckling of Laminated Composite Plates and Shell Panels*, AFWAL-TR-85-3069, Air Force Wright Aeronautical Laboratories, Wright-Patterson Air Force Base, OH (1985).

7.40. Hatcher, D. and Tuttle, M., "Measurement of Critical Buckling Loads and Mode Shapes of Composite Panels," in H. H. Chung and Y. W. Kwon (eds.), *Recent*

Advances in Structural Mechanics, PVP-Vol. **225**/NE-Vol. **7,** 21–26, American Society of Mechanical Engineers, New York (1991).

7.41. Jones, R. M., *Mechanics of Composite Materials,* Hemisphere Publishing Co., New York, 264–270 (1975).

7.42. Dropek, R. K., "Numerical Design and Analysis of Structures, " in T. J. Reinhart (ed.), *Engineered Materials Handbook Volume 1: Composites,* 463–478, ASM International, Materials Park, OH (1987).

7.43. Peters, S. T., Humphrey, W. D., and Foral, R. F., *Filament Winding Composite Structure Fabrication,* 5–45, Society for Advancement of Materials and Process Engineering, Covina, CA (1991).

CHAPTER
8

ANALYSIS OF VISCOELASTIC AND DYNAMIC BEHAVIOR

8.1 INTRODUCTION

In the analyses of Chaps. 1 to 7 it has been assumed that the applied loads are static in nature and that the composite and its constituents exhibit time-independent linear elastic behavior. However, composite structures are often subjected to dynamic loading caused by vibration or wave propagation. In addition, many composites exhibit time-dependent viscoelastic behavior under load; this is particularly true for composites having polymeric constituents. This chapter contains the basic information needed for the analysis of both viscoelastic and dynamic behavior of composites and their constituents.

The word "viscoelastic" has evolved as a way of describing materials which exhibit characteristics of both viscous fluids and elastic solids. Polymeric materials, which are known to be viscoelastic, may behave like fluids or solids, depending on the time scale and/or the temperature. For example, polycarbonate, a thermoplastic polymer, is a liquid during molding at processing temperatures, but is a glassy solid at service (ambient) temperatures. It will deform like a rubber at temperatures just above the glass transition temperature, T_g. At temperatures below T_g, however, it will deform just as much, and in the same way if the test time is long enough.

We know that ideal Hookean elastic solids are capable of energy storage under load, but not energy dissipation, whereas ideal newtonian fluids under nonhydrostatic stresses are capable of energy dissipation, but not energy storage. Viscoelastic materials, however, are capable of both storage and dissipation of energy under load. Another characteristic of viscoelastic materials is memory. Perfectly elastic solids are said to have only "simple memory" because they remember only the unstrained state and the current strains depend only on the current stresses. Viscoelastic materials have what is often referred to as "fading memory" because they remember the past in such a way that the current strains depend more strongly on the recent stress-time history than on the more distant stress-time history.

There are four important physical manifestations of viscoelastic behavior in structural materials, as illustrated by the various conditions of the uniaxially loaded viscoelastic rod in Fig. 8.1. First, if the rod is subjected to a constant stress, the resulting strain will exhibit time-dependent "creep," as shown in Fig. 8.1(a). The time-dependent creep strains are superimposed on the initial elastic strains. Second, if the rod is subjected to a constant strain or displacement, the resulting stress will exhibit time-dependent "relaxation," as shown in Fig. 8.1(b). That is, the stress relaxes from the initial elastic stress. Third, if the bar is subjected to oscillatory loading, the resulting stress-strain curve will describe a "hysteresis loop," as shown in Fig. 8.1(c). The area enclosed by the hysteresis loop is a measure of the damping, or dissipation, of energy in the material. Fourth, if the bar is loaded at various strain rates, the stress-strain curves will exhibit a strain-rate dependence, as shown in Fig. 8.1(d). That is, the stress corresponding to a given strain depends on the rate of straining. An ideal elastic material exhibits none of the above characteristics.

All structural materials exhibit some degree of viscoelasticity, and the extent of such behavior often depends on environmental conditions such as temperature. For example, while a structural steel or aluminum may be essentially elastic at room temperature, viscoelastic effects become apparent at elevated temperatures approaching half the melting temperature. Polymeric materials are viscoelastic at room temperature, and the viscoelastic effects become stronger as the temperature approaches the glass transition temperature. Recall from Chap. 5 that the glass transition region (Fig. 5.1) is a region of transition between glassy behavior and rubbery behavior and a region characterized by the onset of pronounced viscoelastic behavior.

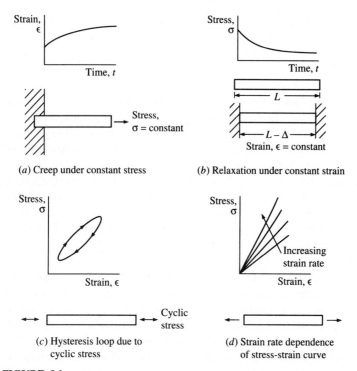

(a) Creep under constant stress

(b) Relaxation under constant strain

(c) Hysteresis loop due to cyclic stress

(d) Strain rate dependence of stress-strain curve

FIGURE 8.1
Physical manifestations of viscoelastic behavior in structural materials, as demonstrated by various types of loading applied to a viscoelastic rod.

Polymers with amorphous microstructures tend to be more viscoelastic than those with crystalline microstructures. As shown in Fig. 8.2, amorphous microstructures consist of three-dimensional arrangements of randomly entangled long-chain polymer molecules which are

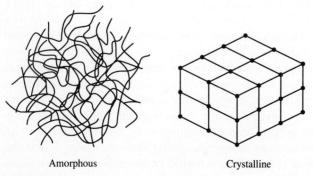

Amorphous   Crystalline

FIGURE 8.2
Amorphous and crystalline microstructures in polymers.

often characterized by analogy to a "bowl of spaghetti." On the other hand, crystalline microstructures consist of regular, ordered crystalline arrays of atoms (Fig. 8.2). Some polymers have both amorphous and crystalline components in their microstructures, and some polymers are purely amorphous.

On the basis of the previous discussion we conclude that viscoelastic behavior of composite materials is more significant for composites having one or more polymeric constituents. Viscoelastic effects in polymer matrix composites are most pronounced in matrix-dominated response to off-axis or shear loading. Viscoelastic deformations and plastic deformations are similar in that both are driven by shear stresses. Indeed, elements of the theory of plasticity are often borrowed for use in the theory of viscoelasticity. For example, it is sometimes assumed in viscoelasticity analysis that the dilatational response to hydrostatic stresses is elastic, but that the distortional response to shear stresses is viscoelastic.

In this chapter we will be concerned with the development of stress-strain relationships for linear viscoelastic materials and composites made of those materials. These stress-strain relationships take on special forms for creep, relaxation, and sinusoidal oscillation. Following the use of certain integral transforms, the viscoelastic stress-strain relationships turn out to be analogous to Hookean elastic stress-strain relationships, leading to the so-called Elastic-Viscoelastic Correspondence Principle.

Dynamic loading is usually categorized as being either impulsive or oscillatory. Dynamic response consists of either a propagating wave or a vibration, depending on the elapsed time and the relative magnitudes of the wavelength of the response and the characteristic structural dimension. Both types of excitation usually cause wave propagation initially. Wave propagation will continue if the response wavelength is much shorter than the characteristic structural dimension, otherwise standing waves (i.e., vibrations) will be set up as the waves begin to reflect back from the boundaries. Wave propagation in composites may involve complex reflection and/or refraction effects at fiber/matrix interfaces or ply interfaces, complicating matters further.

The dynamic response of composites may also be complicated by their anisotropic behavior. For example, the speed of a propagating wave in an isotropic material is independent of orientation, whereas the wave speed in an anisotropic composite depends on the direction of propagation. Anisotropic coupling effects often lead to complex waves or modes of vibration. For example, an isotropic beam which is subjected to an oscillatory bending moment will respond in pure flexural modes of vibration, but a nonsymmetric laminate may respond in a coupled bending-twisting mode or some other complex mode. In this chapter, however, only the analyses for vibrations and wave propagation in specially orthotropic composites or laminates without coupling will be considered.

Damping, which is one of the manifestations of viscoelastic behavior, is obviously important for noise and vibration control. Composites generally have better damping than conventional metallic structural materials, especially if the composite has one or more polymeric constituents. It will be shown that the complex modulus notation and the Elastic-Viscoelastic Correspondence Principle from the viscoelasticity theory are particularly useful in the development of analytical models for predicting the damping behavior of composites.

Finally, it will be shown in this chapter that the effective modulus theory, which was introduced in Chaps. 2 and 3, is indispensable in both viscoelastic and dynamic analysis of composites. Under certain restrictions, the concept of an effective modulus or effective compliance will be used to extend various viscoelastic analyses and dynamic analyses of homogeneous materials to the corresponding analyses of heterogeneous composites.

8.2 LINEAR VISCOELASTIC BEHAVIOR OF COMPOSITES

A linear elastic solid exhibits a linearity between stress and strain, and this linear relationship is independent of time. A linear viscoelastic solid also exhibits a linearity between stress and strain, but the linear relationship depends on the time history of the input. The mathematical criteria for linear viscoelastic behavior are similar to those for linear behavior of any system. Following the notation of Schapery [8.1], the criteria can be stated as follows:

Let the response R to an input I be written as $R = R\{I\}$, where $R\{I\}$ denotes that the current value of R is a function of the time history of the input I. For linear viscoelastic behavior the response $R\{I\}$ must satisfy both of the following conditions:

1. Proportionality: i.e., $R\{cI\} = cR\{I\}$, where c is a constant, and

2. Superposition: i.e., $R\{I_a + I_b\} = R\{I_a\} + R\{I_b\}$, where I_a and I_b may be the same or different time histories.

Any response not satisfying these criteria would be a nonlinear response. These criteria form the basis of the stress-strain relationship known as the Boltzmann superposition integral, which is developed in the next section.

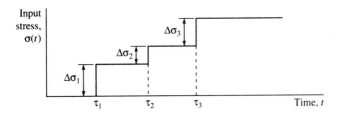

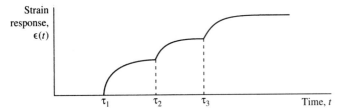

FIGURE 8.3
Input stress and strain response in one-dimensional loading of a linear viscoelastic material for illustration of the Boltzmann Superposition Principle.

8.2.1 Boltzmann Superposition Integrals for Creep and Relaxation

The stress-strain relationships for a linear viscoelastic material can be developed by using the Boltzmann Superposition Principle [8.2]. If the material is at a constant temperature and is "nonaging," then the response at any time t due to an input at time $t = \tau$ is a function of the input and the elapsed time $(t - \tau)$ only. Aging is a time-dependent change in the material which is different from viscoelastic creep or relaxation. Both temperature and aging effects will be considered later in Sec. 8.2.6.

Consider the one-dimensional isothermal loading of a nonaging, isotropic, homogeneous linear viscoelastic material by the stresses $\Delta\sigma_1$, $\Delta\sigma_2$, and $\Delta\sigma_3$ at times τ_1, τ_2, and τ_3, respectively, as shown in Fig. 8.3. According to the Boltzmann Superposition Principle, the strain response is linearly proportional to the input stress, but the proportionality factor is a function of the elapsed time since the input stress. Thus, for the stress-time history in Fig. 8.3 the total strain response at any time $t > \tau_3$ is given by

$$\epsilon(t) = \Delta\sigma_1 S(t - \tau_1) + \Delta\sigma_2 S(t - \tau_2) + \Delta\sigma_3 S(t - \tau_3) \qquad (8.1)$$

where $S(t)$ is the creep compliance, which is zero for $t < 0$. For input stresses having arbitrary time histories Eq. (8.1) can be generalized as the Boltzmann superposition integral, or hereditary law

$$\epsilon(t) = \int_{-\infty}^{t} S(t - \tau) \frac{d\sigma(\tau)}{d\tau} d\tau \qquad (8.2)$$

Alternatively, the stress resulting from arbitrary strain inputs may be given by

$$\sigma(t) = \int_{-\infty}^{t} C(t - \tau) \frac{d\epsilon(\tau)}{d\tau} d\tau \qquad (8.3)$$

where $C(t)$ is the relaxation modulus, which is zero for $t < 0$.

Equation (8.2) can be extended to the more general case of a homogeneous, anisotropic, linear viscoelastic material with multiaxial inputs and responses by using the contracted notation and writing

$$\epsilon_i(t) = \int_{-\infty}^{t} S_{ij}(t - \tau) \frac{d\sigma_j(\tau)}{d\tau} d\tau \qquad (8.4)$$

where $i, j = 1, 2, \ldots, 6$
           $S_{ij}(t) =$ creep compliances

For the specific case of the homogeneous, linear viscoelastic, specially orthotropic lamina in plane stress Eqs. (8.4) become

$$\epsilon_1(t) = \int_{-\infty}^{t} S_{11}(t - \tau) \frac{d\sigma_1(\tau)}{d\tau} d\tau + \int_{-\infty}^{t} S_{12}(t - \tau) \frac{d\sigma_2(\tau)}{d\tau} d\tau$$

$$\epsilon_2(t) = \int_{-\infty}^{t} S_{12}(t - \tau) \frac{d\sigma_1(\tau)}{d\tau} d\tau + \int_{-\infty}^{t} S_{22}(t - \tau) \frac{d\sigma_2(\tau)}{d\tau} d\tau \qquad (8.5)$$

$$\gamma_{12}(t) = \int_{-\infty}^{t} S_{66}(t - \tau) \frac{d\tau_{12}(\tau)}{d\tau} d\tau$$

Similarly, Eq. (8.3) can be generalized to the form

$$\sigma_i(t) = \int_{-\infty}^{t} C_{ij}(t - \tau) \frac{d\epsilon_j(\tau)}{d\tau} d\tau \qquad (8.6)$$

where the $C_{ij}(t)$ are the relaxation moduli. Note that Eqs. (8.4) and (8.6) are analogous to the generalized Hooke's law for linear elastic materials given by Eqs. (2.5) and (2.3), respectively, and that Eqs. (8.5) are analogous to the Hooke's law for the specially orthotropic lamina given by Eqs. (2.24). Thus, the creep compliances, $S_{ij}(t)$, for the viscoelastic material are analogous to the elastic compliances, S_{ij}, and the viscoelastic relaxation moduli, $C_{ij}(t)$, are analogous to the elastic stiffnesses, C_{ij}.

In order to apply the stress-strain relationships in Eqs. (8.4) to (8.6) to *heterogeneous*, anisotropic, linear viscoelastic composites, we again make use of the "effective modulus theory" that was introduced in Chaps. 2 and 3. Recall that in order to apply the stress-strain relationships at a point in a homogeneous material [i.e., Eqs. (2.3) and (2.5)] to the case of a heterogeneous composite, we replaced the stresses and strains at a point with the volume-averaged stresses and strains [Eqs.

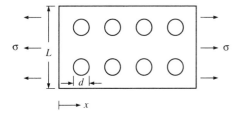

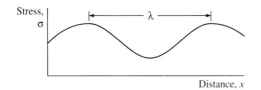

FIGURE 8.4
Critical dimensions which are used in the criteria for the application of the effective modulus theory.

(2.7) and (2.8)] and also replaced the elastic moduli of the heterogeneous composite by effective moduli of an equivalent homogeneous material [Eqs. (2.9) and (2.10)]. Recall also that the criterion for the use of the effective modulus theory was that the scale of the inhomogeneity, d, had to be much smaller than the characteristic structural dimension, L, over which the averaging is done. Since this chapter also deals with dynamic behavior, however, it is appropriate to add another criterion related to dynamic effects. That is, the scale of the inhomogeneity, d, must also be much smaller than the characteristic wavelength, λ, of the dynamic stress distribution (Fig. 8.4). Thus, the criteria for the use of the effective modulus theory in dynamic loading of viscoelastic composites are $d \ll L$ and $d \ll \lambda$. Practically speaking, the second criterion only becomes important when dealing with the propagation of high-frequency waves which have very short wavelengths. On the other hand, the wavelengths associated with typical mechanical vibrations will almost always be sufficiently large so as to satisfy $d \ll \lambda$. The book by Christensen [8.3] gives a more detailed discussion of the effective modulus theory.

Thus, Eqs. (8.4) to (8.6) are valid for heterogeneous, anisotropic, linear viscoelastic composites if at an arbitrary time, t, we simply replace the stresses and strains at a point with the volume-averaged stresses and strains, replace the creep compliances with the effective creep compliances, and replace the relaxation moduli with the effective relaxation moduli. Thus, the effective creep compliance matrix for the specially orthotropic lamina in plane stress is given by

$$S_{ij}(t) = \begin{bmatrix} S_{11}(t) & S_{12}(t) & 0 \\ S_{21}(t) & S_{22}(t) & 0 \\ 0 & 0 & S_{66}(t) \end{bmatrix} \qquad (8.7)$$

For the generally orthotropic lamina we have

$$\bar{S}_{ij}(t) = \begin{bmatrix} \bar{S}_{11}(t) & \bar{S}_{12}(t) & \bar{S}_{16}(t) \\ \bar{S}_{21}(t) & \bar{S}_{22}(t) & \bar{S}_{26}(t) \\ \bar{S}_{16}(t) & \bar{S}_{26}(t) & \bar{S}_{66}(t) \end{bmatrix} \tag{8.8}$$

where the $\bar{S}_{ij}(t)$ are the transformed effective creep compliances. Halpin and Pagano [8.4] have shown that the $\bar{S}_{ij}(t)$ are related to the $S_{ij}(t)$ by the transformations

$$\bar{S}_{11}(t) = S_{11}(t)c^4 + [2S_{12}(t) + S_{66}(t)]c^2 s^2 + S_{22}(t)s^4$$

$$\bar{S}_{12}(t) = S_{12}(t)(s^4 + c^4) + [S_{11}(t) + S_{22}(t) - S_{66}(t)]s^2 c^2$$

$$\bar{S}_{22}(t) = S_{11}(t)s^4 + [2S_{12}(t) + S_{66}(t)]s^2 c^2 + S_{22}(t)c^4$$

$$\bar{S}_{66}(t) = 2[2S_{11}(t) + 2S_{22}(t) - 4S_{12}(t) - S_{66}(t)]c^2 s^2 + S_{66}(t)(s^4 + c^4) \tag{8.9}$$

$$\bar{S}_{16}(t) = [2S_{11}(t) - 2S_{12}(t) - S_{66}(t)]sc^3 - [2S_{22}(t) - 2S_{12}(t) - S_{66}(t)]s^3 c$$

$$\bar{S}_{26}(t) = [2S_{11}(t) - 2S_{12}(t) - S_{66}(t)]s^3 c - [2S_{22}(t) - 2S_{12}(t) - S_{66}(t)]sc^3$$

where $s = \sin\theta$, $c = \cos\theta$, and the angle θ has been previously defined in Fig. 2.6. Note that these equations are entirely analogous to the corresponding elastic compliance transformation equations. Further justification for such direct correspondence between elastic and viscoelastic equations is provided by the Elastic-Viscoelastic Correspondence Principle, which is discussed later.

Recall that for the elastic case strain energy considerations led to the symmetry conditions $S_{ij} = S_{ji}$ and $C_{ij} = C_{ji}$. For the viscoelastic case Schapery [8.1] has used thermodynamic arguments to show that if $S_{ij}(t) = S_{ji}(t)$ for the constituent materials, then the same is true for the composite. Halpin and Pagano [8.4] and others have presented experimental evidence that for transversely isotropic composites under plane stress $S_{12}(t) = S_{21}(t)$. In both elastic and viscoelastic cases further reductions in the number of independent moduli or compliances depend on material property symmetry and the coordinate system used.

Example 8.1. A specially orthotropic, linear viscoelastic composite lamina is subjected to the shear stress-time history shown in Fig. 8.5. If the

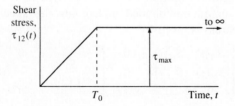

FIGURE 8.5
Shear stress time history for Example 8.1.

effective shear creep compliance is given by

$$S_{66}(t) = A + Bt \qquad \text{when } t \geq 0$$

and $\qquad\qquad S_{66}(t) = 0 \qquad\qquad \text{when } t < 0$

where A and B are material constants and t is time, find the expressions for the creep strain at $t < T_0$ and $t > T_0$.

Solution. The creep strain is given by Eq. (8.4):

$$\epsilon_i(t) = \int_{-\infty}^{t} S_{ij}(t - \tau) \frac{d\sigma_j(\tau)}{d\tau} d\tau$$

which, for the case of $t < T_0$, reduces to

$$\epsilon_6(t) = \gamma_{12}(t) = \int_0^t [A + B(t - \tau)] \frac{\tau_{max}}{T_0} d\tau$$

$$= \frac{A\tau_{max}}{T_0} t + \frac{B\tau_{max}}{2T_0} t^2$$

For $t > T_0$ we have

$$\epsilon_6(t) = \gamma_{12}(t) = \int_0^{T_0} [A + B(t - \tau)] \frac{\tau_{max}}{T_0} d\tau$$

$$+ \int_{T_0}^t (0) \, d\tau = A\tau_{max} + B\tau_{max}t - \frac{B\tau_{max}T_0}{2}$$

8.2.2 Differential Equations and Spring-Dashpot Models

Although the Boltzmann superposition integral is a valid mathematical expression of the stress-strain relationship for a linear viscoelastic material, it does not lend itself easily to the use of physical models which help us to understand viscoelastic behavior better. In this section Laplace transforms will be used to convert the Boltzmann superposition integral to an ordinary differential equation involving time derivatives of stress and strain. Physical models for viscoelastic behavior can be easily interpreted by using differential equations.

The Laplace transform, $\mathscr{L}[f(t)]$ or $\bar{f}(s)$, of a function $f(t)$ is defined by

$$\mathscr{L}[f(t)] = \bar{f}(s) = \int_0^\infty f(t)e^{-st} \, dt \qquad (8.10)$$

where s is the Laplace parameter. For purposes of illustration we now take the Laplace transform of the one-dimensional Boltzmann superposition integral given by Eq. (8.3). The Laplace transform of both sides of the equation is given by

$$\mathscr{L}[\sigma(t)] = \bar{\sigma}(s) = \mathscr{L}\left[\int_{-\infty}^t C(t - \tau) \frac{d\epsilon(\tau)}{d\tau} d\tau \right] \qquad (8.11)$$

Noting that the right-hand side of Eq. (8.11) is in the form of a convolution integral [8.5], we can also write

$$\bar{C}(s)\frac{d\bar{\epsilon}(s)}{d\tau} = \mathscr{L}\left[\int_{-\infty}^{t} C(t-\tau)\frac{d\epsilon(\tau)}{d\tau}d\tau\right] \tag{8.12}$$

Taking the inverse Laplace transform of Eq. (8.12), we find that

$$\mathscr{L}^{-1}\left[\bar{C}(s)\frac{d\bar{\epsilon}(s)}{d\tau}\right] = \int_{-\infty}^{t} C(t-\tau)\frac{d\epsilon(\tau)}{d\tau}d\tau \tag{8.13}$$

Thus, Eq. (8.11) can be written as

$$\bar{\sigma}(s) = \mathscr{L}\left[\mathscr{L}^{-1}\left(\bar{C}(s)\frac{d\bar{\epsilon}(s)}{d\tau}\right)\right] = \bar{C}(s)\frac{d\bar{\epsilon}(s)}{d\tau} \tag{8.14}$$

But from the properties of Laplace transforms of derivatives [8.5],

$$\mathscr{L}\left[\frac{d\epsilon(\tau)}{d\tau}\right] = \frac{d\bar{\epsilon}(s)}{d\tau} = s\bar{\epsilon}(s) - \epsilon(0) \tag{8.15}$$

where $\epsilon(0)$ is the initial strain. If we neglect the initial conditions, Eq. (8.14) becomes

$$\bar{\sigma}(s) = s\bar{C}(s)\bar{\epsilon}(s) \tag{8.16}$$

If we perform similar operations on Eq. (8.2), we find that

$$\bar{\epsilon}(s) = s\bar{S}(s)\bar{\sigma}(s) \tag{8.17}$$

Note that Eqs. (8.16) and (8.17) are now of the same form as Hooke's law for linear elastic materials, except that the Laplace transforms of the stresses and strains are linearly related, and the proportionality constants are the Laplace transform of the creep compliance and the Laplace transform of the relaxation modulus. This is another example of the correspondence between the equations for elastic and viscoelastic materials and is another building block in the Elastic-Viscoelastic Correspondence Principle, which will be discussed later. Note also that according to Eqs. (8.16) and (8.17), the Laplace transform of the creep compliance and the Laplace transform of the relaxation modulus must be related by

$$\bar{S}(s) = \frac{1}{s^2\bar{C}(s)} \tag{8.18}$$

However, the corresponding time domain properties are not generally related by a simple inverse relationship. That is, in general,

$$S(t) \neq \frac{1}{C(t)} \tag{8.19}$$

Equation (8.17) can also be written as a ratio of two polynomials in the

Laplace parameter s as follows:

$$\bar{\epsilon}(s) = s\bar{S}(s)\bar{\sigma}(s) = \frac{Q(s)}{P(s)}\bar{\sigma}(s) \tag{8.20}$$

where
$$P(s) = a_0 + a_1 s + a_2 s^2 + \cdots + a_n s^n$$
$$Q(s) = b_0 + b_1 s + b_2 s^2 + \cdots + b_n s^n$$

Thus, we can write
$$P(s)\bar{\epsilon}(s) = Q(s)\bar{\sigma}(s) \tag{8.21}$$

But if we neglect the initial conditions, the Laplace transform of the nth derivative of a function $f(t)$ is

$$\mathscr{L}\left[\frac{d^n f(t)}{dt^n}\right] = s^n \bar{f}(s) \tag{8.22}$$

Making use of Eq. (8.22) and taking the inverse Laplace transform of Eq. (8.21), we find that

$$a_n \frac{d^n \epsilon}{dt^n} + \cdots + a_2 \frac{d^2 \epsilon}{dt^2} + a_1 \frac{d\epsilon}{dt} + a_0 \epsilon = b_0 \sigma + b_1 \frac{d\sigma}{dt} + b_2 \frac{d^2\sigma}{dt^2} + \cdots + b_n \frac{d^n \sigma}{dt^n} \tag{8.23}$$

Thus, linear viscoelastic behavior may also be described by an ordinary differential equation as well as by the Boltzmann superposition integral. Note that the linear elastic material described by Hooke's law is a special case of Eq. (8.23) when all time derivatives of stress and strain vanish (i.e., $a_0\epsilon = b_0\sigma$). Recall that one of the physical manifestations of viscoelastic behavior is the dependence of stress on strain rate; such strain rate effects can be modeled with Eq. (8.23). We now consider several simple physical models of linear viscoelastic behavior which include various time derivatives of stress and strain.

As shown in Figs. 8.6 to 8.8, useful physical models can be constructed from simple elements such as the elastic spring and the

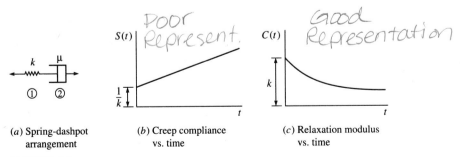

(a) Spring-dashpot arrangement

(b) Creep compliance vs. time

(c) Relaxation modulus vs. time

FIGURE 8.6
Maxwell model, with corresponding creep and relaxation curves.

viscous dashpot, where the spring of modulus k is assumed to follow Hooke's law and the dashpot is assumed to be filled with a newtonian fluid of viscosity μ. Thus, the stress-strain relationship for the elastic spring element is of the form $\epsilon = \sigma/k$, whereas the corresponding equation for the viscous dashpot is $d\epsilon/dt = \sigma/\mu$.

The Maxwell model consists of a spring and a dashpot in series, as shown in Fig. 8.6(a). The total strain across a model of unit length must equal the sum of the strains in the spring and the dashpot, so that

$$\epsilon = \epsilon_1 + \epsilon_2 \tag{8.24}$$

and the strain rate across the model is then

$$\frac{d\epsilon}{dt} = \frac{d\epsilon_1}{dt} + \frac{d\epsilon_2}{dt} = \frac{1}{k}\frac{d\sigma}{dt} + \frac{\sigma}{\mu} \tag{8.25}$$

Note that Eq. (8.25) is just a special case of Eq. (8.23) with only first derivatives of stress and strain. For creep at constant stress $\sigma = \sigma_0$ Eq. (8.25) reduces to

$$\frac{d\epsilon}{dt} = \frac{\sigma_0}{\mu} \tag{8.26}$$

Integrating Eq. (8.26) once, we find that

$$\epsilon(t) = \frac{\sigma_0}{\mu}t + C_1 \tag{8.27}$$

where the constant of integration, C_1, is found from the initial condition $\epsilon(0) = C_1 = \sigma_0/k$. Thus, the creep strain for the Maxwell model is given by

$$\epsilon(t) = \frac{\sigma_0}{\mu}t + \frac{\sigma_0}{k} \tag{8.28}$$

and the corresponding creep compliance is given by

$$S(t) = \frac{\epsilon(t)}{\sigma_0} = \frac{t}{\mu} + \frac{1}{k} \tag{8.29}$$

A plot of the creep compliance vs. time according to Eq. (8.29) is shown in Fig. 8.6(b). The type of creep behavior that is actually observed in experiments is more like that shown in Fig. 8.3, however. Thus, the Maxwell model does not adequately describe creep.

For relaxation at constant strain $\epsilon = \epsilon_0$ the Maxwell model stress-strain relationship in Eq. (8.25) becomes

$$0 = \frac{1}{k}\frac{d\sigma}{dt} + \frac{\sigma}{\mu} \tag{8.30}$$

Integrating Eq. (8.30) once, we find that

$$\ln \sigma = -\frac{k}{\mu}t + C_2 \qquad (8.31)$$

where the constant of integration, C_2, is found from the initial condition $\sigma(0) = \sigma_0$. The resulting stress relaxation function is

$$\sigma(t) = \sigma_0 e^{-kt/\mu} = \sigma_0 e^{-t/\lambda} \qquad (8.32)$$

where $\lambda = \mu/k$ is the relaxation time, or the time required for the stress to relax to $1/e$, or 37 percent of its initial value. The relaxation time is therefore a measure of the internal time scale of the material. The corresponding relaxation modulus is

$$C(t) = \frac{\sigma(t)}{\epsilon_0} = \frac{\sigma_0}{\epsilon_0} e^{-t/\lambda} = ke^{-t/\lambda} \qquad (8.33)$$

Figure 8.6(c) shows the relaxation modulus vs. time from Eq. (8.33), which is in general agreement with the type of relaxation observed experimentally. Thus, the Maxwell model appears to describe adequately the relaxation phenomenon, but not the creep response.

Figure 8.7(a) shows the Kelvin-Voigt model, which consists of a spring and a dashpot in parallel. Using the appropriate equations for a parallel arrangement and following a procedure similar to the one just outlined, it can be shown that the differential equation describing the behavior of the Kelvin-Voigt model is given by

$$\sigma = k\epsilon + \mu \frac{d\epsilon}{dt} \qquad (8.34)$$

Equation (8.34) is seen to be another special case of Eq. (8.23) with only first derivatves of strain. It can also be shown that the creep compliance for the Kelvin-Voigt model is given by

$$S(t) = \frac{1}{k}[1 - e^{-t/\rho}] \qquad (8.35)$$

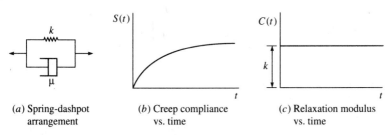

(a) Spring-dashpot   (b) Creep compliance   (c) Relaxation modulus
    arrangement   vs. time   vs. time

FIGURE 8.7
Kelvin-Voigt model, with corresponding creep and relaxation curves.

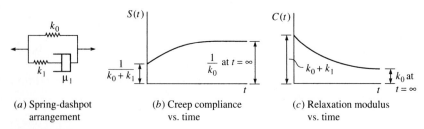

(a) Spring-dashpot
arrangement

(b) Creep compliance
vs. time

(c) Relaxation modulus
vs. time

FIGURE 8.8
Standard linear solid, or Zener model, with corresponding creep and relaxation curves.

where $\rho = \mu/k$ is now referred to as the retardation time. Similarly, the relaxation modulus is given by

$$C(t) = k \tag{8.36}$$

Equations (8.35) and (8.36) are plotted in Figs. 8.7(b) and 8.7(c), respectively. The creep compliance curve agrees with experimental observation, except that the initial elastic response is missing. On the other hand, the relaxation modulus has not been observed to be constant, as shown in Fig. 8.7(c). Thus, like the Maxwell model, the Kelvin-Voigt model does not adequately describe all features of experimentally observed creep and relaxation.

One obvious way to improve the the spring-dashpot model is to add more elements. One such improved model, shown in Fig. 8.8(a), is referred to as the standard linear solid, or Zener model. It can be shown that the differential equation for the Zener model is given by

$$\sigma + \frac{\mu_1}{k_1}\frac{d\sigma}{dt} = k_0\epsilon + \frac{\mu_1}{k_1}(k_0 + k_1)\frac{d\epsilon}{dt} \tag{8.37}$$

where the parameters k_0, k_1, and μ_1 are defined in Fig. 8.8(a). Equation (8.37) is obviously another special case of the general differential equation (8.23). It is also interesting to note that the Zener model shown in Fig. 8.8(a) is just a Maxwell model in parallel with a spring. The creep compliance for the Zener model is given by

$$S(t) = \frac{1}{k_0}\left[1 - \frac{k_1}{k_0 + k_1}e^{-t/\rho_1}\right] \tag{8.38}$$

where $\rho_1 = \frac{\mu_1}{k_0 k_1}(k_0 + k_1)$ is the retardation time.

As shown in Fig. 8.8(b), the shape of the creep compliance curve from Eq. (8.38) matches the expected shape based on experimental observations. The relaxation modulus for the Zener model is given by

$$C(t) = k_0 + k_1 e^{-t/\lambda_1} \tag{8.39}$$

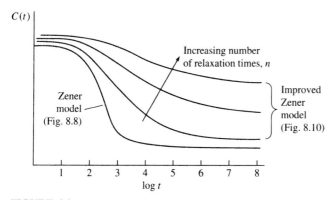

FIGURE 8.9
Effect of increasing number of relaxation times on relaxation curve of Zener model.

where $\lambda_1 = \mu_1/k_1$ is the relaxation time. Note that λ_1 is just the relaxation time for the Maxwell model consisting of μ_1 and k_1. Figure 8.8(c) shows the predicted relaxation modulus curve from Eq. (8.39), and, again, the general shape of the curve appears to be similar to what is experimentally observed.

Although the Zener model is the simplest spring-dashpot model which correctly describes all expected features of experimentally observed creep and relaxation behavior in linear viscoelastic materials, it still is not completely adequate. This remaining inadequacy is best described by plotting the relaxation modulus vs. the logarithm of time, as shown in Fig. 8.9. Practically speaking, complete relaxation for the Zener model occurs in less than a decade in time, but relaxation for real polymers happens over a much longer time scale. For example, the glass-to-rubber transition, which is only one region of polymer viscoelastic behavior, takes about six to eight decades in time to complete [8.6]. This extended relaxation period for polymers is due to the existence of a distribution of relaxation times. By using an improved Zener model such as the parallel arrangement shown in Fig. 8.10, we can introduce such a distribution of relaxation times, λ_i, which makes it possible to extend the range of relaxation to more realistic values. This form of the improved Zener model consists of n Maxwell elements in parallel with the elastic

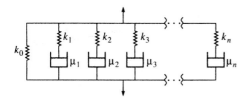

FIGURE 8.10
Improved Zener model, parallel arrangement.

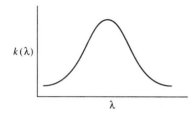

$k(\lambda)$

λ

FIGURE 8.11
Continuous distribution of relaxation times, or relaxation spectrum, $k(\lambda)$, for improved Zener model of Fig. 8.10 with an infinite number of elements.

spring, k_0. It can be easily shown that the relaxation modulus for this improved Zener model is given by

$$C(t) = k_0 + \sum_{i=1}^{n} k_i e^{-t/\lambda_i} \qquad (8.40)$$

where $\lambda_i = \mu_i/k_i$ is the relaxation time for ith Maxwell element.

As shown in Fig. 8.9, the effect of increasing n and the corresponding number of relaxation times is to broaden the range of relaxation. The number of relaxation times needed to describe adequately the viscoelastic behavior of a particular material must be determined experimentally. For an infinite number of elements in the improved Zener model of Fig. 8.10 and a continuous distribution of relaxation times the relaxation modulus can be expressed as [8.7]

$$C(t) = k_0 + \int_0^{\infty} k(\lambda)e^{-t/\lambda} \, d\lambda \qquad (8.41)$$

where $k(\lambda)$ is the distribution of relaxation times, or the relaxation spectrum, which is shown schematically in Fig. 8.11.

By considering an alternative form of an improved Zener model consisting of a spring in series with n Kelvin-Voigt elements, as shown in Fig. 8.12, it can be shown that the corresponding creep compliance expression is

$$S(t) = \frac{1}{k_0} + \sum_{i=1}^{n} \frac{1}{k_i}[1 - e^{-t/\rho_i}] \qquad (8.42)$$

where $\rho_i = \mu_i/k_i$ is the retardation time for ith Kelvin-Voigt element.

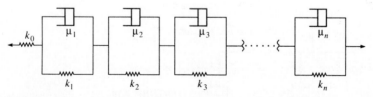

k_0 μ_1 μ_2 μ_3 μ_n

k_1 k_2 k_3 k_n

FIGURE 8.12
Improved Zener model, series arrangement.

Although the above equations have been derived on the basis of simple spring-dashpot models, the generalized relaxation modulus and creep compliance expressions for anisotropic linear viscoelastic composites have the same forms as Eqs. (8.40) and (8.42), respectively. According to Schapery [8.1], if the elastic moduli are positive definite (i.e., always either positive or equal to zero), it can be shown using the thermodynamic theory that the generalized expressions corresponding to Eqs. (8.40) and (8.42) are, respectively,

$$C_{ij}(t) = \sum_{m=1}^{n} C_{ij}^{(m)} e^{-t/\lambda_m} + C_{ij} \tag{8.43}$$

and
$$S_{ij}(t) = \sum_{m=1}^{n} s_{ij}^{(m)} [1 - e^{-t/\rho_m}] + S_{ij} \tag{8.44}$$

where $\quad i, j = 1, 2, \ldots, 6$

C_{ij}, S_{ij} = elastic moduli and compliances, respectively

λ_m, ρ_m = relaxation times and retardation times, respectively

$C_{ij}^{(m)}, S_{ij}^{(m)}$ = coefficients corresponding to λ_m and ρ_m, respectively

As with the simple spring-dashpot models, the numerical values of the parameters on the right-hand side of Eqs. (8.43) and (8.44) must be determined experimentally.

The relaxation times and retardation times are strongly dependent on temperature, and such temperature dependence is the basis of the time-temperature superposition method, which will be discussed later. It is assumed here that the materials are "thermorheologically simple." That is, all the relaxation times, λ_i, and the retardation times, ρ_i, are assumed to have the same temperature dependence. A similar argument holds for the effect of aging, which will also be discussed later.

Example 8.2. For the problem in Example 8.1 the effective shear compliance is to be approximated by a Kelvin-Voigt model of the form

$$S_{66}(t) = \frac{1}{k}(1 - e^{-t/\lambda}) \qquad \text{when } t \geq 0$$

and $\qquad S_{66}(t) = 0 \qquad \text{when } t < 0$

Determine the creep strain at $t < T_0$ and $t > T_0$.

Solution. For the case of $t < T_0$ Eq. (8.4) reduces to

$$\gamma_{12}(t) = \int_0^t \frac{1}{k}[1 - e^{-(t-\tau)/\lambda}] \frac{\tau_{max}}{T_0} d\tau = \frac{\tau_{max}}{kT_0}[t - \lambda(1 - e^{-t/\lambda})]$$

and for $t > T_0$ we have

$$\gamma_{12}(t) = \int_0^{t_0} \frac{1}{k}[1 - e^{-(t-\tau)/\lambda}]\frac{\tau_{max}}{T_0}d\tau + (0)$$

$$= \frac{\tau_{max}}{kT_0}[T_0 - \lambda e^{-t/\lambda}(e^{T_0/\lambda} - 1)]$$

8.2.3 Quasi-Elastic Analysis

From the previous section, it should be clear that the generalized Boltzmann superposition integrals in Eqs. (8.4) and (8.6) can be Laplace-transformed to yield equations of the form

$$\bar{\epsilon}_i(s) = s\bar{S}_{ij}(s)\bar{\sigma}_j(s) \tag{8.45}$$

and

$$\bar{\sigma}_i(s) = s\bar{C}_{ij}(s)\bar{\epsilon}_j(s) \tag{8.46}$$

These equations are of the same form as the corresponding elastic stress-strain relationships and arc presumably easier to work with than the integral equations. In a practical analysis or design problem involving the use of these equations, however, the problem solution in the Laplace domain would then have to be inverse-transformed to get the desired time domain result, and this can present difficulties. Schapery [8.1] has presented several approximate methods for performing such inversions. If the input stresses or strains are constant, however, there is no need for inverse transforms and the time domain equations turn out to be very simple. Schapery refers to this as a "quasi-elastic analysis," and the equations used in such an analysis will be developed in the remainder of this section.

Consider a generalized creep problem with time-varying stresses $\sigma_j(t)$ given by

$$\sigma_j(t) = \sigma_j' H(t) \tag{8.47}$$

where $j = 1, 2, \ldots, 6$, the σ_j' are constant stresses, and $H(t)$ is the unit step function, or Heaviside function, shown in Fig. 8.13(a) and defined as

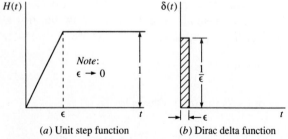

(a) Unit step function   (b) Dirac delta function

FIGURE 8.13
Unit step function and Dirac delta function.

follows [8.2]:

$$\lim_{\epsilon \to 0} H(t) \begin{cases} = 0 & \text{for } t \leq 0 \\ = t/\epsilon & \text{for } 0 \leq t \leq \epsilon \\ = 1 & \text{for } t \geq \epsilon \end{cases} \tag{8.48}$$

The unit step function can be easily shifted along the time axis by an amount ξ by writing the function as $H(t - \xi)$. Substituting the stresses from Eq. (8.47) in the Boltzmann superposition integral, Eq. (8.4), we find that the resulting strains are given by

$$\epsilon_i(t) = \int_{-\infty}^{t} S_{ij}(t - \tau)\sigma_j' \frac{dH(\tau)}{d\tau} d\tau \tag{8.49}$$

but according to Eqs. (8.48), the derivative of the step function must be

$$\frac{dH(t)}{dt} = \delta(t) \begin{cases} = 0 & \text{for } t \leq 0 \\ = 1/\epsilon & \text{for } 0 \leq t \leq \epsilon \\ = 0 & \text{for } t \geq 0 \end{cases} \tag{8.50}$$

where the parameter ϵ can be made arbitrarily small, the derivative in Eq. (8.50) is taken before $\epsilon \to 0$, and $\delta(t)$ is the Dirac delta function shown in Fig. 8.13(b). Thus, the integral in Eq. (8.49) can be written

$$\epsilon_i(t) = \left\{ \int_{-\infty}^{t} S_{ij}(t - \tau)\delta(\tau) \, d\tau \right\} \sigma_j' \tag{8.51}$$

where the constants σ_j' have been moved outside the integral. According to the properties of convolution integrals [8.5], we can also write

$$\epsilon_i(t) = \left\{ \int_{-\infty}^{t} S_{ij}(\tau)\delta(t - \tau) \, d\tau \right\} \sigma_j' \tag{8.52}$$

This integral can be broken down and rewritten as follows:

$$\epsilon_i(t) = \left\{ \int_{-\infty}^{t-\epsilon} (0) \, d\tau + \int_{t-\epsilon}^{t} S_{ij}(t)\delta(t - \tau) \, d\tau \right\} \sigma_j' \tag{8.53}$$

where the $S_{ij}(\tau)$ evaluated over the interval $t - \epsilon \leq \tau \leq t$ can be approximated as $S_{ij}(t)$ since ϵ is very small. The $S_{ij}(t)$ can now be moved outside the integral, leaving the integral of the Dirac delta function, which is defined as [8.5]

$$\int_{t-\epsilon}^{t} \delta(t - \tau) \, d\tau = 1 \tag{8.54}$$

Thus, the final result is

$$\epsilon_i(t) = S_{ij}(t)\sigma_j' \tag{8.55}$$

The form of this equation suggests that we can solve for creep strains under constant stresses, σ_j', by simply replacing the elastic compliances, S_{ij}, in Hooke's law [Eqs. (2.5)] with the corresponding viscoelastic creep compliances, $S_{ij}(t)$. Similarly, it can be shown that if the constant strain inputs

$$\epsilon_j(t) = \epsilon_j' H(t) \tag{8.56}$$

are substituted in Eqs. (8.6), the resulting stresses must be

$$\sigma_i(t) = C_{ij}(t)\epsilon_j' \tag{8.57}$$

Thus, the stress relaxation under constant strains can be found by replacing the elastic moduli, C_{ij}, in Hooke's law [Eqs. (2.3)] with the corresponding viscoelastic relaxation moduli, $C_{ij}(t)$. Equations (8.55) and (8.57) form the basis of the so-called "quasi-elastic analysis" and obviously eliminate the need for Laplace transform analysis in the stress-strain relationships. It should be emphasized again, however, that Eqs. (8.55) and (8.57) are only valid for constant or near constant inputs. Such equations give additional hints of a direct correspondence between the equations for linear elastic systems and those for linear viscoelastic systems, and this correspondence will be discussed in more detail later.

Example 8.3. The filament wound pressure vessel described in Example 2.1 is constructed of a viscoelastic composite which has creep compliances that can be modeled by using one term series representations of the form shown in Eqs. (8.44). Assuming that the internal pressure, p, is constant, determine the creep strains along the principal material directions in the wall of the vessel.

Solution. Since the internal pressure, p, is constant, the stresses in the wall of the vessel are all constant, and we can use a quasi-elastic analysis to predict the creep strains. From Eqs. (8.55), we find that the creep strains along the principal material directions are given by

$$\epsilon_1(t) = S_{11}(t)\sigma_1 + S_{12}(t)\sigma_2$$
$$\epsilon_2(t) = S_{12}(t)\sigma_1 + S_{22}(t)\sigma_2$$

and
$$\epsilon_6(t) = \gamma_{12}(t) = S_{66}(t)\tau_{12}$$

From Example 2.1, the stresses along the principal material directions were found to be

$$\sigma_1 = 20.5p \text{ MPa}$$
$$\sigma_2 = 17.0p \text{ MPa}$$
$$\sigma_6 = \sigma_{12} = \tau_{12} = 6.0p \text{ MPa}$$

Substituting these stresses and the creep compliances from Eqs. (8.44) in the above expressions for the strains, we find that

$$\epsilon_1(t) = (S_{11}^{(1)}[1 - e^{-t/\rho_1}] + S_{11})(20.5p)$$
$$+ (S_{12}^{(1)}[1 - e^{-t/\rho_1}] + S_{12})(17.0p)$$
$$\epsilon_2(t) = (S_{12}^{(1)}[1 - e^{-t/\rho_1}] + S_{12})(20.5p)$$
$$+ (S_{22}^{(1)}[1 - e^{-t/\rho_1}] + S_{22})(17.0p)$$
$$\gamma_{12}(t) = (S_{66}^{(1)}[1 - e^{-t/\rho_1}] + S_{66})(6.0p)$$

8.2.4 Sinusoidal Oscillations and Complex Modulus Notation

In the previous section it was shown that when the inputs are constant, the Boltzmann superposition integrals are reduced to simple algebraic equations which resemble the linear elastic Hooke's law. In this section an analogous simplification will be demonstrated for the case of stresses or strains which vary sinusoidally with time. The results will make it much easier to analyze sinusoidal vibrations of viscoelastic composites. The general procedure here follows that presented by Fung [8.2].

Consider the case where the stresses vary sinusoidally with frequency ω. Using the contracted notation and complex exponentials, such stresses can be written as

$$\tilde{\sigma}_n(t) = A_n e^{i\omega t} \qquad (8.58)$$

where $n = 1, 2, \ldots, 6$
i = imaginary operator is $(-1)^{1/2}$
A_n = complex stress amplitudes

and the superscript tilde $(\sim)$ refers to a sinusoidally varying quantity.

Substituting Eq. (8.58) in Eq. (8.4), we find that the resulting sinusoidally varying strains are given by

$$\tilde{\epsilon}_m(t) = \int_{-\infty}^{t} S_{mn}(t - \tau)i\omega A_n e^{i\omega\tau} \, d\tau \qquad (8.59)$$

where $m, n = 1, 2, \ldots, 6$.

It is now convenient to define a new variable $\xi = t - \tau$, so that

$$\tilde{\epsilon}_m(t) = \int_0^\infty S_{mn}(\xi)e^{-i\omega\xi}i\omega A_n e^{i\omega t} \, d\xi \qquad (8.60)$$

The terms not involving functions of ξ may be moved outside the integral, and since $S_{mn}(t)$ for $t < 0$, the lower limit on the integral can be changed to $-\infty$, so that

$$\tilde{\epsilon}_m(t) = i\omega A_n e^{i\omega t} \int_{-\infty}^{\infty} S_{mn}(\xi) e^{-i\omega \xi} \, d\xi \tag{8.61}$$

The integral in Eq. (8.61) is just the Fourier transform of the creep compliances, $\mathscr{F}[S_{mn}(\xi)]$, or $S_{mn}(\omega)$, which is written as

$$\mathscr{F}[S_{mn}(\xi)] = S_{mn}(\omega) = \int_{-\infty}^{\infty} S_{mn}(\xi) e^{-i\omega \xi} \, d\xi \tag{8.62}$$

Thus, the stress-strain relationship reduces to

$$\tilde{\epsilon}_m(t) = i\omega S_{mn}(\omega) A_n e^{i\omega t} = i\omega S_{mn}(\omega) \tilde{\sigma}_n(t) \tag{8.63}$$

In order to get this equation to resemble Hooke's law more closely, we simply define the frequency domain complex compliances as follows:

$$S_{mn}^*(\omega) = i\omega S_{mn}(\omega) \tag{8.64}$$

so that Eq. (8.63) becomes

$$\tilde{\epsilon}_m(t) = S_{mn}^*(\omega) \tilde{\sigma}_n(t) \tag{8.65}$$

Thus, in linear viscoelastic materials the sinusoidally varying stresses are related to the sinusoidally varying strains by complex compliances in the same way that static stresses and strains are related by elastic compliances in the linear elastic material. In addition, the time domain creep compliances are related to frequency domain complex compliances by the Fourier transforms. It is important to note, however, that the complex compliance is not simply equal to the Fourier transform of the corresponding creep compliance. According to Eq. (8.64), the complex compliance, $S_{mn}^*(\omega)$, is equal to a factor of $i\omega$ times $S_{mn}(\omega)$, and $S_{mn}(\omega)$ is the Fourier transform of the creep compliance, $S_{mn}(t)$.

Alternatively, if we substitute sinusoidally varying strains in Eq. (8.5), we find that the sinusoidally varying stresses are

$$\tilde{\sigma}_m(t) = C_{mn}^*(\omega) \tilde{\epsilon}_n(t) \tag{8.66}$$

where the complex moduli are defined by

$$C_{mn}^*(\omega) = i\omega C_{mn}(\omega) \tag{8.67}$$

and the $C_{mn}(\omega)$ are the Fourier transforms of the corresponding relaxation moduli, $C_{mn}(t)$. Alternatively, Eqs. (8.65) and (8.66) may be written in matrix form as

$$\{\tilde{\epsilon}(t)\} = [S^*(\omega)]\{\tilde{\sigma}(t)\} \tag{8.68}$$

and

$$\{\tilde{\sigma}(t)\} = [C^*(\omega)]\{\tilde{\epsilon}(t)\} \tag{8.69}$$

respectively, where the complex compliance matrix and the complex modulus matrix must be related by $[S^*(\omega)] = [C^*(\omega)]^{-1}$.

The complex modulus notation not only has a mathematical basis in viscoelasticity theory, but it also has a straightforward physical interpretation. Since the complex modulus is a complex variable, we can write it in terms of its real and imaginary parts as follows:

$$C^*_{mn}(\omega) = C'_{mn}(\omega) + iC''_{mn}(\omega)$$
$$= C'_{mn}(\omega)[1 + i\eta_{mn}(\omega)]$$
$$= |C^*_{mn}(\omega)|\, e^{-i\delta_{mn}(\omega)} \qquad (8.70)$$

[no summation on m and n in Eq. (8.70)]

where $C'_{mn}(\omega)$ = storage modulus
$C''_{mn}(\omega)$ = loss modulus

$$\eta_{mn}(\omega) = \text{loss factor} = \tan[\delta_{mn}(\omega)] = \frac{C''_{mn}(\omega)}{C'_{mn}(\omega)}$$

$\delta_{mn}(\omega)$ = phase lag between $\tilde{\sigma}_m(t)$ and $\tilde{\epsilon}_n(t)$

Thus, the real part of the complex modulus is associated with elastic energy storage, whereas the imaginary part is associated with energy dissipation, or damping. A physical interpretation of the one-dimensional forms of these equations may be given with the aid of the rotating vector diagram in Fig. 8.14. The stress and strain vectors are both assumed to be rotating with angular velocity, ω, and the physical oscillation is generated by either the horizontal or vertical projection of the vectors. The complex exponential representations of the rotating stress and strain vectors in the diagram are

$$\tilde{\sigma}(t) = \sigma e^{i(\omega t + \delta)}$$
$$\tilde{\epsilon}(t) = \epsilon e^{i\omega t} \qquad (8.71)$$

and

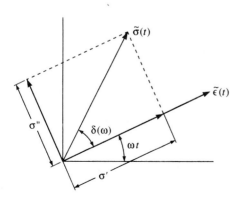

FIGURE 8.14
Rotating vector diagram for physical interpretation of the complex modulus.

so that the one-dimensional complex modulus is defined as

$$C^*(\omega) = \frac{\tilde{\sigma}(t)}{\tilde{\epsilon}(t)} = \frac{\sigma e^{i\delta}}{\epsilon} = \frac{\sigma}{\epsilon}(\cos \delta + i \sin \delta)$$

$$= \frac{\sigma'}{\epsilon} + i\frac{\sigma''}{\epsilon} = C'(\omega) + iC''(\omega)$$

$$= C'(\omega)[1 + i\eta(\omega)] \qquad (8.72)$$

It is seen that the strain lags the stress by the phase angle, δ; the storage modulus, $C'(\omega)$, is the in-phase component of the stress, σ', divided by the strain, ϵ; the loss modulus, $C''(\omega)$, is the out-of-phase component of stress, σ'', divided by the strain, ϵ; and the loss factor, $\eta(\omega)$, is the tangent of the phase angle, δ. Experimental determination of the complex modulus involves the measurement of the storage modulus, $C'(\omega)$, and the loss factor, $\eta(\omega)$, as a function of frequency, ω; and several techniques for doing this will be described in Chap. 10.

The inverse Fourier transform of the parameter $S_{mn}(\omega)$ is the creep compliance, $S_{mn}(t)$, as given by

$$\mathscr{F}^{-1}[S_{mn}(\omega)] = S_{mn}(t) = \frac{1}{2\pi}\int_{-\infty}^{\infty} S(\omega)e^{i\omega t}\, d\omega \qquad (8.73)$$

where $\mathscr{F}^{-1}$ is the inverse Fourier transform operator. Equations (8.62) and (8.73) form the so-called Fourier transform pair, which make it possible to transform back and forth between the time domain and the frequency domain [8.8]. Since experimental frequency data are usually expressed in units of cycles per second, or Hz, it is convenient to define the frequency as $f = \omega/2\pi$ (Hz), so that the Fourier transform pair now becomes symmetric in form:

$$\mathscr{F}[S_{mn}(t)] = S_{mn}(f) = \int_{-\infty}^{\infty} S_{mn}(t)e^{-i2\pi ft}\, dt \qquad (8.74)$$

and $\qquad \mathscr{F}^{-1}[S_{mn}(f)] = S_{mn}(t) = \int_{-\infty}^{\infty} S_{mn}(f)e^{i2\pi ft}\, df \qquad (8.75)$

It can be shown that the time domain relaxation modulus and the corresponding frequency domain complex modulus are related by a similar Fourier transform pair. As a further indication of the usefulness of such equations, inverse Fourier transforms have been used to estimate time domain creep behavior of composites from frequency domain complex modulus data obtained from vibration tests of the same materials [8.9].

Example 8.4. The composite pressure vessel described in Examples 2.1 and 8.3 has an internal pressure, p, that varies sinusoidally with time, as shown

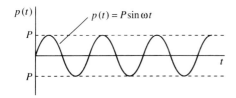

FIGURE 8.15
Sinusoidally varying pressure for Example 8.4.

in Fig. 8.15. If the complex compliances of the composite material are given by

$$S_{mn}^*(\omega) = S_{mn}'(\omega) + i S_{mn}''(\omega)$$

determine all the time-dependent strains associated with the principal material axes.

Solution. From Example 2.1 and Fig. 8.15, the stresses along the 12 directions are

$$\tilde{\sigma}_1(t) = 20.5p = 20.5P \sin \omega t \text{ (MPa)}$$
$$\tilde{\sigma}_2(t) = 17.0p = 17.0P \sin \omega t \text{ (MPa)}$$
$$\tilde{\sigma}_6(t) = \tilde{\tau}_{12}(t) = 6.0p = 6.0P \sin \omega t \text{ (MPa)}$$

The corresponding strains from Eqs. (8.65) are

$$\tilde{\epsilon}_1(t) = S_{11}^*(\omega)\tilde{\sigma}_1(t) + S_{12}^*(\omega)\tilde{\sigma}_2(t) + (0)\tilde{\sigma}_6(t)$$
$$= [S_{11}'(\omega) + i S_{11}''(\omega)]20.5P \sin \omega t + [S_{12}'(\omega) + i S_{12}''(\omega)]17.0P \sin \omega t$$
$$\tilde{\epsilon}_2(t) = S_{12}^*(\omega)\tilde{\sigma}_1(t) + S_{22}^*(\omega)\tilde{\sigma}_2(t) + (0)\tilde{\sigma}_6(t)$$
$$= [S_{12}'(\omega) + i S_{12}''(\omega)]20.5P \sin \omega t + [S_{22}'(\omega) + i S_{22}''(\omega)]17.0P \sin \omega t$$
$$\tilde{\epsilon}_6(t) = \tilde{\gamma}_{12}(t) = S_{66}^*(\omega)\tilde{\sigma}_6(t) = S_{66}^*(\omega)\tilde{\tau}_{12}(t)$$
$$= [S_{66}'(\omega) + i S_{66}''(\omega)]6.0P \sin \omega t$$

8.2.5 Elastic-Viscoelastic Correspondence Principle

In the previous sections of this chapter we have seen a number of examples where the form of the stress-strain relationships for linear viscoelastic materials is the same as that for linear elastic materials. Such analogies between the equations for elastic and viscoelastic analysis have led to the formal recognition of an "Elastic-Viscoelastic Correspondence Principle." The correspondence principle for isotropic materials was apparently introduced by Lee [8.10], whereas the application to anisotropic materials was proposed by Biot [8.11]. The specific application of the correspondence principle to the viscoelastic analysis of anisotropic composites has been discussed in detail by Schapery [8.1, 8.12] and Christensen [8.3].

A summary of the correspondences between elastic and viscoelastic

TABLE 8.1
Elastic-viscoelastic correspondence in stress-strain relationships

Material and input	Stresses	Strains	Properties	Equation
Linear elastic				
Input stresses	σ_j	ϵ_i	S_{ij}	(2.5)
Input strains	σ_i	ϵ_j	C_{ij}	(2.3)
Linear viscoelastic				
Generalized creep	$\bar{\sigma}_j(s)$	$\bar{\epsilon}_i(s)$	$s\bar{S}_{ij}(s)$	(8.45)
Constant stress creep	$\bar{\sigma}_j'$	$\epsilon_i(t)$	$S_{ij}(t)$	(8.55)
Generalized relaxation	$\bar{\sigma}_i(s)$	$\bar{\epsilon}_j(s)$	$s\bar{C}_{ij}(s)$	(8.46)
Constant strain relaxation	$\sigma_i(t)$	ϵ_j'	$C_{ij}(t)$	(8.57)
Sinusoidal stress input	$\tilde{\sigma}_j(t)$	$\tilde{\epsilon}_i(t)$	$S_{ij}^*(\omega)$	(8.65)
Sinusoidal strain input	$\tilde{\sigma}_i(t)$	$\tilde{\epsilon}_j(t)$	$C_{ij}^*(\omega)$	(8.66)

$i, j = 1, 2, \ldots, 6$

stress-strain relationships is given in Table 8.1. The implication of this table is that if we have the necessary equations for a linear elastic solution to a problem, we simply make the corresponding substitutions in the equations to get the corresponding linear viscoelastic solution. Although Table 8.1 is only concerned with the correspondences in the stress-strain relationships, there are obviously other equations involved in a complete solution to an elasticity problem. The correspondences in the equilibrium equations, the strain-displacement relations, the boundary conditions, and the variational methods of elastic analysis are beyond the scope of this book, but detailed discussions of these matters are given by Schapery [8.1, 8.12] and Christensen [8.3, 8.5].

One of the most important implications of the correspondence principle is that analytical models for predicting elastic properties of composites at both the micromechanical and the macromechanical levels can be easily converted for prediction of the corresponding viscoelastic properties. For example, the rule of mixtures for predicting the longitudinal modulus of a unidirectional composite can now be converted for viscoelastic relaxation problems by rewriting Eq. (3.22) as

$$E_1(t) = E_{f1}(t)v_f + E_m(t)v_m \tag{8.76}$$

where $E_1(t)$ = longitudinal relaxation modulus of composite
$E_{f1}(t)$ = longitudinal relaxation modulus of fiber
$E_m(t)$ = relaxation modulus of isotropic matrix
v_f = fiber volume fraction
v_m = matrix volume fraction

The relative viscoelasticity of fiber and matrix materials may make

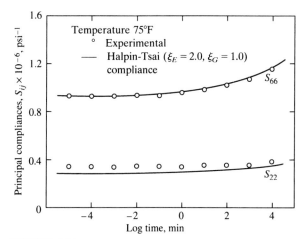

FIGURE 8.16
Measured and predicted creep compliances for glass/epoxy composite. (*From Beckwith* [8.13].)

further simplification possible. In most polymer matrix composites the time dependency of the matrix material would be much more significant than that of the fiber, so the fiber modulus could be assumed to be elastic, and the time dependency of $E_1(t)$ would be governed by $E_m(t)$ alone. The results of a similar analysis of the creep compliances $S_{22}(t)$ and $S_{66}(t)$ for a glass/epoxy composite are shown in Fig. 8.16 from Ref. [8.13]. From these results, it appears that the compliances can be accurately predicted by using the viscoelastic properties of the epoxy matrix in the corresponding viscoelastic forms of the Halpin-Tsai equations [Eqs. (3.57) and (3.58)].

At the macromechanical level equations such as the laminate force-deformation relationships can be converted to viscoelastic form using the correspondence principle. For example, the creep strains in a symmetric laminate under constant in-plane loading can be analyzed by employing the correspondence principle and a quasi-elastic analysis to rewrite Eqs. (7.58) as

$$\begin{Bmatrix} \epsilon_x^0(t) \\ \epsilon_y^0(t) \\ \gamma_{xy}^0(t) \end{Bmatrix} = \begin{bmatrix} A_{11}'(t) & A_{12}'(t) & A_{16}'(t) \\ A_{12}'(t) & A_{22}'(t) & A_{26}'(t) \\ A_{16}'(t) & A_{26}'(t) & A_{66}'(t) \end{bmatrix} \begin{Bmatrix} N_x \\ N_y \\ N_{xy} \end{Bmatrix} \tag{8.77}$$

where $\quad A_{ij}'(t) =$ laminate creep compliances
$N_x, N_y, N_{xy} =$ constant loads

Sims and Halpin [8.14] have used these equations, along with uniaxial creep tests, to determine the creep compliances of glass/epoxy laminates

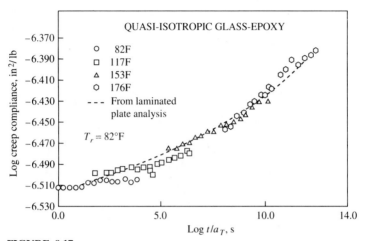

FIGURE 8.17
Predicted and measured creep compliance for a quasi-isotropic glass/epoxy laminate. (*From Sims and Halpin [8.14]. Copyright ASTM. Reprinted with permission.*)

for comparison with predictions. For example, the compliance $A'_{11}(t)$ was determined by applying a constant load N_x and by measuring the creep strain, $\epsilon_x^0(t)$, then using the equation

$$A'_{11}(t) = \frac{\epsilon_x^0(t)}{N_x} \qquad (8.78)$$

These measured values were compared with predicted values from a combined micromechanics-macromechanics analysis which was based on the use of the correspondence principle, the Halpin-Tsai equations, and the Classical Lamination Theory. The agreement between measurements and predictions is excellent, as shown in Fig. 8.17.

When the correspondence principle is used for problems involving sinusoidally varying stresses and strains in viscoelastic composites, we must be particularly careful to make sure that the criteria for using the effective modulus theory are met. These restrictions are discussed in more detail, and applications of the correspondence principle to the prediction of complex moduli of particle and fiber composites are given in papers by Hashin [8.15, 8.16]. For example, assuming that these criteria have been met, micromechanics equations such as Eq. (8.76) can be modified for the case of sinusoidal oscillations as

$$E_1^*(\omega) = E_{f1}^*(\omega)v_f + E_m^*(\omega)v_m \qquad (8.79)$$

where $E_1^*(\omega)$ = longitudinal complex modulus of composite
$E_{f1}^*(\omega)$ = longitudinal complex modulus of fiber
$E_m^*(\omega)$ = complex modulus of isotropic matrix

By setting the real parts of both sides of Eq. (8.79) equal, we find the composite longitudinal storage modulus to be

$$E_1'(\omega) = E_{f1}'(\omega)v_f + E_m'(\omega)v_m \tag{8.80}$$

where $E_1'(\omega)$ = longitudinal storage modulus of composite
$E_{f1}'(\omega)$ = longitudinal storage modulus of fiber
$E_m'(\omega)$ = storage modulus of isotropic matrix

Similarly, by setting the imaginary parts of both sides of Eq. (8.79) equal, we find that the composite longitudinal loss modulus is

$$E_1''(\omega) = E_{f1}''(\omega)v_f + E_m''(\omega)v_m \tag{8.81}$$

where $E_1''(\omega)$ = longitudinal loss modulus of composite
$E_{f1}''(\omega)$ = longitudinal loss modulus of fiber
$E_m''(\omega)$ = loss modulus of isotropic matrix

The composite longitudinal loss factor is found by dividing Eq. (8.81) by Eq. (8.80):

$$\eta_1(\omega) = \frac{E_1''(\omega)}{E_1'(\omega)} = \frac{E_{f1}''(\omega)v_f + E_m''(\omega)v_m}{E_{f1}'(\omega)v_f + E_m'(\omega)v_m} \tag{8.82}$$

The complex forms of the other lamina properties can be determined in a similar fashion. In studies of the complex moduli of aligned discontinuous fiber composites Suarez et al. [8.17] used the complex forms of Eqs. (6.17), (3.40), and (3.57) to determine $E_1^*(\omega)$, $v_{12}^*(\omega)$, $E_2^*(\omega)$, and $G_{12}^*(\omega)$, respectively. These properties were then substituted into the complex form of Eq. (2.39) to obtain the off-axis complex modulus, $E_x^*(\omega)$. The predicted off-axis storage moduli and loss factors for various fiber orientations are compared with experimental data for a continuous fiber graphite/epoxy composite in Fig. 8.18, and the agreement is seen to be quite reasonable. Similar results were obtained for discontinuous fiber composites, but the fiber length effect is dominated by the fiber orientation effect, except for fiber orientations of $\theta \simeq 0°$. It is also interesting to note that there is an optimum fiber orientation for maximizing the loss factor. Thus damping is another design variable in composite structures.

For oscillatory loading of symmetric viscoelastic laminates Eqs. (8.77) can be rewritten, so that the sinusoidally varying strains are related to the sinusoidally varying loads by

$$\begin{Bmatrix} \bar{\epsilon}_x^0(t) \\ \bar{\epsilon}_y^0(t) \\ \bar{\gamma}_{xy}^0(t) \end{Bmatrix} = \begin{bmatrix} A_{11}'^*(\omega) & A_{12}'^*(\omega) & A_{16}'^*(\omega) \\ A_{12}'^*(\omega) & A_{22}'^*(\omega) & A_{26}'^*(\omega) \\ A_{16}'^*(\omega) & A_{26}'^*(\omega) & A_{66}'^*(\omega) \end{bmatrix} \begin{Bmatrix} \tilde{N}_x(t) \\ \tilde{N}_y(t) \\ \tilde{N}_{xy}(t) \end{Bmatrix} \tag{8.83}$$

where the $A_{ij}'^*(\omega)$ are the laminate complex extensional compliances.

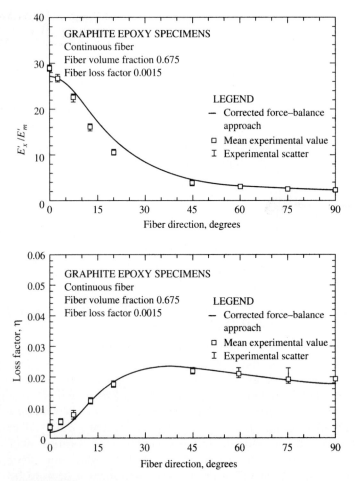

FIGURE 8.18
Predicted and measured off-axis storage modulus ratio, E_x'/E_m' and loss factor, η, of graphite/epoxy for various fiber orientations. (*From Suarez et al.* [8.17].)

The laminate stiffnesses can also be written in complex form [i.e., the $A_{ij}^*(\omega)$, $B_{ij}^*(\omega)$, and $D_{ij}^*(\omega)$], and the resulting equations have been used by Sun et al. [8.18] and others in studies of damping in laminates. Damping in composites will be discussed in more detail later in this chapter.

8.2.6 Temperature and Aging Effects

In the previous sections of this chapter the effects of temperature and aging on viscoelastic behavior have not been taken into account. We now consider these effects, as well as the corresponding methods of analysis.

It is convenient to discuss first the effects of temperature. In Sec. 8.2.2 a thermorheologically simple material was defined as having relaxation times, λ_i, and retardation times, ρ_i, which all have the same temperature dependence. Considering only the temperature dependence, the relaxation times at different temperatures can then be related by the equation

$$\lambda_i(T) = a_T \lambda_i(T_r) \qquad (8.84)$$

where $\lambda_i(T) = i$th relaxation time at temperature T
$\lambda_i(T_r) = i$th relaxation time at reference temperature, T_r
a_T = temperature-dependent shift factor

A similar equation can be used to express the temperature dependence of the retardation times. The effect of increasing temperature is to reduce the relaxation and retardation times and to speed up the relaxation and creep processes. This "speeding up" of the viscoelastic response can also be thought of as a process operating in "reduced time" [8.19]. For the purpose of illustration, we now consider the effect of the temperature-dependent relaxation times on the relaxation modulus by using the Zener single relaxation model in Fig. 8.8. The relaxation modulus at time, t, and temperature, T, is determined by modifying Eq. (8.39) as

$$C(t, T) = k_0 + k_1 e^{-t/\lambda_1(T)} \qquad (8.85)$$

whereas the relaxation modulus at time, t, and reference temperature, T_r, is

$$C(t, T_r) = k_0 + k_1 e^{-t/\lambda_1(T_r)} \qquad (8.86)$$

If we let the time at the reference temperature T_r be the "reduced time,"

$$\xi = t/a_T \qquad (8.87)$$

then Eq. (8.86) becomes

$$C(\xi, T_r) = k_0 + k_1 e^{-t/a_T\lambda_1(T_r)}$$
$$= k_0 + k_1 e^{-t/\lambda_1(T)}$$
$$= C(t, T) \qquad (8.88)$$

Thus, *the effect of changing temperature on the relaxation modulus is the same as the effect of a corresponding change in the time scale,* and this is the basis of the well-known time-temperature superposition (TTS) principle, or the method of reduced variables [8.20].

One of the most useful applications of TTS is to extend the time range of short-term creep or relaxation test data by taking such data at various temperatures and then shifting the data along the time axis to

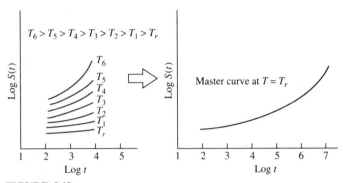

FIGURE 8.19
Shifting of creep data at various temperatures to generate a master curve at a reference temperature.

form a "master curve" at a reference temperature, as shown in Fig. 8.19. However, the usefulness of the method depends on the ability to determine the shift factor, a_T. When the temperature, T, is greater than the glass transition temperature, T_g, the shift factor can be determined empirically with the well-known Williams-Landel-Ferry (WLF) equation [8.20]

$$\log a_T = \frac{-c_1(T - T_r)}{c_2 + (T - T_r)} \qquad (8.89)$$

where c_1 and c_2 are material constants which must be determined from the experimental data. It has been found that when T_r is approximately 50°C above T_g the values $c_1 = 8.86$ and $c_2 = 101.6$ are valid for a variety of polymers.

TTS has been successfully applied to many polymers at temperatures above T_g, but it is a different matter below the glass transition. Although TTS has been shown to be suitable for short-term creep or relaxation data at those temperatures, it does not produce valid results when applied to long-term test data. The reason is that a process called "physical aging" occurs in a polymer below T_g, and this aging process changes the viscoelastic response of the material during a long-term creep test. In a short-term test, because the test duration is much less than the aging time, no significant aging occurs during the test. Physical aging in polymers is associated with a slow loss of free volume which has been trapped in the polymer microstructure after quenching below T_g [8.21]. As the polymer gives up free volume, the polymer chain mobility decreases and the relaxation times increase, thus reducing the speed of the relaxation or creep [8.6]. Pioneering work on aging of polymers has been done by Struik [8.21, 8.22], and more recent work has been reported by Janas and McCullough [8.23] and Ogale and McCullough [8.24]. Still more recently, Sullivan [8.25] has shown that physical aging significantly affects the creep behavior of polymer matrix composites.

Since aging time, t_a, and temperature, T, both affect the relaxation times, Sullivan [8.25] has suggested that new shift factor, $a(T, t_a)$, be defined by modifying Eq. (8.84) as

$$a = a(T, t_a) = \frac{\lambda_i(T, t_a)}{\lambda_i(T_r, t_{ar})} \qquad (8.90)$$

where $\lambda_i(T, t_a)$ = ith relaxation time at temperature, T, and aging time, t_a

$\lambda_i(T_r, t_{ar})$ = ith relaxation time at reference temperature, T_r, and reference aging time, t_{ar}

Struik [8.21] proposes that the TTS relationship for creep compliance be modified to include aging time effects by writing

$$S(t, T, t_a) = B(T)S(at, T_r, t_{ar}) \qquad (8.91)$$

where $B(T)$ = temperature-dependent vertical shift factor

$S(t, T, t_a)$ = creep compliance at time, t, temperature, T, and aging time, t_a

$S(at, T_r, t_{ar})$ = creep compliance at shifted time, at, reference temperature, T_r, and reference aging time, t_{ar}

Note that Eq. (8.91) is analogous to the TTS relationship for relaxation in Eq. (8.88), and a modified equation similar to Eq. (8.91) can be written for relaxation. This new shift factor may be related to $\bar{a}_T$, the temperature shift factor below T_g, and a_{ta}, the shift factor for aging time, by the equation [8.25]

$$\log a = \log \bar{a}_T + \log a_{ta} \qquad (8.92)$$

Figure 8.20 shows Sullivan's data on the effect of aging time on the shear creep compliance $S_{66}(t)$ of a glass/vinyl ester composite [8.25]. Clearly, the creep rate decreases with increased aging time, indicating an increase in the relaxation times and a slowing of the creep process. Support for the conclusion that TTS works well for short-term creep at constant age is provided by additional data from Sullivan [8.25] in Figs. 8.21 and 8.22. Figure 8.21 shows the short-term (or momentary) creep at various temperatures and "constant age," where the creep testing time is limited to no more than 10 percent of the aging time used in preconditioning the specimens. Figure 8.22 shows the corresponding momentary master curve at a reference temperature of 60°C. Both horizontal and vertical shifting of the momentary creep data was necessary to obtain the

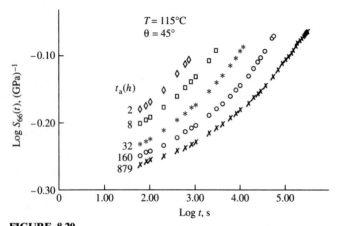

FIGURE 8.20
Effect of aging time, t_a, on shear creep compliance of 45° off-axis glass/vinyl ester composite at a test temperature of 115°C. (*From Sullivan* [8.25]. Reprinted *by permission of Elsevier Science Publishers, Ltd.*)

master curve [8.25]. The difference between long-term creep curves and the master curve from momentary creep data is shown in Fig. 8.23. Again, the conclusion is that aging slows down the creep process and that TTS does not work for long-term creep. Also shown in Fig. 8.23 are predicted long-term creep curves based on effective time theory [8.25], which is not discussed here.

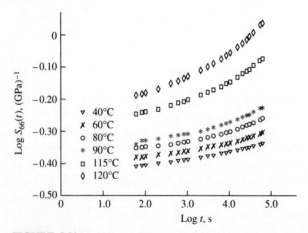

FIGURE 8.21
Momentary shear creep compliance data for glass/vinyl ester composite at various temperatures and constant aging time, $t_a = 166$ h. (*From Sullivan* [8.25]. *Reprinted by permission of Elsevier Science Publishers, Ltd.*)

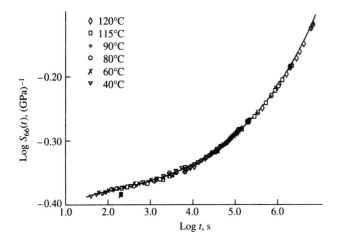

FIGURE 8.22
Momentary master curve for glass/vinyl ester composite at $t_a = 166$ h, $T_r = 60°C$, based on the test data from Fig. 8.21. (*From Sullivan* [8.25]. *Reprinted by permission of Elsevier Science Publishers, Ltd.*)

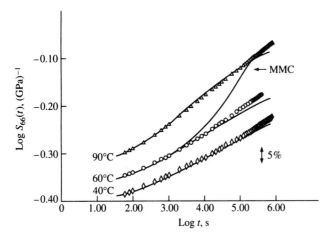

FIGURE 8.23
Long-term shear creep compliance and a momenetary master curve (MMC) for glass/vinyl ester composite, $t_a = 1$ h. Also shown are predicted curves from the effective time theory, which is not discussed here. (*From Sullivan* [8.25]. *Reprinted by permission of Elsevier Science Publishers, Ltd.*)

Example 8.5. The momentary master curve for the shear creep compliance, $S_{66}(t)$, of a unidirectional glass/vinyl ester composite at a reference temperature of 60°C and a reference aging time of 166 h is shown on a log-log scale in Fig. 8.22. (a) Neglecting aging effects, determine the time required to reach a compliance of 0.63 $(GPa)^{-1}$ at a temperature of 60°C, and (b) neglecting vertical shifting, estimate the time required to reach the same compliance of 0.63 $(GPa)^{-1}$ at a temperature of 100°C. From experimental data, it is known that for this material the WLF parameters are $c_1 = -1.01$ and $c_2 = -89.2$.

Solution.
(a) From Fig. 8.22 at a value of $\log S_{66}(t) = \log (0.63) = -0.2$, the corresponding value from the curve is $\log t = 6$, and so $t = 10^6$ s at $T = 60°C$.
(b) From Eq. (8.91), we have

$$S_{66}(t, T, t_a) = B(T)S_{66}(at, T_r, t_{ar})$$

Since we are neglecting vertical shifting, $B(T) = 1$. Since the data are for a constant aging time, we have $a = a_T$ and Eq. (8.91) becomes

$$S_{66}(t, T) = S_{66}(at, T_r) = S_{66}(a_T t, T_r)$$

where the shift factor, a_T, is found from the WLF equation

$$\log a_T = \frac{-c_1(T - T_r)}{c_2 + (T - T_r)} = \frac{-(-1.01)(100 - 60)}{-89.2 + (100 - 60)} = -0.8211$$

or $a_T = 0.151$, which means that $a_T t = 0.151(10^6) = 1.51 \times 10^5$ s. Thus, the creep compliance curve at 100°C is shifted to the left of the curve at the reference temperature of 60°C, and it takes only 15 percent as much time to reach the compliance of 0.63 $(GPa)^{-1}$ at 100°C as it does at 60°C.

8.3 DYNAMIC BEHAVIOR OF COMPOSITES

In this section the basic concepts of dynamic behavior of composites will be introduced by discussing wave propagation, vibration, and damping of specially orthotropic composites without coupling. Only one-dimensional wave propagation without dispersion, reflection, or refraction will be considered, as three-dimensional wave propagation, wave dispersion, and reflection/refraction effects are beyond the scope of this book. For detailed discussions of these topics the reader is referred to publications by Christensen [8.3], Hearmon [8.26], Achenbach [8.27], Ross and Sierakowski [8.28], and Moon [8.29]. Longitudinal vibrations of composite bars and flexural vibrations of composite beams and plates without coupling will also be considered. Vibrations of laminates with coupling and laminated plate boundary conditions other than simply supported will not be considered. These topics are discussed in detail in books by Whitney [8.30] and Vinson and Sierakowski [8.31]. The use of the Elastic-Viscoelastic Correspondence Principle and a strain energy method to analyze damping in composites will also be discussed.

The basic premise of all analyses presented in this section is that the criteria for valid use of the effective modulus theory have been met. That is, the scale of the inhomogeneity is assumed to be much smaller than the characteristic structural dimension and the characteristic wavelength of the dynamic stress distribution. Thus, all heterogeneous composite material properties are assumed to be effective properties of equivalent homogeneous materials. If the wavelength is not long in comparison with the scale of the inhomogeneity in the material, the wave shape is distorted as it travels through the material, and this is referred to as dispersion. Dispersion in composites has been discussed in several previous publications [8.3, 8.27, 8.29].

8.3.1 Longitudinal Wave Propagation and Vibrations in Specially Orthotropic Composite Bars

As shown in any vibrations book [8.32], longitudinal wave propagation and vibration in a homogeneous, isotropic, linear elastic bar (Fig. 8.24) are governed by the one-dimensional wave equation

$$\frac{\partial}{\partial x}\left(AE\frac{\partial u}{\partial x}\right) = \rho A\,\frac{\partial^2 u}{\partial t^2} \tag{8.93}$$

where x = distance from end of bar
   t = time
   $u = u(x, t)$ is the longitudinal displacement of a cross section in bar at a distance x and time t
   $A = A(x)$ is the cross-sectional area of bar
   ρ = mass density of bar
   $E = E(x)$ is the modulus of elasticity of bar

It is assumed that the displacement $u(x, t)$ is uniform across a given cross section. Using effective modulus theory for a heterogeneous, specially orthotropic, linear elastic composite bar, we simply replace the properties ρ and E with the corresponding effective properties of an equivalent homogeneous material. The effective modulus E then depends on the orientation of fibers relative to the axis of the bar. For fibers oriented along the x direction, $E = E_1$; for fibers oriented along the transverse direction, $E = E_2$; and for a specially orthotropic laminate we use the

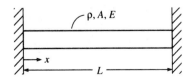

FIGURE 8.24
Bar of density ρ, cross-sectional area A, and length L.

effective laminate engineering constant $E = E_x$. If the area and the modulus are not functions of position, Eq. (8.93) reduces to

$$c^2 \frac{\partial^2 u}{\partial x^2} = \frac{\partial^2 u}{\partial t^2} \tag{8.94}$$

where $c = \left(\dfrac{E}{\rho}\right)^{1/2}$ is the wave speed.

The most common solutions to the one-dimensional wave equation are of the d'Alembert type or the separation of variables type. The d'Alembert solution is of the form

$$u(x, t) = p(x + ct) + q(x - ct) \tag{8.95}$$

The function $p(x + ct)$ represents a wave traveling to the left with velocity c. That is, a point located at $\xi = x + ct$ moves to the left with velocity c if ξ is a constant, since $x = \xi - ct$. Similarly, $q(x - ct)$ represents a wave traveling to the right with velocity c. For a sine wave we have

$$u(x, t) = A \sin \frac{2\pi}{\lambda}(x + ct) + A \sin \frac{2\pi}{\lambda}(x - ct) \tag{8.96}$$

where λ is the wavelength. Note that this is the wavelength, λ, that must be greater than the scale of the inhomogeneity, d, in order for the effective modulus theory to be valid. Alternatively, we can write Eq. (8.96) as

$$u(x, t) = A \sin(2\pi kx + \omega t) + A \sin(2\pi kx - \omega t) \tag{8.97}$$

where $k = 1/\lambda$ is the number, the number of waves per unit distance
$\omega = 2\pi c/\lambda$ is the frequency of wave

Using trigonometric identities, we find that

$$u(x, t) = 2A \sin 2\pi kx \cos \omega t \tag{8.98}$$

which represents a standing wave of profile $2A \sin 2\pi kx$, which oscillates with frequency ω. Generally, the combined wave motion in opposite directions is caused by reflections from the boundaries. Thus, wave propagation without reflection will not lead to a standing wave (or vibration).

A separation of variables solution is found by letting

$$u(x, t) = U(x)F(t) \tag{8.99}$$

where $U(x)$ is a function of x alone and $F(t)$ is a function of t alone. Substituting this solution in Eq. (8.94) and separating variables, we obtain

$$c^2 \frac{1}{U}\frac{d^2 U}{dx^2} = \frac{1}{F}\frac{d^2 F}{dt^2} \tag{8.100}$$

The left-hand side of Eq. (8.100) is a function of x alone and the right-hand side is a function of t alone; therefore, each side must be equal to a constant. If we let this constant be, say, $-\omega^2$, then Eq. (8.100) gives the two ordinary differential equations

$$\frac{d^2F}{dt^2} + \omega^2 F = 0 \qquad (8.101a)$$

$$\frac{d^2U}{dx^2} + \left(\frac{\omega}{c}\right)^2 U = 0 \qquad (8.101b)$$

and the solutions to these equations are of the form

$$F(t) = A_1 \sin \omega t + B_1 \cos \omega t \qquad (8.102)$$

$$U(x) = A_2 \sin \frac{\omega}{c} x + B_2 \cos \frac{\omega}{c} x \qquad (8.103)$$

where A_1 and B_1 depend on the initial conditions and A_2 and B_2 depend on the boundary conditions. For a bar which is fixed on both ends (Fig. 8.24) the substitution of the boundary conditons $u(0, t) = u(L, t) = 0$ leads to the conclusion that $B_2 = 0$ and

$$\sin \frac{\omega}{c} L = 0 \qquad (8.104)$$

Equation (8.104) is the eigenvalue equation, which has an infinite number of solutions, ω_n, such that

$$\frac{\omega_n L}{c} = n\pi \qquad (8.105)$$

where $\quad n =$ mode number $= 1, 2, 3, \ldots, \infty$
$\quad\quad \omega_n =$ eigenvalues, or natural frequencies (rad/s) $= 2\pi f_n$
$\quad\quad f_n =$ natural frequencies (Hz)

Thus,

$$f_n = \frac{nc}{2L} = \frac{n}{2L} \left(\frac{E}{\rho}\right)^{1/2} \qquad (8.106)$$

For the nth mode of vibration the displacements are then

$$u_n(x, t) = (A' \sin \omega_n t + B' \cos \omega_n t) \sin \frac{n\pi x}{L} \qquad (8.107)$$

where $A' = A_1 A_2$ and $B' = B_1 B_2$. The mode shape for the nth mode is given by the eigenfunction

$$U_n(x) = \sin \frac{n\pi x}{L} \qquad (8.108)$$

and the general solution is the superposition of all modal responses

$$u(x, t) = \sum_{n=1}^{\infty} (A' \sin \omega_n t + B' \cos \omega_n t) \sin \frac{n\pi x}{L} \qquad (8.109)$$

Mode shapes, natural frequencies, and wavelengths for the first three modes of the fixed-fixed bar are shown in Fig. 8.25. The most important point here is that as the mode number increases, the wavelength decreases and the use of effective modulus theory becomes more questionable. In general, the wavelengths associated with typical mechanical vibration frequencies of structures in the audio frequency range will satisfy the criteria $d \ll \lambda$. The wavelengths associated with ultrasonic wave propagation may be short enough to cause concern about the use of effective modulus theory, however.

The equations developed in this section are instructive not only from the point of view of the limitations of effective modulus theory, but for material characterization as well. The two basic approaches to

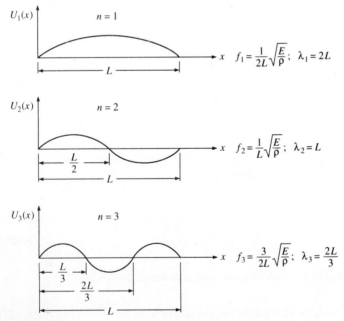

FIGURE 8.25
Mode shapes, natural frequencies, and wave lengths for the first three modes of longitudinal vibration of a bar with both ends fixed (Fig. 8.24).

measurement of dynamic mechanical properties of materials involve the use of either wave propagation experiments or vibration experiments. Assuming that the criteria for the use of effective modulus theory have been met, the effective modulus of a specially orthotropic composite can be determined by measuring the longitudinal wave speed, c, in a specimen of density, ρ, and then solving for $E = c^2\rho$. Alternatively, the nth mode natural frequency, f_n, can be measured in a vibration experiment, and the effective modulus can be found from an equation like Eq. (8.106). Dynamic mechanical testing of composites will be discussed in more detail in Chap. 10.

Finally, the equations presented here can be modified for linear viscoelastic composites in sinusoidal vibration by using the Elastic Viscoelastic Correspondence Principle. This means that the effective modulus E will be replaced by the complex modulus, $E^*(\omega)$. Alternatively, the stress-strain relationship used in deriving the equation of motion could be an equation of the form shown in Eq. (8.23) or a special case of that equation.

8.3.2 Flexural Vibration of Composite Beams

Transverse, or flexural, motion of a homogeneous, isotropic, linear elastic beam (Fig. 8.26) without shear or rotary inertia effects is described by the well-known Bernoulli-Euler equation

$$-\frac{\partial^2}{\partial x^2}\left(EI\frac{\partial^2 w}{\partial x^2}\right) = \rho A \frac{\partial^2 w}{\partial t^2} \tag{8.110}$$

where $I =$ moment of inertia of cross section about centroidal axis of beam
  $w = w(x, t)$ is the transverse displacement of centroidal axis of beam

and $x, t, \rho, A,$ and E are as defined in Eq. (8.93). If the beam is such that EI is constant along the length, Eq. (8.110) reduces to

$$EI\frac{\partial^4 w}{\partial x^4} + \rho A \frac{\partial^2 w}{\partial t^2} = 0 \tag{8.11}$$

Assuming that the criteria for the use of effective modulus theory have

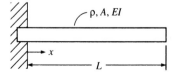

FIGURE 8.26
Cantilever beam for the Bernoulli-Euler beam theory.

been met, these equations can be used for specially orthotropic composites or laminates without coupling if the modulus E is replaced by the effective flexural modulus, E_f. Recall that, depending on the laminate configuration, E_f may be found from equations like Eqs. (7.8), (7.9), (7.68), or (7.69). For laminates with coupling the equations of motion based on the Classical Lamination Theory will be developed in the next section.

As an example of a solution of the Bernoulli-Euler equation, consider a separation of variables solution for harmonic free vibration:

$$w(x, t) = W(x)e^{i\omega t} \qquad (8.112)$$

where ω = frequency
$W(x)$ = mode shape function

Substitution of this solution in Eq. (8.111) yields the equation

$$\frac{d^4W(x)}{dx^4} - k^4W(x) = 0 \qquad (8.113)$$

where $k = \left(\dfrac{\omega^2 \rho A}{EI}\right)^{1/4}$.

The solution for Eq. (8.113) is of the form

$$W(x) = C_1 \sin kx + C_2 \cos kx + C_3 \sinh kx + C_4 \cosh kx \qquad (8.114)$$

where the constants C_1, C_2, C_3, and C_4 depend on the boundary conditions. For example, for a cantilever beam (Fig. 8.26) the four boundary conditions yield the following relationships:

$$W(x) = 0 \quad \text{when } x = 0; \quad \text{therefore, } C_2 = -C_4$$

$$\frac{dW(x)}{dx} = 0 \quad \text{when } w = 0, \quad \text{therefore, } C_1 = -C_3$$

$$\frac{d^2W(x)}{dx^2} = 0 \quad \text{when } x = L;$$

$$\text{therefore, } C_1(\sin kL + \sinh kL) + C_2(\cos kL + \cosh kL) = 0$$

$$\frac{d^3W(x)}{dx^3} = 0 \quad \text{when } x = L;$$

$$\text{therefore, } C_1(\cos kL + \cosh kL) + c_2(\sin kL - \sinh kL) = 0$$

For nontrivial solutions C_1 and C_2 in the last two equations the determinant of the coefficients must be equal to zero and

$$\cos kL \cosh kL + 1 = 0 \qquad (8.115)$$

This is the eigenvalue equation for the cantilever beam, which has an infinite number of solutions, $k_n L$. The subscript n refers to the mode number. The eigenvalues for the first three modes are

$$k_1 L = 1.875, \qquad k_2 L = 4.694, \qquad k_3 L = 7.855 \qquad (8.116)$$

Substituting the eigenvalues in the definition of k [see Eq. (8.113)], rearranging, and using the relationship $\omega = 2\pi f$, we have the frequency equation

$$f_n = \frac{(k_n L)^2}{2\pi L^2}\left(\frac{EI}{\rho A}\right)^{1/2} \qquad (8.117)$$

The mode shape function for the nth mode is then

$$W_n(x) = C_2[\cos k_n x - \cosh k_n x + \sigma_n(\sin k_n x - \sinh k_n x)] \qquad (8.118)$$

where
$$\sigma_n = \frac{\sin k_n L - \sinh k_n L}{\cos k_n L + \cosh k_n L}$$

The mode shapes and frequencies for the first three modes of the cantilever beam are shown in Fig. 8.27. The effect of increasing the mode number and the corresponding reduction in wavelength is again apparent.

If transverse shear and rotary inertia effects are included in the derivation of the equation of motion for transverse vibration of a beam,

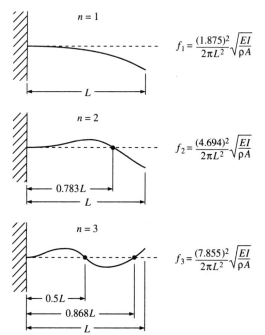

FIGURE 8.27
Mode shapes and natural frequencies for the first three modes of flexural vibration of the cantilever beam in Fig. 8.26.

the result is the well-known Timoshenko beam equation [8.33]

$$EI \frac{\partial^4 w}{\partial x^4} + \rho A \frac{\partial^2 w}{\partial t^2} + \frac{J\rho}{FG} \frac{\partial^4 w}{\partial t^4} - \left(J + \frac{EI\rho}{FG}\right) \frac{\partial^4 w}{\partial x^2 \partial t^2} = 0 \qquad (8.119)$$

where J = rotary inertia per unit length
        F = shape factor for cross section
        G = shear modulus

This equation can also be used for specially orthotropic composites and laminates without coupling by replacing E and G with the effective flexural modulus, E_f, and the effective through-the-thickness shear modulus, respectively, for the composite. For example, for a unidirectional, transversely isotropic composite with the fibers along the beam axis the appropriate shear modulus to use is $G = G_{13} = G_{12}$. If the fibers are oriented in the transverse direction, $G = G_{23}$. Both shear and rotary inertia effects become more important as the mode number increases, and both effects reduce the natural frequencies below the Bernoulli-Euler values. The beam length-to-thickness ratio, L/h, is an important factor in the determination of the shear effect, with decreasing L/h generating increased shear effects. It appears that for highly anisotropic composite beams shear effects may be significant unless L/h is greater than about 100 [8.34]. The transverse shear effect is also strongly dependent on the ratio E/G, which is much greater for composite beams than for isotropic beams. For a typical isotropic metal $E/G \simeq 2.6$, but for composites such as unidirectional graphite/epoxy $E_1/G_{12} \simeq 20$ or higher. Sandwich beams with foam or honeycomb cores have even higher E/G ratios due to the low shear stiffness of the core and are very susceptible to transverse shear effects.

As in the previous section, the equations developed here can be used in dynamic mechanical testing to determine the effective moduli of a composite specimen. The equations can also be converted to linear viscoelastic form by replacing the elastic moduli with the corresponding complex moduli, or by deriving the equation of motion from a viscoelastic stress-strain relationship. More sophisticated analytical models for vibrating composite beams including various effects such as viscoelastic behavior, transverse shear, and bending-twisting coupling have been developed [8.30, 8.31, 8.35–8.37], but these are beyond the scope of this book.

Example 8.6. For a symmetric laminated beam having a rectangular cross section of width b and thickness h, determine (a) the equation of motion for free vibration and (b) the natural frequencies. Assume that the criteria for use of the effective modulus theory have been met.

Solution.

(*a*) Substituting the flexural modulus, E_{fx}, from Eq. (7.68) in the expression for *EI*, we find that

$$EI = E_{fx}I = \frac{12}{h^3 D_{11}'} \frac{bh^3}{12} = \frac{b}{D_{11}'}$$

(Note that *h* is used to denote thickness here since *t* is used for time.) Thus, the Bernoulli-Euler beam equation [Eq. (8.111)] becomes

$$\frac{b}{D_{11}'} \frac{\partial^4 w}{\partial x^4} + \rho A \frac{\partial^2 w}{\partial t^2} = 0$$

(*b*) The natural frequencies are then found from Eq. (8.117):

$$f_n = \frac{(k_n L)^2}{2\pi L^2} \left(\frac{b}{D_{11}' \rho A}\right)^{1/2}$$

where the eigenvalues, k_n, depend on the boundary conditions.

8.3.3 Transverse Vibration of Laminated Plates

Although the equations for vibration of composite beams in the previous section are useful, they are limited to laminates without coupling. The more general equations of motion for transverse vibration of a laminated plate can be derived by modifying the static equilibrium equations which were developed for the analysis of static deflections of laminated plates in Sec. 7.9. For example, according to Newton's second law, Eq. (7.119) must now be modified, so that the summation of forces along the *x* direction in Fig. 7.35 is given by

$$N_x \, dy + \frac{\partial N_x}{\partial x} dx \, dy + N_{xy} \, dx + \frac{\partial N_{xy}}{\partial y} dx \, dy$$

$$- N_x \, dy - N_{xy} \, dx = \rho_0 \, dx \, dy \frac{\partial^2 u^0}{\partial t^2} \quad (8.120)$$

where ρ_0 = mass per unit area of laminate is ρh
ρ = mass density of laminate is the mass per unit volume
h = thickness of laminate (since *t* is used for time here)
$u^0 = u^0(x, y, t)$ is the middle surface displacement in the *x* direction

Equation (8.120) may be simplified as

$$\frac{\partial N_x}{\partial x} + \frac{\partial N_{xy}}{\partial y} = \rho_0 \frac{\partial^2 u^0}{\partial t^2} \quad (8.121)$$

Similarly, the summation of forces along the y direction yields

$$N_y \, dx + \frac{\partial N_x}{\partial y} \, dx \, dy + N_{xy} \, dy + \frac{\partial N_{xy}}{\partial x} \, dx \, dy$$

$$- N_y \, dx - N_{xy} \, dy = \rho_0 \, dx \, dy \frac{\partial^2 v^0}{\partial t^2} \qquad (8.122)$$

or
$$\frac{\partial N_y}{\partial y} + \frac{\partial N_{xy}}{\partial x} = \rho_0 \frac{\partial^2 v^0}{\partial t^2} \qquad (8.123)$$

where $v^0 = v^0(x, y, t)$ is the middle surface displacement in the y direction. The summation of forces along the z direction gives

$$Q_x \, dy + \frac{\partial Q_x}{\partial x} \, dx \, dy + Q_y \, dx + \frac{\partial Q_y}{\partial y} - Q_x \, dy$$

$$- Q_y \, dx + q(x, y) = \rho_0 \frac{\partial^2 w}{\partial t^2} \qquad (8.124)$$

or
$$\frac{\partial Q_x}{\partial x} + \frac{\partial Q_y}{\partial y} + q(x, y) = \rho_0 \frac{\partial^2 w}{\partial t^2} \qquad (8.125)$$

where $w = w(x, y, t)$ is the displacement in the z direction.

For moment equilibrium we consider the moments about the x axis and the y axis while neglecting rotary inertia. Thus, the summation of moments about the x axis gives

$$- M_y \, dx - \frac{\partial M_y}{\partial y} \, dy \, dx - M_{xy} \, dy - \frac{\partial M_{xy}}{\partial x} \, dx \, dy + Q_y \, dx \, dy$$

$$+ \frac{\partial Q_y}{\partial y} \, dy \, dx \, dy + q(x, y) \, dx \, dy \, dy/2 + Q_x \, dy \, dy/2 + \frac{\partial Q_x}{\partial x} \, dx \, dy \, dy/2$$

$$+ M_y \, dx + M_{xy} \, dy - Q_x \, dy \, dy/2 = 0 \qquad (8.126)$$

Simplifying and neglecting products of differentials, we get

$$\frac{\partial M_y}{\partial y} + \frac{\partial M_{xy}}{\partial x} = Q_y \qquad (8.127)$$

A similar summation of moments about the y axis gives

$$\frac{\partial M_x}{\partial x} + \frac{\partial M_{xy}}{\partial y} = Q_x \qquad (8.128)$$

Substitution of Eqs. (8.127) and (8.128) in Eq. (8.125) yields

$$\frac{\partial^2 M_x}{\partial x^2} + 2 \frac{\partial^2 M_{xy}}{\partial x \, \partial y} + \frac{\partial^2 M_y}{\partial y^2} + q(x, y) = \rho_0 \frac{\partial^2 w}{\partial t^2} \qquad (8.129)$$

Equations (8.121), (8.123), and (8.129) are the differential equations of motion of the plate in terms of stress and moment resultants. The corresponding equations of motion in terms of displacements can be derived by substituting the laminate force-deformation equations (7.41), the strain-displacement relations (7.29), and the curvature-displacement equations (7.30) in Eqs. (8.121), (8.123), and (8.129). The resulting equations are

$$
A_{11}\frac{\partial^2 u^0}{\partial x^2} + 2A_{16}\frac{\partial^2 u^0}{\partial x\,\partial y} + A_{66}\frac{\partial^2 u^0}{\partial y^2} + A_{16}\frac{\partial^2 v^0}{\partial x^2} + (A_{12}+A_{66})\frac{\partial^2 v^0}{\partial x\,\partial y}
$$

$$
+ A_{26}\frac{\partial^2 v^0}{\partial y^2} - B_{11}\frac{\partial^3 w}{\partial x^3} - 3B_{16}\frac{\partial^3 w}{\partial x^2\,\partial y} - (B_{12}+2B_{66})\frac{\partial^3 w}{\partial x\,\partial y^2}
$$

$$
- B_{26}\frac{\partial^3 w}{\partial y^3} = \rho_0\frac{\partial^2 u^0}{\partial t^2} \quad (8.130)
$$

$$
A_{16}\frac{\partial^2 u^0}{\partial x^2} + (A_{12}+A_{66})\frac{\partial^2 u^0}{\partial x\,\partial y} + A_{26}\frac{\partial^2 u^0}{\partial y^2} + A_{66}\frac{\partial^2 v^0}{\partial x^2}
$$

$$
+ 2A_{26}\frac{\partial^2 v^0}{\partial x\,\partial y} + A_{22}\frac{\partial^2 v^0}{\partial y^2} - B_{16}\frac{\partial^3 w}{\partial x^3} - (B_{12}+2B_{66})\frac{\partial^3 w}{\partial x^2\,\partial y}
$$

$$
- 3B_{26}\frac{\partial^3 w}{\partial x\,\partial y^2} - B_{22}\frac{\partial^3 w}{\partial y^3} = \rho_0\frac{\partial^2 v^0}{\partial t^2} \quad (8.131)
$$

$$
D_{11}\frac{\partial^4 w}{\partial x^4} + 4D_{16}\frac{\partial^4 w}{\partial x^3\,\partial y} + 2(D_{12}+2D_{66})\frac{\partial^4 w}{\partial x^2\,\partial y^2} + 4D_{26}\frac{\partial^4 w}{\partial x\,\partial y^3}
$$

$$
+ D_{22}\frac{\partial^4 w}{\partial y^4} - B_{11}\frac{\partial^3 u^0}{\partial x^3} - 3B_{16}\frac{\partial^3 u^0}{\partial x^2\,\partial y} - (B_{12}+2B_{66})\frac{\partial^3 u^0}{\partial x\,\partial y^2}
$$

$$
- B_{26}\frac{\partial^3 u^0}{\partial y^3} - B_{16}\frac{\partial^3 v^0}{\partial x^3} - (B_{12}+2B_{66})\frac{\partial^3 v^0}{\partial x^2\,\partial y} - 3B_{26}\frac{\partial^3 v^0}{\partial x\,\partial y^2}
$$

$$
- B_{22}\frac{\partial^3 v^0}{\partial y^3} + \rho_0\frac{\partial^2 w}{\partial t^2} = q(x, y) \quad (8.132)
$$

As with the static case in Sec. 7.9, the in-plane displacements, u^0 and v^0, are coupled with the transverse displacements, w, when the B_{ij} are present. For symmetric laminates with $B_{ij} = 0$ Eq. (8.132) alone becomes the governing equation for transverse displacements. These governing partial differential equations must be solved subject to the appropriate boundary conditions. As in the static case, when the in-plane displacements are coupled with the transverse displacements, the boundary conditions must be a combination of boundary conditions for a planar theory of elasticity problem and boundary conditions for a plate bending

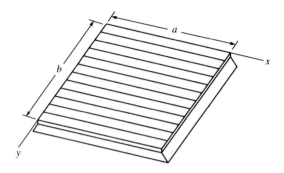

FIGURE 8.28
Simply supported, specially orthotropic plate for free transverse vibration analysis.

problem. In this section we will only discuss transverse vibrations according to Eq. (8.132) with all $B_{ij} = 0$ and the transverse distributed load $q(x, y) = 0$.

Let us now consider the case of free transverse vibration of the rectangular, specially orthotropic plate which is simply supported on all edges, as shown in Fig. 8.28. The discussion here follows the analysis of Whitney [8.30]. For a specially orthotropic plate all $B_{ij} = 0$, $A_{16} = A_{26} = D_{16} = D_{26} = 0$, and Eq. (8.132) becomes

$$D_{11}\frac{\partial^4 w}{\partial x^4} + 2(D_{12} + 2D_{66})\frac{\partial^4 w}{\partial x^2\,\partial y^2} + D_{22}\frac{\partial^4 w}{\partial y^4} + \rho_0\frac{\partial^2 w}{\partial t^2} = 0 \quad (8.133)$$

For free harmonic vibration at frequency, ω, we can assume that

$$w(x, y, t) = W(x, y)e^{i\omega t} \quad (8.134)$$

where $W(x, y)$ is a mode shape function. Substituting Eq. (8.134) in Eq. (8.133), we have

$$D_{11}\frac{\partial^4 W}{\partial x^4} + 2(D_{12} + 2D_{66})\frac{\partial^4 W}{\partial x^2\,\partial y^2} + D_{22}\frac{\partial^4 W}{\partial y^4} - \rho_0\omega^2 W = 0 \quad (8.135)$$

For the simply supported boundary condition the transverse displacements and bending moments must vanish at the edges as in the static case. Thus, from Eqs. (7.135) and (7.136), we have, again,

along $x = 0$ and $x = a$,
$$W(x, y) = 0$$

and
$$M_x = -D_{11}\frac{\partial^2 W}{\partial x^2} - D_{12}\frac{\partial^2 W}{\partial y^2} = 0 \quad (8.136)$$

and along $y = 0$ and $y = b$,
$$W(x, y) = 0$$

and
$$M_y = -D_{12}\frac{\partial^2 W}{\partial x^2} - D_{22}\frac{\partial^2 W}{\partial y^2} = 0 \quad (8.137)$$

It can be shown that the equation of motion and the boundary conditions are satisfied by solutions of the form

$$W(x, y) = A_{mn} \sin \frac{m\pi x}{a} \sin \frac{n\pi y}{b} \qquad (8.138)$$

where m and n are mode indices which refer to the number of half wavelengths along the x and y directions, respectively, for mode mn, and a and b are the plate dimensions along the x and y directions, respectively. Substitution of Eq. (8.138) in Eq. (8.135) yields the frequency equation

$$\omega_{mn}^2 = \frac{\pi^4}{\rho_0 a^4} [D_{11} m^4 + 2(D_{12} + 2D_{66})(mnR)^2 + D_{22}(nR)^4)] \qquad (8.139)$$

where the plate aspect ratio $R = a/b$ and ω_{mn} is the natural frequency for mode mn [8.30]. For the fundamental mode, where $m = n = 1$, the natural frequency is given by

$$\omega_{11}^2 = \frac{\pi^4}{\rho_0} \left[\frac{D_{11}}{a^4} + \frac{2(D_{12} + 2D_{66})}{a^2 b^2} + \frac{D_{22}}{b^4} \right] \qquad (8.140)$$

and the mode shape function is given by

$$W(x, y) = \sin \frac{\pi x}{a} \sin \frac{\pi y}{b} \qquad (8.141)$$

We now consider numerical results given by Whitney [8.30] for frequencies and mode shapes of two square plates. One plate is orthotropic with $D_{11}/D_{22} = 10$ and $(D_{12} + 2D_{66})/D_{22} = 1$; the other is isotropic with $D_{11}/D_{22} = 1$ and $(D_{12} + 2D_{66})/D_{22} = 1$. The four lowest natural frequencies for the two plates are compared in Table 8.2 and the

TABLE 8.2
Predicted natural frequencies for the first four modes of simply supported plates made of specially orthotropic and isotropic materials

Mode	Orthotropic $\omega = k\pi^2/b^2 \sqrt{D_{22}/\rho_0}$			Isotropic $\omega = k\pi^2/b^2 \sqrt{D/\rho_0}$		
	m	n	k	m	n	k
1st	1	1	3.62	1	1	2.0
2nd	1	2	5.68	1	2	5.0
3rd	1	3	10.45	2	1	5.0
4th	2	1	13.0	2	2	8.0

Source: From Whitney [8.30].

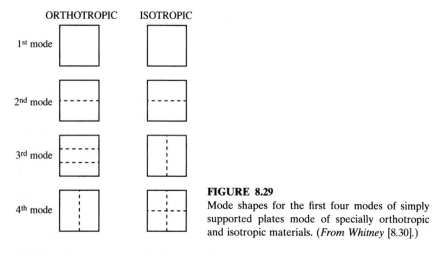

FIGURE 8.29
Mode shapes for the first four modes of simply supported plates mode of specially orthotropic and isotropic materials. (*From Whitney* [8.30].)

corresponding mode shapes are compared in Fig. 8.29. The dotted lines in Fig. 8.29 denote the nodal lines of zero displacement for a particular mode. It is interesting to note that, in order of increasing frequency, the sequence of mode numbers is different for the isotropic and orthotropic plates. Due to the high stiffness of the orthotropic plate along the x direction, its frequencies are higher than the corresponding isotropic plate frequencies. It is also interesting to note that for the isotropic plate $\omega_{12} = \omega_{21}$, but for the orthotropic plate $\omega_{21} > \omega_{12}$.

As with the static case, it is generally not possible to find exact mode shape functions similar to those given by Eq. (8.138) for boundary conditions such as clamped edges or free edges. For such cases approximate solutions must be derived using approaches such as the Rayleigh-Ritz method or the Galerkin method. For more detailed discussions of these methods the reader is referred to books by Whitney [8.30] and Vinson and Sierakowski [8.31].

The equation of motion for a specially orthotropic, laminated beam is found by reducing Eq. (8.133) to the one-dimensional form

$$D_{11} \frac{\partial^4 w}{\partial x^4} + \rho_0 \frac{\partial^2 w}{\partial t^2} = 0 \tag{8.142}$$

If we substitute $\rho_0 = \rho h$, and if we multiply Eq. (8.142) by the beam width, b, we have

$$b D_{11} \frac{\partial^4 w}{\partial x^4} + \rho b h \frac{\partial^2 w}{\partial t^2} = 0 \tag{8.143}$$

For the one-dimensional case $D_{11} = 1/D'_{11}$, and since $bh = A$, we have

$$\frac{b}{D'_{11}} \frac{\partial^4 w}{\partial x^4} + \rho A \frac{\partial^2 w}{\partial t^2} = 0 \tag{8.144}$$

which is the same as the equation that was derived from the beam theory earlier in Example 8.6.

Example 8.7. A unidirectional AS/3501 graphite/epoxy plate is simply supported on all four edges. The plate is 300 mm by 300 mm square, 2 mm thick, and has a mass density of 1.6 mg/mm³. Determine the frequency of the fundamental mode of the plate.

Solution. Using the lamina stiffnesses, Q_{ij}, from Example 7.3 and the thickness of 2 mm in Eqs. (7.40) for a laminate consisting of a single orthotropic lamina, we find the laminate bending stiffnesses to be

$$D_{11} = 92.53 \text{ GPa-mm}^3 \qquad D_{12} = 1.813 \text{ GPa-mm}^3$$
$$D_{22} = 6.03 \text{ GPa-mm}^3 \qquad D_{66} = 4.6 \text{ GPa-mm}^3$$

The mass per unit area is

$$\rho_0 = \rho h = (1.6 \text{ mg/mm}^3)(2 \text{ mm}) = 3.2 \text{ mg/mm}^2 = 0.0032 \text{ g/mm}^2$$

The fundamental frequency is then found from Eq. (8.140) as

$$\omega_{11}^2 = \frac{\pi^4}{(0.0032)(300)^4}[92.53 + 2(1.813 + 2(4.6)) + 6.03](10^9) = 4.53(10^5) \text{ rad}^2/\text{s}^2$$

or

$$\omega_{11} = 673 \text{ rad/s}$$

(*Note:* GPa-mm³ = 10^9 g-mm²/s² in the above equation.)

8.3.4 Analysis of Damping in Composites

Damping is simply the dissipation of energy during dynamic deformation. As structures and machines are pushed to higher and higher levels of precision and performance, and as the control of noise and vibration becomes more of a societal concern, it is becoming essential to take damping into account in the design process. In conventional metallic structures it is commonly accepted that much of the damping comes from friction in structural joints or from add-on surface damping treatments because the damping in the metal itself is typically very low. On the other hand, polymer composites have generated increased interest in the development of highly damped, lightweight, structural composites because of their good damping characteristics and the inherent design flexibility which allows trade-offs between such properties as damping and stiffness. The purpose of this section is to give a brief overview of the analysis of linear viscoelastic damping in composites. Dynamic mechanical testing of composites, which includes experimental determination of damping, will be discussed in Chap. 10. More detailed treatments of damping in composites are given in publications by Gibson [8.38–8.40], Bert [8.41], Adams [8.42], Chaturvedi [8.43], and Kinra and Wolfenden [8.44].

As described in Sec. 8.1, damping is one of the important physical manifestations of viscoelastic behavior in dynamically loaded structural materials, and the stress-strain hysteresis loop in Fig. 8.1(c) is typical of damped response under cyclic loading. Viscoelastic behavior of fiber and/or matrix materials is not the only mechanism for structural damping in composite materials although it does appear to be the dominant mechanism in undamaged polymer composites vibrating at small amplitudes. Other damping mechanisms include thermoelastic damping due to cyclic heat flow, coulomb friction due to slip in unbonded regions of the fiber/matrix interface, and energy dissipation at sites of cracks and/or delaminations [8.38]. Thermoelastic damping is generally more important for metal composites than for polymer composites. Damping due to poor interface bonding, cracks, and/or delaminations cannot be relied upon in the design of structures, but the measurement of such damping may be the basis of a valuable nondestructive evaluation methodology [8.39].

In order to understand linear viscoelastic damping better, it is important to recognize the relationship between the time scale of the applied deformation and the internal time scale of the material. The time scale for cyclic deformation is determined by the oscillation frequency, ω. Recall that the relaxation times, λ_i, or retardation times, ρ_i, are measures of the internal time scale of the material. We will now use the Zener single relaxation model to illustrate how damping depends on the relationship between these two time scales.

For sinusoidal oscillation of the Zener single relaxation model [Fig. 8.8(a)] we can write

$$\sigma = \sigma_0 e^{i\omega t} = (E' + iE'')\epsilon \tag{8.145}$$

where σ = stress
σ_0 = stress amplitude
ϵ = strain
ω = frequency
E' = storage modulus is $E'(\omega)$
E'' = loss modulus is $E''(\omega)$
i = imaginary operator, which is $(-1)^{1/2}$

Substituting Eq. (8.145) in the stress-strain relationship for the Zener model [Eq. (8.37)] and separating into real and imaginary parts, we find that

$$E' = E'(\omega) = \frac{k_0 + (k_0 + k_1)\omega^2\lambda_1^2}{1 + \omega^2\lambda_1^2} \tag{8.146}$$

$$E'' = E''(\omega) = \frac{\omega\lambda_1 k_1}{1 + \omega^2\lambda_1^2} \tag{8.147}$$

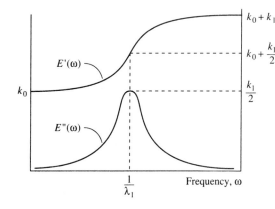

FIGURE 8.30
Variation of storage modulus, $E'(\omega)$, and loss modulus, $E''(\omega)$, with frequency for the Zener single relaxation model.

and

$$\eta = \eta(\omega) = \frac{E''(\omega)}{E'(\omega)} = \frac{\omega \lambda_1 k_1}{k_0 + (k_0 + k_1)\omega^2 \lambda_1^2} \qquad (8.148)$$

where $\lambda_1 = \mu_1 / k_1$ is the relaxation time from Eq. (8.39)
 $\eta = \eta(\omega)$ is the loss factor

The variations of E' and E'' with frequency, ω, are shown schematically in Fig. 8.30. Note that when the frequency is the reciprocal of the relaxation time, $\omega = 1/\lambda_1$, the loss modulus peaks and the storage modulus passes through a transition region. Such damping peaks in the frequency domain are often referred to as "relaxation peaks." The loss factor has a peak at a different frequency, not shown in Fig. 8.30 because the relative position of that peak depends on the numerical values of the parameters. But the important point is that the dissipation of energy, whether characterized by the loss modulus or the loss factor, is maximized when the time scale of the deformation is the same as the internal time scale of the material. If the two time scales are substantially different, the energy dissipation is reduced. For example, notice in Fig. 8.30 that $E'' \rightarrow 0$ as $\omega \rightarrow 0$ and as $\omega \rightarrow \infty$. This behavior is typical for viscoelastic materials, but, as mentioned earlier, the actual transitions occur over a wider range (in this case a wider frequency range) than the single relaxation model produces. Thus, as before, an improved Zener model (Fig. 8.10 or 8.12) with a distribution of relaxation times makes it possible to extend the range of the relaxation to approximate the actual behavior better.

Analytical models have been developed for predicting damping in composites at both the micromechanical and macromechanical levels. Only in certain special cases, such as thermoelastic damping [8.45] or dislocation damping [8.46] in metals, can the damping be predicted from first principles without knowledge of constituent material damping properties. (These damping mechanisms will not be discussed here.) If the damping mechanism is of the linear viscoelastic type, there are two basic approaches to the development of analytical models, both of which are based on the existence of experimental damping data for constituent materials. The two approaches are as follows:

1. The use of the Elastic-Viscoelastic Correspondence Principle in combination with elastic solutions from the mechanics of materials or the elasticity theory
2. The use of a strain energy formulation which relates the total damping in the structure to the damping of each element and the fraction of the total strain energy stored in that element

The basis of the first approach is that linear elastostatic analyses can be converted to vibratory linear viscoelastic analyses by replacing static stresses and strains with the corresponding vibratory stresses and strains, and by replacing the elastic moduli or compliances with the corresponding complex moduli or compliances, respectively. According to this procedure, the elastostatic stress-strain relationships in Eqs. (2.5) would be converted to the viscoelastic vibratory Eqs. (8.65) and Eqs. (2.3) would be converted to Eqs. (8.66), as described in Sec. 8.2.5. The use of this approach to derive the micromechanics equation for the longitudinal loss factor of a unidirectional composite [Eq. (8.82)] has already been demonstrated. The same approach has been used to derive micromechanics equations for the prediction of damping in aligned discontinuous fiber composites having various fiber aspect ratios and fiber orientations [8.17, 8.47] and in randomly oriented short fiber composites [8.48].

The correspondence principle has also been used in combination with the Classical Lamination Theory to develop equations for the laminate loss factors [8.18]. For example, the extensional loss factors for a laminate can be expressed in terms of the real and imaginary parts of the corresponding laminate extensional stiffnesses

$$\eta_{ij}^{(A)} = \frac{A_{ij}''}{A_{ij}'} \tag{8.149}$$

Similar equations can be used to describe laminate coupling and flexural loss factors [8.18]. The major limitation of such analyses is that the Classical Lamination Theory neglects interlaminar stresses, so that interlaminar damping is not included. As shown later in this section, a more general three-dimensional analysis including interlaminar damping may be developed by using a strain energy method.

Although sinusoidally varying stresses and strains were assumed in the development of the complex modulus notation in Sec. 8.2.4, it has been shown that as long as the stiffness and damping show some frequency dependence, the complex modulus notation is also valid for the more general nonsinusoidal case [8.49]. Anomalous analytical results such as noncausal response can occur if the components of the complex modulus are independent of frequency. Composite materials (particularly polymer composites) generally have frequency-dependent complex moduli, however.

The second approach involves the use of a strain energy relationship which was first presented in 1962 by Ungar and Kerwin [8.50]. Ungar and Kerwin found that for an arbitrary system of linear viscoelastic elements the system loss factor can be expressed as a summation of the products of the individual element loss factors and the fraction of the total strain energy stored in each element:

$$\eta = \frac{\sum\limits_{i=1}^{n} \eta_i W_i}{\sum\limits_{i=1}^{n} W_i} \tag{8.150}$$

where η_i = loss factor for the ith element in system
   W_i = strain energy stored in the ith element at maximum vibratory displacement
   n = total number of elements in system

When applying this equation to composite damping analysis, the composite becomes the "system," and the nature of the elements depends on whether the analysis is micromechanical or macromechanical. For example, this equation has been used in combination with mechanics of materials solutions for the strain energy of aligned discontinuous fiber composites [8.51]. In this analysis the damping in the fiber was neglected (i.e., the fiber loss factor $\eta_f = 0$), so that the longitudinal loss factor of the aligned discontinuous fiber composite was approximated by the following form of Eq. (8.150):

$$\eta_1 = \frac{\eta_m W_m}{W_f + W_m} \tag{8.151}$$

where η_m = matrix loss factor
   W_m = strain energy in matrix at maximum vibratory displacement
   W_f = strain energy in fiber at maximum vibratory displacement

The strain energy terms W_f and W_m were determined from mechanics of materials by using the stress distributions from the Cox model [Eqs. (6.14) and (6.19)]. The longitudinal storage modulus, E_1', was also determined from the Cox model [Eq. (6.17)], and the loss modulus was

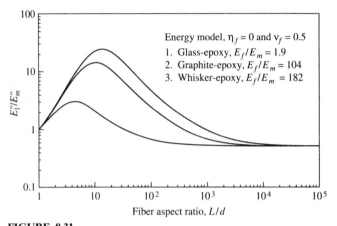

FIGURE 8.31
Variation of loss modulus ratio, E_1''/E_m'', with fiber aspect ratio, L/d, for several aligned-discontinuous fiber composite systems. (*From Gibson et al. [8.51]. Reprinted by permission of Chapman & Hall as Publishers.*)

found from $E_1'' = E_1'\eta_1$. Figure 8.31 shows the variation of the predicted ratio E_1''/E_m'' with fiber length-to-diameter ratio, L/d, for several fiber/matrix combinations. It is seen that each composite has an optimum L/d where the ratio E_1''/E_m'' is maximized, and that both the peak value of E_1''/E_m'' and the optimum L/d shift to higher values as the modulus ratio E_f/E_m increases. This means that the damping, which is primarily due to interfacial shear deformation, is increased when the mismatch between the fiber and the matrix stiffnesses (as determined by E_f/E_m) is increased.

The Ungar-Kerwin equation is ideally suited for finite element implementation in the analysis of complex structures. In the finite element implementation the element index "i" in Eq. (8.150) refers to the element number, n refers to the total number of finite elements, and the strain energy terms, W_i, are determined from the finite element analysis. It appears that the equation was first implemented in finite element form in the so-called "modal strain energy" approach for the analysis of modal damping in complex structures [8.52]. The strain energy/finite element approach has also been used in numerous composite analysis applications at both the micromechanical level [8.53, 8.54] and the laminate level [8.55–8.59]. For example, in studies of the fiber/matrix interphase the finite element models shown in Fig. 8.32 were used in conjunction with the equation

$$\eta = \frac{\eta_f W_f + \eta_m W_m + \eta_i W_i}{W_f + W_m + W_i} \qquad (8.152)$$

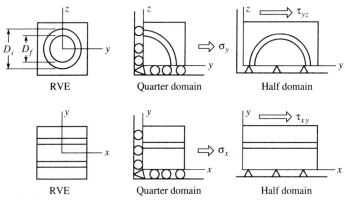

FIGURE 8.32
Models used for strain energy/finite element analysis of effect of interphase on damping of unidirectional graphite/epoxy under different loading conditions. (*From Gibson et al.* [8.54]. *Reprinted by permission of the Society for the Advancement of Material and Process Engineering.*)

where η_f = fiber loss factor
η_i = interphase loss factor
W_i = strain energy in interphase region at maximum vibratory displacement

Typical results for four different loading conditions are shown in Fig. 8.33. It appears that the in-plane shear loss factor, η_{xy}, is the most sensitive of the four loss factors to the size of the interphase region.

Three-dimensional finite element analysis has been used in conjunction with the Ungar-Kerwin equation to study interlaminar damping and the effects of coupling on damping in laminates [8.55–8.57]. In these

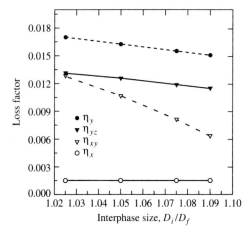

FIGURE 8.33
Predicted effect of interphase size on loss factor for material and loading conditions described in Fig. 8.32. (*From Gibson et al.* [8.54]. *Reprinted by permission of the Society for the Advancement of Material and Process Engineering.*)

studies the laminate loss factor was modeled using the equation

$$\eta = \sum_{k=1}^{N} [\eta_x^{(k)}W_x^{(k)} + \eta_y^{(k)}W_y^{(k)} + \eta_{xy}^{(k)}W_{xy}^{(k)} + \eta_z^{(k)}W_z^{(k)}$$
$$+ \eta_{yz}^{(k)}W_{yz}^{(k)} + \eta_{xz}^{(k)}W_{xz}^{(k)}]/W_t \qquad (8.153)$$

where

k = lamina number

N = total number of laminae

W_t = total strain energy stored in laminate at maximum vibratory displacement

$$= \sum_{k=1}^{N} [W_x^{(k)} + W_y^{(k)} + W_{xy}^{(k)} + W_z^{(k)}$$
$$+ W_{yz}^{(k)} + W_{xz}^{(k)}]$$

x, y, z = global laminate coordinates

$\eta_x^{(k)}, \eta_y^{(k)}, \eta_{xy}^{(k)}$ = in-plane loss factors for the kth lamina

$\eta_z^{(k)}, \eta_{yz}^{(k)}, \eta_{xz}^{(k)}$ = out-of plane loss factors for the kth lamina

$W_x^{(k)}, W_y^{(k)}, W_{xy}^{(k)}$ = in-plane strain energy terms for the kth lamina

$W_z^{(k)}, W_{yz}^{(k)}, W_{xz}^{(k)}$ = out-of plane strain energy terms for the kth lamina

Thus, the decomposition of the total damping into contributions associated with each stress component is a relatively simple task with the strain energy approach. For example, Fig. 8.34 shows the contribution of the different components of interlaminar damping as a function of fiber orientation for angle-ply graphite laminates under uniaxial extension

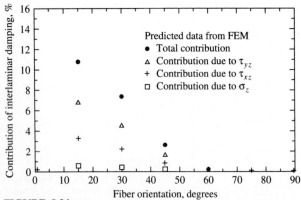

FIGURE 8.34
Contribution of different components of interlaminar damping for various fiber orientations for $[\pm\theta]_s$ graphite/epoxy laminates (with laminate width/thickness = 4 and length/thickness = 6) under uniaxial loading. (*From Hwang and Gibson* [8.5]. *Reprinted by permission of Elsevier Science Publishers, Ltd.*)

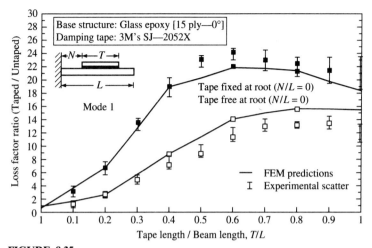

FIGURE 8.35

Measured and predicted damping for unidirectional glass/epoxy beam with constrained viscoelastic layer damping tapes of different lengths and tape end fixity conditions. (*From Mantena et al.* [8.58]. *Copyright AIAA 1990—Used with permission.*)

[8.55]. The finite element model for this work was shown previously in Fig. 7.23. It is seen that the interlaminar damping is maximized at a particular fiber orientation, and that the interlaminar shear stress, τ_{xz}, is the most significant contributor to interlaminar damping in this case. A similar approach was used to study damping in composite beams with constrained viscoelastic layer damping treatments, and Fig. 8.35 shows the effect of constrained viscoelastic layer (damping tape) length on damping for a glass/epoxy beam [8.58]. In this case damping is seen to be strongly dependent on the ratio of damping tape length to beam length and the tape end fixity condition.

Finally, although the loss factor is a convenient measure of damping because of its connection with the complex modulus notation, it is not the only parameter used to describe damping. For materials with small damping ($\eta \ll 1$) other measures of damping which appear in the literature are related to the loss factor as follows [8.41]:

$$\eta = \frac{\psi}{2\pi} = \frac{\Delta}{\pi} = 2\zeta = \frac{1}{Q} \tag{8.154}$$

where ψ = specific damping capacity
Δ = logarithmic decrement
ζ = damping ratio, or damping factor
Q = quality factor

Most of these parameters are associated with the damping of a single degree of freedom vibration model and are used to obtain damping from vibration test data. Such tests will be discussed in more detail in Chap. 10. In summary, damping has become an important consideration in the design of dynamically loaded composite materials and structures. As a result, there is increased interest in the prediction of damping in composites. Several analytical methods for making such predictions have been reviewed, and sample results have been presented. Because of the design flexibility that is inherent in composite materials, the potential for improvement and optimization of damping appears to be much greater than that for conventional structural materials.

Example 8.8. The constituent materials in a unidirectional graphite/epoxy material have the following dynamic mechanical properties at a certain frequency, ω:

$$E'_{f1} = 220 \, \text{GPa} \quad (32 \times 10^6 \, \text{psi}); \quad \eta_{f1} = 0.002; \quad v_f = 0.6$$
$$E'_m = 3.45 \, \text{GPa} \quad (0.5 \times 10^6 \, \text{psi}); \quad \eta_m = 0.02; \quad v_m = 0.4$$

Determine the composite longitudinal loss factor and the percentage of the total longitudinal damping due to each constituent.

Solution. Substituting the above data in Eq. (8.82) from the Elastic-Viscoelastic Correspondence Principle, or using the strain energy approach and Eq. (8.150), we find that the composite longitudinal loss factor is

$$\eta_1 = \frac{E''_{f1} v_f + E''_m v_m}{E'_{f1} v_f + E'_m v_m} = \frac{\eta_{f1} E'_{f1} v_f + \eta_m E'_m v_m}{E'_{f1} v_f + E'_m v_m}$$

$$= \frac{0.002(220)(0.6) + 0.02(3.45)(0.4)}{220(0.6) + 3.45(0.4)}$$

$$= 0.001979 + 0.000207$$

$$= 0.002186$$

Thus, the fiber contributes $(0.001979/0.002186) \times 100 = 90.5\%$ of the damping and the matrix contributes the remaining 9.5 percent. Even though the matrix has a greater loss factor than the fiber, most of the strain energy is stored in the fiber, and this is why the fiber contributes more to the total composite damping. This is not true for the off-axis case, however, as the strain energy in the matrix becomes more significant. For example, the composite transverse loss factor is dominated by the matrix contribution.

PROBLEMS

8.1. For a linear viscoelastic material the creep response under a constant stress is followed by a "recovery response" after the stress is removed at some time, t_0. Using the Boltzmann Superposition Principle, find an expression for the uniaxial recovery compliance, $R(t)$, for times $t > t_0$ in terms of the creep compliance, $S(t)$, the time of stress removal, t_0, and the time, t.

8.2. In general, the creep compliances, $S_{ij}(t)$, and the relaxation moduli, $C_{ij}(t)$,

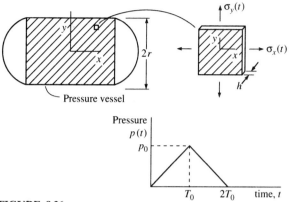

FIGURE 8.36
State of stress in composite pressure vessel and variation of internal pressure with time for Problem 8.4.

are not related by a simple inverse relationship. Show that only when $t \to 0$ and when $t \to \infty$, can we say that

$$[C_{ij}(t)] = [S_{ij}(t)]^{-1}$$

8.3. The shear creep compliance, $S_{66}(t)$, for a unidirectional viscoelastic composite is given by $S_{66}(t) = \gamma_{12}(t)/\tau_{12}$, where $\gamma_{12}(t)$ is the time-dependent shear creep strain and τ_{12} is the constant shear stress. If $S_{66}(t)$ can be approximated by a power law as $S_{66}(t) = at^b$, where a and b are material constants and t is time, determine the "constant loading rate compliance" $U_{66}(t) = \gamma_{12}(t)/\tau_{12}(t)$, where the shear stress is due to a constant loading rate, so that $\gamma_{12}(t) = Kt$, where K is a constant. *Boltzmann superpos.*

8.4. The time-dependent axial stress, $\sigma_x(t)$, and the time-dependent circumferential stress, $\sigma_y(t)$, in the wall of a filament-wound, thin-walled composite pressure vessel shown in Fig. 8.36 are caused by the internal pressure $p(t)$, where t is time. The required dimensions of the vessel are the wall thickness, h, and the mean radius, r. Note that x and y are not the principal material axes, but, rather, are the longitudinal and transverse axes for the vessel. The variation of $p(t)$ with time is also shown in Fig. 8.36. If the creep compliances associated with the x, y axes are given in contracted notation by

$$\bar{S}_{ij}(t) = \bar{E}_{ij} + \bar{F}_{ij}t \qquad i, j = 1, 2, \ldots, 6$$

where the $\bar{E}_{ij}$ and the $\bar{F}_{ij}$ are material constants, determine all the time-dependent strains along the x, y axes for $t > 2T_0$. Answers should be given in terms of p_0, r, h, T_0, t, and the individual $\bar{E}_{ij}$ and $\bar{F}_{ij}$.

8.5. A linear viscoelastic, orthotropic lamina has principal creep compliances which are given in contracted notation by

$$S_{ij}(t) = E_{ij} + F_{ij}t \qquad i, j = 1, 2, \ldots, 6$$

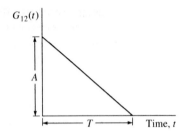

FIGURE 8.37
Variation of shear relaxation modulus, $G_{12}(t)$, with time for Problem 8.10.

where the E_{ij} and the F_{ij} are material constants and t is time. The lamina is subjected to plane stress with constant stresses

$$\sigma_i(t) = \sigma_i' H(t) \qquad i, j = 1, 2, \ldots, 6$$

where $\sigma_i' = $ constants

         $H(t) = $ unit step function

If the failure strains for pure longitudinal, transverse, and shear loading of the lamina are e_L, e_T, and e_{LT}, respectively, find the expressions for the time to failure for each of the three strains.

8.6. Derive the equations for the stress-strain relationship, the creep compliance, and the relaxation modulus for the Kelvin-Voigt model.

8.7. Derive the equations for the stress-strain relationship, the creep compliance, and the relaxation modulus for the Zener model.

8.8. Derive Eq. (8.40).

8.9. Derive Eq. (8.42).

8.10. The shear relaxation modulus, $G_{12}(t)$, of an orthotropic lamina is idealized, as shown in Fig. 8.37. Find the corresponding equations for the shear storage modulus, $G_{12}'(\omega)$, and the shear loss modulus, $G_{12}''(\omega)$, and draw sketches of both parts of the complex modulus in the frequency domain.

8.11. For the Maxwell model in Fig. 8.6, express the storage modulus, $E'(\omega)$; the loss modulus, $E''(\omega)$; and the loss factor, $\eta(\omega)$, in terms of the parameters μ and k and the frequency, ω. Sketch the variation of $E'(\omega)$, $E''(\omega)$, and $\eta(\omega)$ in the frequency domain. It is not necessary to use Fourier transforms here.

8.12. Derive Eqs. (8.146) and (8.147).

8.13. The composite pressure vessel in Problem 8.4 is subjected to an internal pressure which varies sinusoidally with time according to the relationship $p(t) = P_0 \sin \omega t$, and the principal complex compliances are given by

$$S_{mn}^*(\omega) = S_{mn}'(\omega) + i S_{mn}''(\omega) \qquad m, n = 1, 2, \ldots, 6$$

where ω is the frequency. Determine all the time-dependent strains associated with the x, y axes in terms of P_0, r, h, ω, and the individual $S_{mn}'(\omega)$ and $S_{mn}''(\omega)$.

8.14. The polymer matrix material in a linear viscoelastic, unidirectional composite material has a relaxation modulus which can be characterized by the

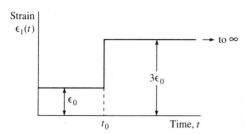

FIGURE 8.38
Composite longitudinal strain-time history for Problem 8.14.

Maxwell model in Fig. 8.6. The fibers are assumed to be linear elastic. If the composite longitudinal strain-time history is as shown in Fig. 8.38, express the composite longitudinal stress as a function of time.

8.15. The matrix material in a linear viscoelastic, unidirectional composite material is to be modeled by using a Maxwell model having parameters k_m and μ_m, while the fiber is to be modeled by using a Kelvin-Voigt model having parameters k_f and μ_f.

(a) Determine the complex extensional moduli of fiber and matrix materials in terms of the Maxwell and Kelvin-Voigt parameters and the frequency, ω.

(b) Determine the complex longitudinal modulus of the unidirectional composite. Assume that the fiber and matrix materials are isotropic. It is not necessary to use the Fourier transforms.

8.16. The dynamic mechanical behavior of an isotropic polymer matrix material may be characterized by two independent complex moduli such as the complex extensional modulus, $E^*(\omega)$, and the complex shear modulus, $G^*(\omega)$. Based on experimental evidence, however, the imaginary parts of $E^*(\omega)$ and $G^*(\omega)$ are not independent because the material can be assumed to be viscoelastic in shear but elastic in dilatation (i.e., the shear modulus, $G^*(\omega)$, is complex and frequency-dependent, but the bulk modulus, k, is real and frequency-independent). Use this simplifying assumption to develop an expression for the shear loss factor, $\eta_G(\omega)$, in terms of the extensional loss factor, $\eta_E(\omega)$, the extensional storage modulus, $E'(\omega)$, and the bulk modulus, k. Assume all loss factors $\ll 1$.

8.17. A drive shaft in the shape of a hollow tube and made of a linear viscoelastic angle-ply laminate is subjected to a torque, T, as shown in Fig. 8.39. Develop an analytical model for predicting the vibratory shear deformation in the shaft from the vibratory shear force, $\tilde{N}_{xy}(t)$, when the torque T varies sinusoidally with time. The input to the model should include the properties and volume fractions of fiber and matrix materials, lamina orientations, and lamina stacking sequences. That is, the model should include both micromechanical and macromechanical components. No calculations are necessary, but the key equations should be described, all parameters should be defined, and key assumptions should be delineated.

8.18. Longitudinal vibration of an isotropic, particle-reinforced composite bar may be modeled by using the one-dimensional wave equation [Eq. (8.94)] if the material is linear elastic. Derive the equation of motion for longitudinal

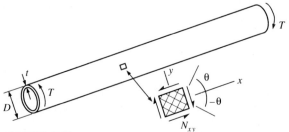

FIGURE 8.39
Composite drive shaft under applied torque for Problem 8.17.

vibration of the bar if it can be assumed to be a Kelvin-Voigt linear viscoelastic material having the stress-strain relationship given by Eq. (8.34).

8.19. Find the separation of variables solution for the longitudinal displacement, $u(x, t)$, of the equation derived in Problem 8.18. Leave the answer in terms of constants which must be determined from the boundary conditions.

8.20. Derive the equation of motion for free transverse vibration of a simply supported, specially orthotropic plate which is subjected to in-plane loads per unit length N_x and N_y, as shown in Fig. 8.40.

8.21. For the plate described in Problem 8.20, find the equation for the plate natural frequencies and determine the effects of positive (tensile) and negative (compressive) in-plane loads N_x and N_y on the natural frequencies.

8.22. If the plate described in Problem 8.20 is clamped on all edges, investigate solutions of the form

$$W(x, y) = A_{mn}\left(1 - \cos\frac{2\pi x}{a}\right)\left(1 - \cos\frac{2\pi y}{b}\right)$$

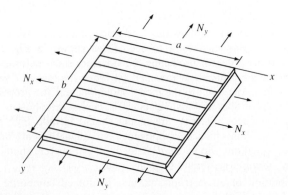

FIGURE 8.40
Simply supported, specially orthotropic plate under in-plane loads for Problem 8.20.

Does this solution satisfy the boundary conditions? Can it be used to find the natural frequencies? Explain your answers.

REFERENCES

8.1. Schapery, R. A., "Viscoelastic Behavior and Analysis of Composite Materials," in G. P. Sendeckyj (ed.), *Composite Materials Volume 2: Mechanics of Composite Materials*, 85–168, Academic Press, New York (1974).

8.2. Fung, Y. C., *Foundations of Solid Mechanics*, Prentice-Hall, Inc. Englewood Cliffs, NJ (1965).

8.3. Christensen, R. M., *Mechanics of Composite Materials*, John Wiley & Sons, New York (1979).

8.4. Halpin, J. C. and Pagano, N. J., "Observations on Linear Anisotropic Viscoelasticity," *Journal of Composite Materials*, **2**(1), 68–80 (1968).

8.5. Christensen, R. M., *Theory of Viscoelasticity: An Introduction*, 2d ed., Academic Press, New York (1982).

8.6. Sullivan, J. L., "Polymer Viscoelasticity", Unpublished notes, Ford Motor Co., Dearborn, MI (1992).

8.7. McCrum, N. G., Buckley, C. P., and Bucknall, C. B., *Principles of Polymer Engineering*, Oxford University Press, Oxford, England (1988).

8.8. Bracewell, R. N., *The Fourier Transform and Its Application*, 2d ed., McGraw-Hill, Inc., New York (1978).

8.9. Gibson, R. F., Hwang, S. J., and Sheppard, C. H., "Characterization of Creep in Polymer Composites by the Use of Frequency-Time Transformations," *Journal of Composite Materials*, **24**, 441–453 (1990).

8.10. Lee, E. H., "Stress Analysis in Viscoelastic Bodies," *Quarterly of Applied Mathematics*, **13**, 183–190 (1955).

8.11. Biot, M. A., "Linear Thermodynamics and the Mechanics of Solids," in *Proceedings of the Third U.S. National Congress of Applied Mechanics*, 1–18 (1958).

8.12. Schapery, R. A., "Stress Analysis of Viscoelastic Composite Materials," *Journal of Composite Materials*, **1**, 228–267 (1967).

8.13. Beckwith, S. W., *Viscoelastic Characterization of a Nonlinear Glass/Epoxy Composite Including the Effects of Damage*, Ph.D. Dissertation, Texas A&M University, College Station, TX (1974).

8.14. Sims, D. F. and Halpin, J. C., "Methods for Determining the Elastic and Viscoelastic Response of Composite Materials," *Composite Materials: Testing and Design (Third Conference)*, ASTM STP **546**, 46–66, American Society for Testing and Materials, Philadelphia, PA (1974).

8.15. Hashin, Z., "Complex Moduli of Viscoelastic Composites I: General Theory and Application to Particulate Composites," *International Journal of Solids and Structures*, **6**, 539–552 (1970).

8.16. Hashin, Z., "Complex Moduli of Viscoelastic Composites II: Fiber Reinforced Materials," *International Journal of Solids and Structures*, **6**, 797–807 (1970).

8.17. Suarez, S. A., Gibson, R. F., Sun, C. T., and Chaturvedi, S. K., "The Influence of Fiber Length and Fiber Orientation on Damping and Stiffness of Polymer Composite Materials, *Experimental Mechanics*, **26**(2), 175–184 (1986).

8.18. Sun, C. T., Wu, J. K., and Gibson, R. F., "Prediction of Material Damping of Laminated Polymer Matrix Composites," *Journal of Materials Science*, **22**, 1006–1012 (1987).

8.19. Findley, W. N., Lai, J. S., and Onaran, K., *Creep and Relaxation of Nonlinear Viscoelastic Materials*. Dover Publications, New York (1976).

8.20. Ferry, J. D., *Viscoelastic Properties of Polymers*, 2d ed., John Wiley & Sons, New York (1970).

8.21. Struik, L. C. E., "Physical Aging in Plastics and other Glassy Materials," *Polymer Engineering and Science*, **17**, 165–173 (1977).

8.22. Struik, L. C. E., *Physical Aging in Amorphous Polymers and Other Materials*, Elsevier, Amsterdam (1978).

8.23. Janas, V. F. and McCullough, R. L., "The Effects of Physical Aging on the Viscoelastic Behavior of a Thermoset Polyester," *Composites Science and Technology*, **30**, 99–118 (1987).

8.24. Ogale, A. A. and McCullough, R. L., "Physical Aging of Polyether Ether Ketone," *Composites Science and Technology*, **30**, 137–148 (1987).

8.25. Sullivan, J. L., "Creep and Physical Aging of Composites," *Composites Science and Technology*, **39**, 207–232 (1990).

8.26. Hearmon, R. F. S., *An Introduction to Applied Anisotropic Elasticity*, Oxford University Press, Oxford, England (1961).

8.27. Achenbach, J. D., "Waves and Vibrations in Directionally Reinforced Composites," in G. P. Sendeckyj (ed.), *Composite Materials Volume 2: Mechanics of Composite Materials*, Academic Press, New York (1974).

8.28. Ross, C. A. and Sierakowski, R. L., "Elastic Waves in Fiber Reinforced Materials," *The Shock and Vibration Digest*, **7**(1), 1–12 (1975).

8.29. Moon, F. C., "Wave Propagation and Impact in Composite Materials," in C. C. Chamis, (ed.), *Composite Materials Volume 7*, Academic Press, New York (1974).

8.30. Whitney, J. M., *Structural Analysis of Laminated Anisotropic Plates*, Technomic Publishing Co., Lancaster, PA (1987).

8.31. Vinson, J. R. and Sierakowski, R. L., *The Behavior of Structures Composed of Composite Materials*, Martinus Nijhoff Publishers, Dordrecht, The Netherlands (1986).

8.32. Meirovitch, L., *Elements of Vibration Analysis*, 2d ed., McGraw-Hill, Inc., New York (1986).

8.33. Timoshenko, S. P., Young, D. H., and Weaver, W., Jr., *Vibration Problems in Engineering*, John Wiley & Sons, New York (1974).

8.34. Dudek, T. J., "Young's and Shear Moduli of Unidirectional Composites by a Resonant Beam Method," *Journal of Composite Materials*, **4**, 232–241 (1970).

8.35. Ni, R. G. and Adams, R. D., "The Damping and Dynamic Moduli of Symmetric Laminated Beams—Theoretical and Experimental Results," *Journal of Composite Materials*, **18**, 104–121 (1984).

8.36. Huang, T. C. and Huang, C. C., "Free Vibrations of Viscoelastic Timoshenko Beam," *Journal of Applied Mechanics*, **38**, Series E (2), 515–521 (1971).

8.37. Nakao, T. Okano, T., and Asano, I., "Theoretical and Experimental Analysis of Flexural Vibration of the Viscoelastic Timoshenko Beam," *Journal of Applied Mechanics*, **52**(3), 728–731 (1985).

8.38. Gibson, R. F., "Damping Characteristics of Composite Materials and Structures," *Journal of Engineering Materials and Performance*, **1**(1), 11–20 (1992).

8.39. Gibson, R. F., "Dynamic Mechanical Properties of Advanced Composite Materials and Structures: A Review," *The Shock and Vibration Digest*, **19**(7), 13–22 (1987).

8.40. Gibson, R. F., "Dynamic Mechanical Properties of Advanced Composite Materials and Structures: A Review of Recent Research," *The Shock and Vibration Digest*, **22**(8), 3–12 (1990).

8.41. Bert, C. W., "Composite Materials: A Survey of the Damping Capacity of Fiber Reinforced Composites," in P. J. Torvik, (ed.), *Damping Applications for Vibration Control*, AMD Vol. **38**, 53–63, American Society of Mechanical Engineers, New York (1980).

8.42. Adams, R. D., "Damping Properties Analysis of Composites", in T. J. Reinhart (ed.),

Engineered Materials Handbook Volume 1: Composites, 206–217, ASM International, Materials Park, OH (1987).

8.43. Chaturvedi, S. K., "Damping of Polymer Matrix Composite Materials," in S. Lee, (ed.), *Encyclopedia of Composites*, VCH Publishing Co., New York (1989).

8.44. Kinra, V. K. and Wolfenden, A. (eds.), *M3D: Mechanics and Mechanisms of Material Damping*, ASTM/STP **1169**, American Society for Testing and Materials, Philadelphia, PA (1992).

8.45. Zener, C., *Elasticity and Anelasticity of Metals*, The University of Chicago Press, Chicago (1948).

8.46. Granato, A. V. and Lucke, K., "Application of Dislocation Theory to Internal Friction Phenomena at High Frequencies," *Journal of Applied Physics*, **27**(7), 789–805 (1956).

8.47. Sun, C. T., Chaturvedi, S. K., and Gibson, R. F., "Internal Material Damping of Polymer Matrix Composites under Off-Axis Loading," *Journal of Materials Science*, **20**, 2575–2585 (1985).

8.48. Sun, C. T., Wu, J. K., and Gibson, R. F., "Prediction of Material Damping in Randomly Oriented Short Fiber Polymer Matrix Composites," *Journal of Reinforced Plastics and Composites*, **4**, 262–272 (1985).

8.49. Nashif, A. D., Jones, D. I. G., and Henderson, J. P., *Vibration Damping*, John Wiley & Sons, New York (1985).

8.50. Ungar, E. E. and Kerwin, E. M., Jr., "Loss Factors of Viscoelastic Systems in Terms of Strain Energy," *Journal of the Acoustical Society of America*, **34**(2), 954–958 (1962).

8.51. Gibson, R. F., Chaturvedi, S. K., and Sun, C. T., "Complex Moduli of Aligned Discontinuous Fiber Reinforced Polymer Composites," *Journal of Materials Science*, **17**, 3499–3509 (1982).

8.52. Johnson, C. D., and Kienholz, D. A., "Finite Element Prediction of Damping in Structures with Constrained Viscoelastic Layers," *AIAA Journal*, **20**(9), 1284–1290 (1982).

8.53. Hwang, S. J. and Gibson, R. F., "Micromechanical Modeling of Damping in Discontinuous Fiber Composites Using a Strain Energy/Finite Element Approach," *Journal of Engineering Materials and Technology*, **109**, 47–52 (1987).

8.54. Gibson, R. F., Hwang, S. J., and Kwak, H., "Micromechanical Modeling of Damping in Composites Including Interphase Effects," in *How Concept Becomes Reality— Proceedings of 36th International SAMPE Symposium*, Vol. **1**, 592–606, Society for the Advancement of Material and Process Engineering, Covina, CA (1991).

8.55. Hwang, S. J. and Gibson, R. F., "The Effects of Three-Dimensional States of Stress on Damping of Laminated Composites," *Composites Science and Technology*, **41**, 379–393 (1991).

8.56. Hwang, S. J. and Gibson, R. F., "Contribution of Interlaminar Stresses to Damping in Thick Laminated Composites under Uniaxial Extension," *Composite Structures*, **20**, 29–35 (1992).

8.57. Hwang, S. J., Gibson, R. F., and Singh, J., "Decomposition of Coupling Effects on Damping of Laminated Composites under Flexural Vibration," *Composites Science and Technology*, **43**, 159–169 (1992).

8.58. Mantena, P. R., Gibson, R. F., and Hwang, S. J., "Optimal Constrained Viscoelastic Tape Lengths for Maximizing Damping in Laminated Composites," *AIAA Journal*, **29**(10), 1678–1685 (1991).

8.59. Hwang, S. J. and Gibson, R. F., "The Use of Strain Energy-Based Finite Element Techniques in the Analysis of Various Aspects of Damping of Composite Materials and Structures," *Journal of Composite Materials*, **26**(17), 2585–2605 (1992).

CHAPTER
9
ANALYSIS OF FRACTURE

9.1 INTRODUCTION

Except for a brief discussion in Sec. 7.8.2, the previous chapters of this book have not considered the analysis of the effects of notches, cracks, delaminations, or other discontinuities in composites. For example, the conventional strength analyses outlined in Chap. 4 involved the use of gross "effective lamina strengths" in various semiempirical failure criteria without regard for specific micromechanical failure modes which are related to such discontinuities. While such procedures, along with the use of empirical "safety factors," may produce a satisfactory design for static loading, failures may still occur due to the growth of cracks or delaminations under dynamic loading. The purpose of this chapter is to give an introduction to the analysis of fracture of composites due to cracks, notches, and delaminations.

First, the prediction of the strength of composites with through-thickness cracks and notches is considered by using both fracture mechanics and stress fracture approaches. Next, the use of fracture mechanics in the analysis of interlaminar fracture will be discussed. Each of these topics is the subject of many publications. Thus, only brief

338

introductions to the subjects are given here, along with key references where more detailed analyses may be found. Each of these topics is also the subject of considerable current research, and the reader is encouraged to consult technical journals for the results of the most recent research. A regularly published listing of the tables of contents of technical journals where such articles appear may be particularly useful [9.1]. The Special Technical Publication (STP) series by the American Society for Testing and Materials is another source of recent research findings [9.2–9.7]. The application of fracture mechanics to composites is the subject of a book [9.8], as is delamination in composites [9.9].

9.2 FRACTURE MECHANICS ANALYSES OF THROUGH-THICKNESS CRACKS

Much of the early work on fracture in composites involved investigations of the applicability of linear elastic fracture mechanics, which had been originally developed for the analysis of through-thickness cracks in homogeneous, isotropic metals. The origin of fracture mechanics can be traced back to the seminal work of Griffith [9.10], who explained the discrepancy between the measured and predicted strength of glass by considering the stability of a small crack. The stability criterion was developed by using an energy balance on the crack.

Consider the through-thickness crack in the uniaxially loaded homogeneous, isotropic, linear elastic plate of infinite width shown in Fig. 9.1. Griffith reasoned that the strain energy of the cracked plate would be less than the corresponding strain energy of the uncracked plate, and from a stress analysis, he estimated that the strain energy released by the creation of the crack under plane stress conditions would be

$$U_r = \frac{\pi \sigma^2 a^2 t}{E} \tag{9.1}$$

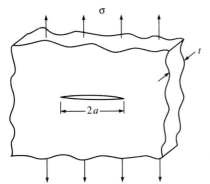

FIGURE 9.1
The Griffith crack: A through-thickness crack in a uniaxially stressed plate of infinite width.

where U_r = strain energy released
 σ = applied stress
 a = half-crack length
 t = plate thickness
 E = modulus of elasticity of the plate

The corresponding expression in Griffith's original paper was later found to be in error, and Eq. (9.1) is consistent with the corrected expression in more recent publications [9.11, 9.12]. In addition, Griffith's energy terms were given on a per unit thickness basis. Equation (9.1) is also consistent with the strain energy released by relaxation of an elliptical zone having major and minor axes of lengths 4a and 2a, respectively, where the minor axis is coincident with the crack and the major axis is perpendicular to the crack. The volume of such an ellipse is

or

$$V = \pi(2a)(a)(t) = 2\pi a^2 t \tag{9.2}$$

Since the plate was assumed to be uniformly stressed before the introduction of the crack, the strain energy released due to relaxation of the elliptical volume around the crack is

$$U_r = \frac{1}{2}\frac{\sigma^2}{E}V = \frac{\pi\sigma^2 a^2 t}{E} \tag{9.3}$$

Griffith also assumed that the creation of new crack surfaces required the absorption of an amount of energy given by

$$U_s = 4at\gamma_s \tag{9.4}$$

where U_s = energy absorbed by creation of new crack surfaces
 γ_s = surface energy per unit area

As the crack grows, if the rate at which energy is absorbed by creating new surfaces is greater than the rate at which strain energy is released, then

$$\frac{\partial U_s}{\partial a} > \frac{\partial U_r}{\partial a} \tag{9.5}$$

and crack growth is stable. If the strain energy is released at a greater rate than it can be absorbed, then

$$\frac{\partial U_r}{\partial a} > \frac{\partial U_s}{\partial a} \tag{9.6}$$

and crack growth is unstable. The threshold of stability, or the condition of neutral equilibrium, is therefore given by

$$\frac{\partial U_r}{\partial a} = \frac{\partial U_s}{\partial a} \tag{9.7}$$

or
$$\frac{\pi\sigma^2 a}{E} = 2\gamma_s \qquad (9.8)$$

Thus, the critical stress, σ_c, for self-sustaining extension of the crack in plane stress is

$$\sigma_c = \sqrt{\frac{2E\gamma_s}{\pi a}} \qquad (9.9)$$

Alternatively, the critical flaw size for plane stress at stress level σ is

$$a_c = \frac{2E\gamma_s}{\pi\sigma^2} \qquad (9.10)$$

It is interesting to note that when we rearrange Eq. (9.8) as

$$\sigma\sqrt{\pi a} = \sqrt{2E\gamma_s} \qquad (9.11)$$

the terms on the left-hand side depend only on loading and geometry, whereas the terms on the right-hand side depend only on material properties. Thus, when the stress reaches the critical fracture stress, σ_c, the left-hand side becomes $\sigma_c\sqrt{\pi a}$. The term $\sigma_c\sqrt{\pi a}$ is now referred to as the fracture toughness, K_c. This is a very important concept, which we will return to later.

The application of the Griffith-type analysis to composites presents some difficulties, but, fortunately, many of these problems have been solved over the years since Griffith's work. For example, for metals and many polymers the energy absorbed in crack extension is actually greater than the surface energy. Recognizing this, both Irwin [9.13] and Orowan [9.14] modified the Griffith analysis to include energy absorption due to plastic deformation at the crack tip. In this analysis the factor $2\gamma_s$ on the right-hand side of Eq. (9.8) and in all subsequent equations is replaced by the factor $2(\gamma_s + \gamma_p)$, where γ_p is the energy of plastic deformation. The solutions of several other problems encountered in the development of composite fracture mechanics have been made possible by the use of several different analytical techniques. Two of these techniques, now referred to as the "stress intensity factor" approach and the "strain energy release rate" approach, will be discussed in the following sections.

9.2.1 Stress Intensity Factor Approach

The Griffith analysis was originally developed for homogeneous, isotropic materials. Using effective modulus theory, we can replace the heterogeneous, anisotropic composite with an equivalent homogeneous, anisotropic material. It turns out that by considering the stress distribution around the crack tip, we can develop another interpretation of the Griffith analysis which can be applied equally well to homogeneous isotropic or anisotropic materials and to states of stress other than the

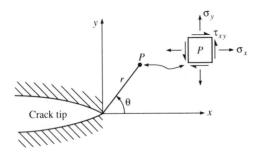

FIGURE 9.2
Stresses at the tip of a crack under plane stress.

simple uniaxial stress that Griffith used. Referring to the plane stress condition in the vicinity of the uniaxially loaded crack in Fig. 9.2, Westergaard [9.15] used a complex stress function approach to show that the stresses for the isotropic case at a point P defined by polar coordinates r, θ can be expressed as

$$\sigma_x = \frac{K_I}{\sqrt{2\pi r}} f_1(\theta) \tag{9.12}$$

$$\sigma_y = \frac{K_I}{\sqrt{2\pi r}} f_2(\theta) \tag{9.13}$$

$$\tau_{xy} = \frac{K_I}{\sqrt{2\pi r}} f_3(\theta) \tag{9.14}$$

where K_I is the stress intensity factor for the crack opening mode, as defined by

$$K_I = \sigma \sqrt{\pi a} \tag{9.15}$$

and the $f_i(\theta)$ are trigonometric functions of the angle θ. Irwin [9.16] recognized that the term $\sigma \sqrt{\pi a}$ controls the magnitudes of the stresses at a point r, θ near the crack tip. Returning to the discussion following Eq. (9.11), we see that the critical value of the stress intensity factor, K_{Ic}, corresponding to the critical stress, σ_c, is the fracture toughness. That is,

$$K_{Ic} = \sigma_c \sqrt{\pi a} \tag{9.16}$$

The fracture toughness, K_{Ic}, is a material property which can be determined experimentally, as shown later. Thus, if the fracture toughness of the material is known, the fracture mechanics analysis can be used in two ways, depending on whether the applied stress or the crack size is known. If the applied stress, σ, is known, equations like Eq. (9.15) can be used to find the critical crack size, a_c, which will lead to unstable and catastrophic crack growth. Knowing the critical crack size, we can specify inspection of the component in question to make sure that there are no cracks of that size. On the other hand, if the crack size, a, is known, then

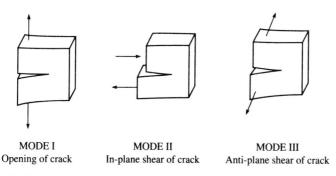

MODE I	MODE II	MODE III
Opening of crack	In-plane shear of crack	Anti-plane shear of crack

FIGURE 9.3
The three basic modes of crack deformation.

equations like Eq. (9.15) can be used to find the critical stress, σ_c, which will lead to unstable and catastrophic crack growth. Loading on the component in question would then be specified so as not to exceed this stress.

The reader is cautioned that the stress intensity factor is defined as $k_1 = \sigma \sqrt{a}$ in some publications. This definition corresponds to the cancellation of $\sqrt{\pi}$ in both the numerator and denominator of Eq. (9.12) to (9.14), so that the denominator corresponding to k_1 would be $\sqrt{2r}$ instead of $\sqrt{2\pi r}$, and thus $K_I = k_1 \sqrt{\pi}$.

Expressions for stress distributions for other types of loading and crack geometries in isotropic materials lead to expressions that are similar to Eqs. (9.12) to (9.14), and the corresponding stress intensity factors can be found in the same way [9.17]. Other important results such as finite width correction factors (recall that the Griffith analysis is for a crack in an infinite width plate) have been tabulated in Ref. [9.17]. The three basic modes of crack deformation are shown in Fig. 9.3. Thus, for the crack opening mode in the above example (mode I) we have the stress intensity factor K_I. For the in-plane shear mode (mode II) we have the stress intensity factor K_{II}, and for the antiplane shear mode (mode III) we have K_{III}. For example, for the cases of pure shear loading in modes II and III we have

$$K_{II} = \tau \sqrt{\pi a} \quad \text{and} \quad K_{III} = \tau \sqrt{\pi a} \tag{9.17}$$

where the shear stress, τ, is different for modes II and III, as shown in Fig. 9.3.

Although the stress analyses for the corresponding anisotropic material cases are more difficult and the expressions are more complicated, the stress intensity factors for certain loading conditions and crack geometries are the same as those for the isotropic case. For example, Lekhnitskii [9.18] has used a stress function approach to show that if the crack shown in Figs. 9.1 and 9.2 lies in an anisotropic material for which the xy plane is a plane of material property symmetry, then the stresses are given by

$$\sigma_x = \frac{K_I}{\sqrt{2\pi r}} F_1(\theta, s_1, s_2) \tag{9.18}$$

$$\sigma_y = \frac{K_I}{\sqrt{2\pi r}} F_2(\theta, s_1, s_2) \tag{9.19}$$

and $$\tau_{xy} = \frac{K_I}{\sqrt{2\pi r}} F_3(\theta, s_1, s_2) \tag{9.20}$$

where the functions $F_i(\theta, s_1, s_2)$ include not only trigonometric functions of the angle, θ, but also s_1 and s_2, which are complex roots of the characteristic equation corresponding to a differential equation in the stress function [9.18]. As pointed out by Wu [9.19], the magnitudes of the stresses at point r, θ in an isotropic material [see Eqs. (9.12) to (9.14)] are completely determined by the stress intensity factors, but in the aniso-tropic case [Eqs. (9.18) to (9.20)] these magnitudes also depend on s_1 and s_2. Wu [9.19] has also shown, however, that if the crack lies along a principal material direction in the anisotropic material, then the stress intensity factors given by Eqs. (9.15) and (9.17) are still valid for their respective loading conditions shown in Fig. 9.3.

Several experimental investigations have shown that the concept of a critical stress intensity factor can be used to describe the fracture behavior of through-thickness cracked unidirectional composites and laminates. Wu [9.19] reasoned that if the fracture toughness, K_{Ic}, is a material constant, then by considering the logarithm of Eq. (9.16), the slope of the log σ_c vs. log a_c plot must be -0.5. Wu's experimental results for unidirectional E-glass/epoxy showed good agreement with this prediction. Konish et al. [9.20] showed that the critical stress intensity factors for 0°, 90°, 45° $[\pm 45°]_s$, and $[0°/\pm 45°/90°]_s$ graphite/epoxy laminates could be determined by using the same fracture toughness test method that had been developed for metals. Parhizgar et al. [9.21] showed both analytically and experimentally that the fracture toughness of unidirectional E-glass/epoxy composites is a constant material pro-perty which does not depend on crack length but which does depend on fiber orientation.

The fracture toughness, K_{Ic}, has been found to be an essentially constant material property for a variety of randomly oriented short fiber composites, as shown in papers by Alexander et al. [9.22] and Sun and Sierakowski [9.23]. Although the random fiber orientation in such materials allows one to use the numerous tabulated solutions for stress intensity factors of isotropic materials [9.17], it appears that the simple crack growth assumed in the Griffith-type analysis does not always occur in these materials. As an alternative to crack growth, the concept of a damage zone ahead of the crack tip in short fiber composites has been proposed by Gaggar and Broutman [9.24].

9.2.2 Strain Energy Release Rate Approach

One of the major drawbacks of the stress intensity factor approach is that a stress analysis of the crack tip region is required. While such analyses have been done for a variety of loading conditions and crack geometries for isotropic materials [9.17], the corresponding analyses for anisotropic materials have only been done for relatively few cases because of mathematical difficulties. A very useful alternative to the stress intensity factor approach is referred to as the "strain energy release rate" approach. The strain energy release rate has an easily understood physical interpretation which is equally valid for either isotropic or anisotropic materials, and it turns out that this rate is also related to the stress intensity factor. The strain energy release rate approach has proved to be a powerful tool in both experimental and computational studies of crack growth.

The derivation of the strain energy release rate presented here follows that of Irwin [9.25], as explained by Corten [9.26]. We first consider a through-thickness cracked linear elastic plate under a uniaxial load, as shown in Fig. 9.4(a). An increase in the load, P, from the unloaded condition causes a linearly proportional change in the displacement, u, at the point of application of the load, as shown in the

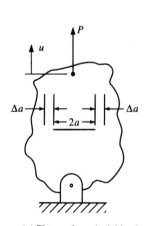

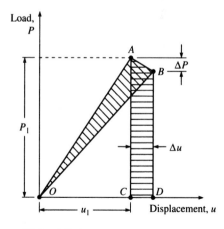

(*a*) Plate under uniaxial load   (*b*) Load-displacement curve

FIGURE 9.4
Loaded plate and corresponding load-displacement curve used for strain energy release rate analysis.

load-displacement plot in Fig. 9.4(b). We now assume that once the load reaches the value P_1 and the corresponding displacement reaches u_1, the crack extends a small increment, Δa. The crack extension causes the load to drop by an amount ΔP and the displacement to increase by an amount Δu. Just before the crack extension occurs, the potential energy, U, stored in the plate is given by the triangular area OAC in Fig. 9.4(b). The potential energy, ΔU, released by the crack extension is given by the triangular area OAB. During the incremental displacement Δu the increment of work done on the plate is ΔW, or the area $ABDC$. For this mode I crack deformation the strain energy release rate, G_I (do not confuse with the shear modulus, G), or the rate of change of the strain energy with respect to the crack extension area, A, is defined by [9.26]

$$G_I = \lim_{\Delta A \to 0} \frac{\Delta W - \Delta U}{\Delta A} = \frac{dW}{dA} - \frac{dU}{dA} \tag{9.21}$$

The system compliance, s, is given by

$$s = \frac{u}{P} \tag{9.22}$$

(Note that this is the system compliance, s, not the material compliance, S, defined earlier as being a ratio of strain to stress.) Thus, the potential energy of the plate in Fig. 9.4(a) is

$$U = \frac{1}{2} Pu = \frac{1}{2} sP^2 \tag{9.23}$$

so that

$$\frac{dU}{dA} = sP \frac{\partial P}{\partial A} + \frac{1}{2} P^2 \frac{\partial s}{\partial A} \tag{9.24}$$

The incremental work done during the crack extension is approximately

$$\Delta W = P(\Delta u) \tag{9.25}$$

so that

$$\frac{dW}{dA} = \lim_{\Delta A \to 0} \frac{\Delta W}{\Delta A} = \lim_{\Delta A \to 0} P \frac{\Delta u}{\Delta A} = P \frac{du}{dA} = P \frac{d}{dA} (sP)$$

$$= Ps \frac{\partial P}{\partial A} + P^2 \frac{\partial s}{\partial A} \tag{9.26}$$

Substitution of Eqs. (9.24) and (9.26) in Eq. (9.21) gives

$$G_I = \frac{P^2}{2} \frac{\partial s}{\partial A} \tag{9.27}$$

For a plate of constant thickness, t, $\partial A = t \, \partial a$ and

$$G_1 = \frac{P^2}{2t} \frac{\partial s}{\partial a} \qquad (9.28)$$

Thus, we can determine G_1 by plotting the compliance as a function of crack length and finding the slope of the curve, ds/da, corresponding to the value of the load, P. The critical strain energy release rate, G_{1c}, for this mode I crack deformation corresponds to the values P_c and $(ds/da)_c$ at fracture. That is,

$$G_{1c} = \frac{P_c^2}{2t} \left(\frac{\partial s}{\partial a} \right)_c \qquad (9.29)$$

From the point of view of the experimentalist, the obvious advantage of Eq. (9.29) is that knowledge of material properties or crack stress distributions is not needed since all the parameters can be determined from measurements on a test specimen. Note also that the method applies to either isotropic or anisotropic materials. As shown later in Sec. 9.4, Eq. (9.21) has been used extensively for both measurement and calculation of the strain energy release rate for mode I delamination in laminates. Measurements of the strain energy release rate based on these equations will be discussed in Chap. 10.

Another major advantage of the strain energy release rate is that it is related to the stress intensity factor. As shown by Irwin [9.13], for mode I crack deformation in isotropic materials under plane stress

$$K_1^2 = G_1 E \qquad (9.30)$$

so that the critical stress intensity factor, or fracture toughness, K_{1c}, is related to the critical strain energy release rate, G_{1c}, by

$$K_{1c}^2 = G_{1c} E \qquad (9.31)$$

This relationship has been used to determine the K_{1c} of composites from measurements of the G_{1c} [9.23] and to find G_{1c} from measurements of K_{1c} [9.20].

Cruse [9.27] has been shown that for a through-thickness mode I crack in an orthotropic laminate having N angle-ply components and having strain compatibility among the plies ahead of the crack the critical strain energy release rate, G_{1c}, for the laminate is related to the corresponding lamina properties by a simple rule of mixtures of the form

$$G_{1c} = \frac{\sum\limits_{i=1}^{N} G_{1ci} t_i}{t} \qquad (9.32)$$

where G_{Ic} = critical strain energy release rate for the laminate
  G_{Ici} = critical strain energy release rate for the ith angle-ply component
  t = total laminate thickness
  t_i = thickness of the ith angle-ply component

The predictions from this equation were found to show good agreement with experimental results for graphite/epoxy laminates [9.27].

The strain energy release rate has also proved to be useful in the characterization of the crack growth rate under cyclic loading. Interest in the possible relationship between fatigue crack growth rate and the strain energy release rate was prompted by the previous work of Paris and Erdogan [9.28], which showed that the mode I crack growth rate, da/dN, in many metals and polymers can be characterized by the equation

$$\frac{da}{dN} = B(\Delta K)^m \tag{9.33}$$

where N = number of cycles of repetitive loading
  ΔK = stress intensity factor range = $K_{Imax} - K_{Imin}$
   = $(\sigma_{max} - \sigma_{min})\sqrt{\pi a}$ for mode I crack growth
  σ_{max} = maximum stress
  σ_{min} = minimum stress
  B, m = experimentally determined empirical factors for a given material, loading conditions, and environment

Equation (9.33) has also found limited use in composites. For example, Kunz and Beaumont [9.29] observed that transverse crack growth in unidirectional graphite/epoxy composites under cyclic compressive loading could be described by such an equation. Fatigue damage in composites cannot always be described in terms of self-similar crack growth, however. More often than not, fatigue damage is a very complex condition involving mixed modes of failure, and the analytical determination of the stress intensity factor for such a condition may be nearly impossible. Thus, the strain energy release rate range, ΔG, may be a more convenient parameter to use than the stress intensity factor range, ΔK. For example, Spearing et al. [9.30] have modeled fatigue damage growth in notched graphite/epoxy laminates by using an equation formed by combining Eqs. (9.30) and (9.33):

$$\frac{da}{dN} = C(\Delta G)^{m/2} \tag{9.34}$$

where $C = BE^{m/2}$

Example 9.1. A quasi-isotropic graphite/epoxy laminate has a fracture toughness $K_{1c} = 30$ MPa m$^{1/2}$ and a tensile strength of 500 MPa. As shown in Fig. 9.5, a 25-mm-wide structural element made from this material has an edge crack of length $a = 3$ mm. If the element is subjected to a uniaxial stress, σ, determine the critical value of the stress which would cause unstable propagation of the crack. Compare this stress with the tensile strength of the material, which does not take cracks into account.

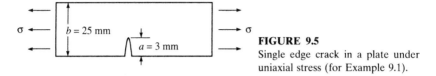

FIGURE 9.5
Single edge crack in a plate under uniaxial stress (for Example 9.1).

Solution. From tabulated solutions [9.17], the stress intensity factor for the single edge crack in Fig. 9.5 is

$$K_1 = \sigma\sqrt{\pi a}\,f(a/b)$$

where the function $f(a/b)$ is given by the empirical formula [9.17]

$$f(a/b) = 1.12 - 0.231(a/b) + 10.55(a/b)^2$$
$$- 21.72(a/b)^3 + 30.39(a/b)^4$$

which is said to be accurate within 0.5 percent when $a/b \le 0.6$. For this case, $a/b = 3/25 = 0.12$ and $f(a/b) = 1.213$. The critical stress is then

$$\sigma_c = \frac{K_{1c}}{\sqrt{\pi a}\,f(a/b)} = \frac{30}{\sqrt{\pi(0.003)}\,(1.213)} = 255 \text{ MPa}$$

Comparing this stress with the tensile strength of 500 MPa, we see that, in this case the cracked element can sustain only about 50 percent of the stress that an uncracked element could withstand.

9.3 STRESS FRACTURE CRITERIA FOR THROUGH-THICKNESS NOTCHES

Although fracture mechanics concepts have been successfully used in some cases to analyze the effects of through-thickness cracks and notches in composite laminates, Whitney and Nuismer [9.31, 9.32] questioned the need for such an approach and then proceeded to develop a simpler approach which is perhaps more useful to designers. As pointed out previously, the use of fracture mechanics in such applications has always been in question because the self-similar crack growth that occurs in metals does not always occur in composite laminates. Additional motivation for the work of Whitney and Nuismer was provided by the need to understand better experimental results which showed that larger holes in laminates under tension cause greater strength reductions than do smaller holes. In a previous attempt to explain this effect Waddoups et al. [9.33]

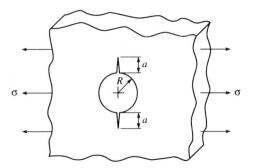

FIGURE 9.6
Uniaxially stressed plate with an edge-cracked hole.

had employed a fracture mechanics analysis of a hole in an isotropic plate with two symmetrically placed cracks extending from either side of the hole, as shown in Fig. 9.6. The stress intensity factor for a mode I crack having this geometry was derived using the previous solution of Bowie [9.34] as

$$K_1 = \sigma\sqrt{\pi a}\, f(a/R) \tag{9.35}$$

While the function $f(a/R)$ has been tabulated for the isotropic case [9.17], it has not been determined for the anisotropic case. Thus, although the analysis of Waddoups et al. [9.33] predicted the experimentally observed trends regarding the effect of hole size, the effects of anisotropy were obviously not considered. In addition, no physical interpretation was given for the cracks at the edge of the hole (i.e., such cracks were used in the analysis but were not necessarily present in the experiments which showed the hole size effect).

Whitney and Nuismer [9.31, 9.32] reasoned that the hole size effect could also be explained by observing the differences in the stress distributions near the hole for large and small holes. For example, the theory of elasticity solutions [9.35] for the normal stress distribution, σ_y, along the x axis near a hole in an infinite isotropic plate under uniform tensile stress are shown in Fig. 9.7 for small ($R = 0.1$ in) and large ($R = 1.0$ in) holes. The stress distribution for the smaller hole obviously has a sharper concentration near the hole than does the stress distribution for the larger hole. Whitney and Nuismer observed that since the plate with the smaller hole would be more capable of redistributing high stresses near the hole than would the plate with the larger hole, the plate with the smaller hole would be stronger. This observation led to the development of two failure criteria which were based on solutions for the normal stress, σ_y, along the x axis near circular holes (Fig. 9.7) and center cracks (Fig. 9.8) in infinite orthotropic plates. The Whitney-Nuismer criteria [9.32] are now summarized.

The hole of radius R in Fig. 9.7 is assumed to be in an infinite orthotropic plate which is under uniform stress, σ, at infinity. The

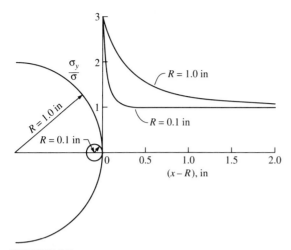

FIGURE 9.7
Normal stress distribution for a circular hole in an infinite isotropic plate. (*From Nuismer and Whitney* [9.32]. *Copyright ASTM. Reprinted with permission.*)

normal stress, $\sigma_y(x, 0)$, along the x axis near the hole is approximately

$$\sigma_y(x, 0) = \frac{\sigma}{2}\left\{2 + \left(\frac{R}{x}\right)^2 + 3\left(\frac{R}{x}\right)^4 - (K_T^\infty - 3)\left[5\left(\frac{R}{x}\right)^6 - 7\left(\frac{R}{x}\right)^8\right]\right\} \quad (9.36)$$

where $x > R$ and the orthotropic stress concentration factor, K_T^∞, for an

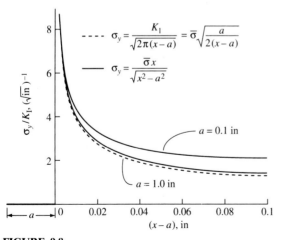

FIGURE 9.8
Normal stress distribution for a center crack in an infinite anisotropic plate. (*From Nuismer and Whitney* [9.32]. *Copyright ASTM. Reprinted with permission.*)

infinite width plate is given by Lekhnitskii [9.36] as

$$K_T^\infty = 1 + \sqrt{\frac{2}{A_{22}}\left(\sqrt{A_{11}A_{22}} - A_{12} + \frac{A_{11}A_{22} - A_{12}^2}{2A_{66}}\right)} \qquad (9.37)$$

where the A_{ij} are the laminate extensional stiffnesses from the Classical Lamination Theory and the subscript 1 denotes the direction parallel to the applied stress, σ.

The first failure criterion proposed by Whitney and Nuismer, referred to as the "point stress criterion," is based on the assumption that failure occurs when the stress σ_y at some fixed distance, d_0, away from the edge of the hole reaches the unnotched tensile strength of the material, σ_0. This criterion is given by

$$\sigma_y(R + d_0, 0) = \sigma_0 \qquad (9.38)$$

By combining Eqs. (9.36) and (9.38), we find that the ratio of notched to unnotched strength is

$$\frac{\sigma_N^\infty}{\sigma_0} = \frac{2}{2 + \xi_1^2 + 3\xi_1^4 - (K_T^\infty - 3)(5\xi_1^6 - 7\xi_1^8)} \qquad (9.39)$$

where
$$\xi_1 = \frac{R}{R + d_0}$$

and the notched tensile strength, σ_N^∞, of the infinite width laminate is equal to the applied stress, σ, at failure. Whitney and Nuismer noted that for very large holes $\xi_1 \rightarrow 1$, and the classical stress concentration result, $\sigma_N^\infty/\sigma_0 = 1/K_T^\infty$, is recovered. As $\xi_1 \rightarrow 0$, however, $\sigma_N^\infty/\sigma_0 \rightarrow 1$, as expected.

The second failure criterion proposed by Whitney and Nuismer, referred to as the "average stress criterion," is based on the assumption that failure occurs when the average value of σ_y over some fixed distance, a_0, from the edge of the hole reaches the unnotched tensile strength of the material, σ_0. This criterion is given by

$$\frac{1}{a_0}\int_R^{R+a_0} \sigma_y(x, 0)\, dx = \sigma_0 \qquad (9.40)$$

By combining Eqs. (9.36) and (9.40), we find that the ratio of notched to unnotched strength is

$$\frac{\sigma_N^\infty}{\sigma_0} = \frac{2(1 - \xi_2)}{2 - \xi_2^2 - \xi_2^4 + (K_T^\infty - 3)(\xi_2^6 - \xi_2^8)} \qquad (9.41)$$

where
$$\xi_2 = \frac{R}{R + a_0}$$

and σ_N^∞ is again the notched tensile strength of the infinite width laminate. As in the point stress criterion, the expected limits are recovered for the cases when $\xi_2 \to 1$ and $\xi_2 \to 0$.

Whitney and Nuismer also applied the point stress criterion and the average stress criterion to the case of the center crack of length $2a$ in an infinite anisotropic plate under uniform tensile stress, σ, as shown in Fig. 9.8. They used Lekhnitskii's [9.36] exact elasticity solution for the normal stress, σ_y, along the x axis near the edge of the crack, which is given by

$$\sigma_y(x, 0) = \frac{\sigma x}{\sqrt{x^2 - a^2}} = \frac{K_I x}{\sqrt{\pi a(x^2 - a^2)}} \tag{9.42}$$

where $x > a$ and $K_I = \sigma\sqrt{\pi a}$ is the mode I stress intensity factor. Substitution of this stress distribution in the point stress failure criterion given by Eq. (9.38) leads to the expression

$$\frac{\sigma_N^\infty}{\sigma_0} = \sqrt{1 - \xi_3^2} \tag{9.43}$$

where
$$\xi_3 = \frac{a}{a + d_0}$$

Substitution of the stress distribution from Eq. (9.42) in the average stress criterion given by Eq. (9.40) yields

$$\frac{\sigma_N^\infty}{\sigma_0} = \sqrt{\frac{1 - \xi_4}{1 + \xi_4}} \tag{9.44}$$

where
$$\xi_4 = \frac{a}{a + a_0}$$

Whitney and Nuismer then reasoned that the effect of crack size on the measured fracture toughness of the notched laminate could be better understood by defining a parameter

$$K_Q = \sigma_N^\infty \sqrt{\pi a} \tag{9.45}$$

which is the fracture toughness corresponding to the notched tensile strength of the infinite width laminate. Substitution of Eq. (9.43) in Eq. (9.45) yields

$$K_Q = \sigma_0 \sqrt{\pi a(1 - \xi_3^2)} \tag{9.46}$$

for the point stress criterion. Similarly, substitution of Eq. (9.44) in Eq.

(9.45) yields

$$K_Q = \sigma_0 \sqrt{\frac{\pi a(1 - \xi_4)}{1 + \xi_4}} \tag{9.47}$$

for the average stress criterion. For vanishly small crack lengths, a, the numerical values of both Eqs. (9.46) and (9.47) approach the limit $K_Q = 0$. For large crack lengths K_Q asymptotically approaches

$$K_Q = \sigma_0 \sqrt{2\pi d_0} \tag{9.48}$$

for the point stress criterion and

$$K_Q = \sigma_0 \sqrt{\pi a_0 / 2} \tag{9.49}$$

for the average stress criterion.

In order to use these stress fracture criteria, it is necessary to do enough experiments to establish values of d_0 or a_0 which give acceptable predicted values of σ_N^∞. Whitney and Nuismer observed that the applicability of these criteria in design depends to a great extent on whether the distance d_0 or a_0 is constant for all hole or crack sizes in at least a particular laminate of a particular material system. If d_0 or a_0 were constant for all laminates of all material systems, the criteria would be even more useful.

Whitney and Nuismer showed that fixed values of d_0 and a_0 in the criteria gave reasonably good agreement with experimental results for graphite/epoxy and glass/epoxy laminates in two different laminate configurations [9.32]. For example, Fig. 9.9 shows a comparison of the

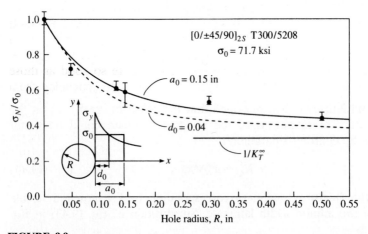

FIGURE 9.9
Comparison of predicted and measured failure stresses for circular holes in $[0/\pm45/90]_{2s}$ T300/5208 graphite/epoxy. (*From Nuismer and Whitney* [9.32]. *Copyright ASTM. Reprinted with permission.*)

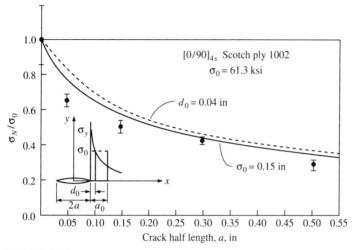

FIGURE 9.10

Comparison of predicted and measured failure stresses for center cracks in $[0/90]_{4s}$ Scotchply 1002 E-glass/epoxy. (*From Nuismer and Whitney* [9.32]. *Copyright ASTM. Reprinted with permission.*)

predictions from the point stress criterion for circular holes [Eq. (9.39)] and the average stress criterion for circular holes [Eq. (9.41)] with experimental data for $[0/\pm45/90]_{2s}$ graphite/epoxy laminates. Similarly, Fig. 9.10 shows a comparison of the predictions from the point stress criterion for center cracks [Eq. (9.43)] and the average stress criterion for center cracks [Eq. (9.44)] with experimental data for $[0/90]_{4s}$ glass/epoxy laminates. Note that the same values of d_0 and a_0 were used for both material systems and laminate configurations, and that both criteria correctly predict the effect of the hole size or crack size on the notched strength. The results for graphite/epoxy are not quite so good as those for glass/epoxy, however. Even though it could not be concluded from this work that d_0 and a_0 are universal constants, the equations can be used with confidence for a particular material system under uniaxial loading. It should also be remembered that these criteria can be used for any through-thickness discontinuity for which the theoretical stress distribution can be found, not just for circular holes or straight cracks. Thus, given the relative simplicity of the equations, the Whitney-Nuismer criteria appear to be of considerable value to designers.

Example 9.2. A large plate made from the quasi-isotropic graphite/epoxy laminate in Example 9.1 has a center crack of length $2a = 6$ mm and is subjected to a uniform uniaxial stress. Compare the predicted fracture strengths of the plate according to the fracture mechanics criterion, the point stress criterion, and the average stress criterion. Use the Whitney-Nuismer values of d_0 and a_0 from Figs. 9.9 and 9.10.

Solution. For the fracture mechanics approach we rearrange Eq. (9.16) as

$$\sigma_c = \frac{K_{Ic}}{\sqrt{\pi a}} = \frac{30}{\sqrt{\pi(0.003)}} = 309 \text{ MPa}$$

For the point stress criterion we use $d_0 = 0.04$ in $= 1.016$ mm and $a = 3$ mm in Eq. (9.43) as

$$\sigma_N^\infty = \sigma_0 \sqrt{1 - \xi_3^2} = (500)\sqrt{1 - [3.0/(3.0 + 1.016)]^2} = 332 \text{ MPa}$$

For the average stress criterion we use $a_0 = 0.15$ in $= 3.81$ mm and $a = 3$ mm in Eq. (9.44) as

$$\sigma_N^\infty = \sigma_0 \sqrt{\frac{1 - \xi_4}{1 + \xi_4}} = (500)\sqrt{\frac{1 - (3.0/(3.0 + 3.81))}{1 + (3.0/(3.0 + 3.81))}} = 312 \text{ MPa}$$

The results from all three analyses are reasonably close, and the fracture mechanics criterion is slightly more conservative than the point stress criterion and the average stress criterion in this case. Clearly, the predicted fracture strengths in all three cases are considerably lower than the unnotched tensile strength of 500 MPa, and we see that the effects of such cracks should not be ignored in design.

9.4 INTERLAMINAR FRACTURE

Delamination, or interlaminar fracture, is a very important failure mode in composite laminates, and research activity regarding the onset and growth of delaminations has continued at a high level for the past decade or so. The mechanics of interlaminar stresses and several mechanics of materials approaches to the prediction of the onset of delamination were discussed previously in Chap. 7. In this section we will discuss the use of fracture mechanics approaches, particularly those involving the use of the strain energy release rate, for the prediction of delamination growth and failure.

Delamination provides one of the few examples of self-similar crack growth in composite laminates. A delamination is in effect a crack separating adjacent laminae, and the plane of the crack lies in the plane of the interface between laminae. Like a crack in a metallic material, a delamination grows in a stable manner until it reaches a critical size, whereupon further growth occurs in an unstable manner. These characteristics make interlaminar fracture a prime candidate for the application of fracture mechanics analysis. On the other hand, as pointed out in Chap. 7, interlaminar stresses are part of a complex three-dimensional state of stress that leads to delamination. While such a complex state of stress at the crack tip inhibits the effective use of the stress intensity factor approach, it makes the problem ideally suited for the strain energy release rate approach.

One of the first reports on the use of the strain energy release rate approach in the analysis of delamination was apparently that of Roderick et al. [9.37], who correlated strain energy release rates with the rates of cyclic debonding between metal panels and composite reinforcement using an equation similar to Eq. (9.34). Shortly thereafter, in a critical review of the applications of fracture mechanics in composites, Kanninen et al. [9.38] noted that the strain energy release rate had seen little application to composites. This observation led to the use of the strain energy release rate by Rybicki et al. [9.39] in an analytical and experimental study of free-edge delamination in boron/epoxy laminates. Rather than using Eq. (9.21) to calculate the strain energy release rate, Rybicki et al. [9.39] employed a finite element formulation of a crack closure integral technique. Such techniques are beyond the scope of this book and will not be discussed further here.

Wang [9.40] conducted experimental and analytical studies of delamination growth in unidirectional glass/epoxy composite specimens. As shown in Fig. 9.11, delamination crack initiators were introduced in the specimens by cutting across several surface plies with a razor blade. The specimens were then subjected to cyclic tension-tension fatigue loading while the length of the delamination, l_d, was measured. Figure 9.12 shows typical data on delamination crack length vs. the number of load cycles, N, at different stress levels. The delamination growth rate, dl_d/dN, at any number N is the tangent of the curve at that value of N. It is particularly important to note in Fig. 9.12 that at a critical number of loading cycles, N_c, corresponding to a critical delamination size for a given stress level, the delamination growth becomes unstable and rapid crack propagation occurs. Such experiments provided further proof of the similarity between crack growth in metals and delamination growth in composite laminates and justified the use of the principles of fracture mechanics in the analysis of delamination.

Wang [9.40] used a hybrid stress finite element method to determine the stress intensity factors K_I and K_{II} for the mixed mode crack growth, which were then correlated with the delamination growth rate by equations similar to Eq. (9.33). In this case, due to the mixed mode delamination, the relationships for the two crack deformation modes are

$$\frac{dl_d}{dN} \sim (\Delta K_I)^a \tag{9.50}$$

for mode I crack opening and

$$\frac{dl_d}{dN} \sim (\Delta K_{II})^b \tag{9.51}$$

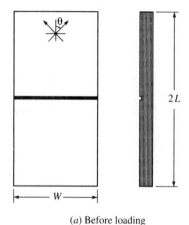

(*a*) Before loading

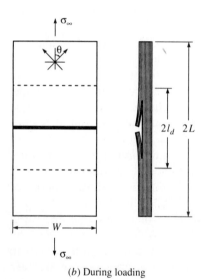

(*b*) During loading

FIGURE 9.11
Specimen for delamination crack growth study ($2L = 152.4$ mm, $W = 25.4$ mm). (*From Wang* [9.40]. *Copyright ASTM. Reprinted with permission.*)

for mode II crack shearing, where a and b are empirically determined exponents. Equations (9.50) and (9.51), when plotted on a log-log plot, should form a straight line. The validity of these equations is confirmed by plotting the experimental data on a tridimensional log-log plot, as shown in Fig. 9.13. The data in Fig. 9.13 were found to follow a general relationship of the form

$$\frac{\log\left(dl_d/dN\right)}{\alpha_1} = \frac{\log\left(\Delta K_I\right) + C_1}{\alpha_2} = \frac{\log\left(\Delta K_{II}\right) + C_2}{\alpha_3} \quad (9.52)$$

where the α_i ($i = 1, 2, 3$) are the directional cosines of the line

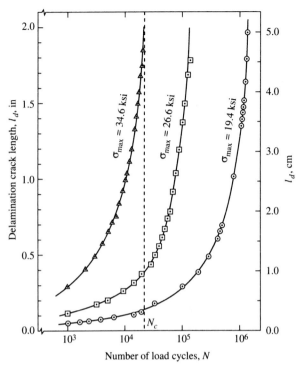

FIGURE 9.12
Delamination crack growth during fatigue in unidirectional glass/epoxy. (*From Wang* [9.40]. *Copyright ASTM. Reprinted with permission.*)

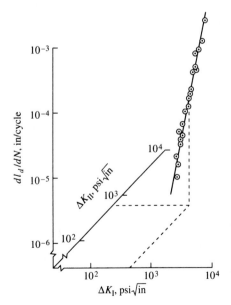

FIGURE 9.13
Fatigue delamination crack growth rate, dl_d/dN, as a function of mixed mode stress intensity factor ranges ΔK_I and ΔK_{II} for unidirectional glass/epoxy. (*From Wang* [9.40]. *Copyright ASTM. Reprinted with permission.*)

$dl_d/dN = f(\Delta K_I, \Delta K_{II})$ with respect to the three axes, respectively, and C_1 and C_2 are constants associated with the opening and shearing modes, respectively.

Both mechanics of materials and fracture mechanics analyses were used by O'Brien [9.41] to study the onset and growth of edge delaminations (see previous Fig. 7.31) in graphite/epoxy laminates. O'Brien's mechanics of materials approach was discussed previously in Chap. 7. A laminate stacking sequence of $[\pm 30/\pm 30/90/\overline{90}]_s$ was selected so that edge delamination growth in tensile specimens would readily occur under cyclic loading, and delamination growth was monitored nondestructively. The strain energy release rate, G, associated with delamination growth was determined from two different analyses, only one of which will be discussed here. One method involved the use of the general equation for the strain energy release rate, Eq. (9.21). The work done during crack extension, W, was ignored, so that

$$G = -\frac{dU}{dA} \qquad (9.53)$$

The subscript I on G has been dropped here because the edge delamination growth is of the mixed mode type and the strain energy release rate may have components due to G_I, G_{II}, and G_{III}. Superposition of the strain energy release rates for different modes will be discussed later. Expressing the strain energy in terms of the strain energy density, $E\epsilon^2/2$, and the volume, V, Eq. (9.53) becomes

$$G = -V\frac{\epsilon^2}{2}\frac{dE}{dA} \qquad (9.54)$$

where ϵ = nominal longitudinal strain

E = longitudinal Young's modulus of a laminate partially delaminated along one or more interfaces

In this case $dA = 2L\,da$ and $V = 2bLt$, where a, b, and t were defined previously in Fig. 7.31 and L is the length of the laminate. Substituting these definitions in Eq. (9.54), along with the definition of E from Eq. (7.115), O'Brien found that

$$G = \frac{\epsilon^2 t}{2}(E_x - E_{td}) \qquad (9.55)$$

where E_x and E_{td} were defined previously along with Eq. (7.115). Thus, the strain energy release rate is independent of delamination size and depends only on E_x and E_{td} (which are determined by the laminate lay-up and the location of the delaminated interfaces), the strain, ϵ, and the thickness, t. The critical strain, ϵ_c, at the onset of delamination was measured for the $[\pm 30/\pm 30/90/\overline{90}]_s$ laminates and used in Eq. (9.55) to

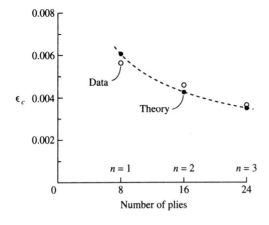

FIGURE 9.14
Edge delamination onset prediction compared with experimental data for $[+45_n/-45_n/0_n/90_n]_s$ graphite/epoxy, where $n = 1, 2, 3$. (*From O'Brien* [9.41]. *Copyright ASTM. Reprinted with permission.*)

determine the corresponding critical strain energy release rate, G_c. This value of G_c was then used to predict the critical value, ϵ_c, at the onset of delamination in $[+45_n/-45_n/0_n/90_n]_s$ laminates. A comparison of measured and predicted values of ϵ_c for different numbers of plies, n, is shown in Fig. 9.14, and the agreement is seen to be very good.

As previously mentioned, the edge delamination test used by O'Brien [9.41] involved mixed mode crack deformations. He used a finite element implementation of a crack closure technique developed by Rybicki et al. [9.39] to find the components G_I, G_{II}, and G_{III}. The total G was then found from the superposition relationship

$$G = G_I + G_{II} + G_{III} \qquad (9.56)$$

In this case G_{III} turned out to be negligible. Equation (9.56) is valid when the plane of the crack and the plane of crack extension coincide with a principal axis of material property symmetry [9.26].

O'Brien also found excellent correlation between delamination growth rate, da/dN, and the maximum strain energy release rate, G_{max}, by using an equation of the form

$$\frac{da}{dN} = cG_{max}^{\beta} \qquad (9.57)$$

where c and β are empirically determined constants. Figure 9.15 shows a comparison of predictions from this equation with experimental data, and the agreement is excellent.

As described above, the experiments of Wang [9.40] and O'Brien [9.41] involved mixed mode delamination, and the different components of the stress intensity factor or the strain energy release rate corresponding to modes I, II, and III had to be determined separately by using finite element techniques. In order to understand delamination better and,

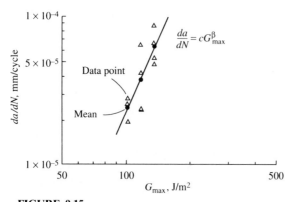

FIGURE 9.15
Power law curve fit for da/dN as a function of G_{max} for $[\pm30/30/90/\overline{90}]_s$ graphite/epoxy. (*From O'Brien* [9.41]. *Copyright ASTM. Reprinted with permission.*)

consequently, the best ways to improve interlaminar fracture toughness, there is an obvious need for delamination experiments which make it possible to isolate a single mode of crack growth. In the following paragraphs the most widely used experiments for single-mode measurement of interlaminar strain energy release rates will be briefly discussed, but details of the techniques will be left for Chap. 10 on mechanical testing of composites.

Mode I delamination has always been of interest because of the obvious weakness of the interlaminar region in through-thickness tension. Perhaps the most widely used mode I interlaminar fracture test method is the double cantilever beam (DCB) test, which was originally developed for studying fracture of adhesively bonded joints, then later adapted for interlaminar fracture of composite laminates [9.42–9.49]. A DCB specimen is shown in Fig. 9.16(*a*). In the DCB test the specimen is loaded transversely as shown in Fig. 9.16(*a*), so that mode I crack opening delamination occurs along the middle plane. The required test data are taken and the delamination G_{Ic} is calculated by using one of several different forms of Eq. (9.21) or Eq. (9.27), as described later in Chap. 10. Typical values of delamination G_{Ic} for several advanced composites, as determined by DCB tests, are tabulated in Table 9.1. The results of some of the attempts to improve the interlaminar fracture toughness are seen in Table 9.1, and these methods will be discussed in more detail later in this section.

Although mode I delamination has received considerable attention in the literature, there is increased interest in mode II delamination because of its apparent relationship to impact damage tolerance of laminates [9.50]. As mentioned in Sec. 7.8.2, transverse impact can cause internal cracks and delaminations which may be difficult to detect. If the

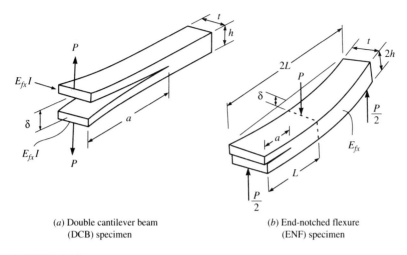

(a) Double cantilever beam
(DCB) specimen

(b) End-notched flexure
(ENF) specimen

FIGURE 9.16
Double cantilever beam and end-notched flexure specimens.

laminate is subsequently subjected to in-plane compressive loading, such cracks and delaminations can lead to buckling and reductions of in-plane compressive strength (Fig. 7.34). There is evidence that the so-called compression after impact (CAI) strength is improved by increasing the mode II critical interlaminar strain energy release rate, G_{IIc} [9.50]. One of the most popular tests for measurement of the critical strain energy release rate for mode II delamination is the end-notched flexure (ENF) test. An ENF specimen is shown in Fig. 9.16(b). The strain energy release rate analysis of the ENF specimen, which has been improved and used by several investigators [9.50–9.55], will be discussed in more detail in the review of test methods in Chap. 10.

Once the capability to measure G_{Ic} and G_{IIc} separately had been developed, it became possible to evaluate various interactive criteria for mixed mode delamination growth. Although there is no universal agreement on which mixed mode delamination growth criterion is the most accurate, one of the simplest and most widely used of these criteria is given by the equation

$$\left(\frac{G_I}{G_{Ic}}\right)^m + \left(\frac{G_{II}}{G_{IIc}}\right)^n = 1 \tag{9.58}$$

where G_I, G_{II} = strain energy release rates for delamination growth in modes I and II, respectively

G_{Ic}, G_{IIc} = critical strain energy release rates for delamination growth in modes I and II, respectively

m, n = empirically determined exponents

TABLE 9.1
Critical interlaminar strain energy release rates, G_{Ic}, for several advanced composites, as determined by double cantilever beam tests

Fiber/matrix combination	Lay-up	G_{Ic} J/m^2 (in-lb/in^2)	Source
T-300/5208 graphite/epoxy	$[0]_{24}$	87.6(0.50)	Wilkins et al. [9.43]
AS-1/3502 graphite/epoxy	$[0]_{24}$	140.1(0.80)	Whitney et al. [9.44]
AS-4/3502 graphite/epoxy	$[0]_{24}$	161.1(0.92)	Whitney et al. [9.44]
T-300/V387A graphite/bismaleimide	$[0]_{24}$	71.8(0.41)	Whitney et al. [9.44]
AS-1/polysulfone graphite/polysulfone	$[0]_{12}$	585.0(3.34)	Whitney et al. [9.44]
T-300/976 graphite/epoxy bidirectional cloth	Woven fabric, 10 plies	282.0(1.61)	Whitney et al. [9.44]
AS-4/3501-6 graphite/epoxy	$[0]_{24}$ †	198–254(1.31–1.45)	Aliyu and Daniel [9.45]
T-300/F-185 graphite/epoxy	$[0]_{24}$ ‡	1880–1500(10.7–8.6)	Daniel et al. [9.46]
AS-4/PEEK graphite/ polyetheretherketone	$[0]_{40}$ §	2890–2410(16.5–13.8)	Leach et al. [9.47]

Source: From Wilkins et al.; Aliyu and Daniel; Daniel et al; and Leach et al.
Copyright ASTM. Reprinted with permission.
Also from Whitney et al. *Copyright Technomic Publishing Company. Reprinted with permission.*

† Range of G_{Ic} is given for crack velocities of 0.05–49.0 mm/s, respectively. Thus, G_{Ic} increases with increasing strain rate for this material. The matrix is Hercules 3501-6, a standard prepreg-type epoxy resin [9.45].

‡ Range of G_{Ic} is given for crack velocities of 0.01–21.0 mm/s, respectively. Thus, G_{Ic} decreases with increasing strain rate for this material. The matrix is Hexcel F-185, which is an elastomer-modified and toughened epoxy [9.46].

§ Range of G_{Ic} is given for stable and unstable crack growth, respectively [9.47].

Good agreement between the predictions from this equation and experimental data has been reported by O'Brien et al. [9.56] and Johnson and Mangalgiri [9.57] when $m = n = 1$. O'Brien et al. [9.56] investigated the use of Eq. (9.58) for graphite/epoxy laminates having various lay-ups, and predictions are compared with experimental data from the edge delamination test [9.41] in Fig. 9.17. Some previous data from Murri and O'Brien [9.58] are included in Fig. 9.17. Johnson and Mangalgiri tested various matrix resins using the DCB, ENF, and several other methods, and comparisons of the predictions of Eq. (9.58) with experimental data

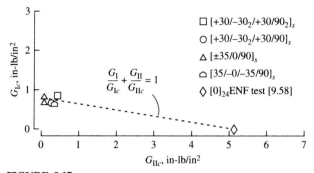

FIGURE 9.17
Comparison of predictions from Eq. (9.58) with mixed mode fracture data for T300/5208 graphite/epoxy laminates. (*From O'Brien et al.* [9.56]. *Copyright ASTM. Reprinted with permission.*)

are shown in Fig. 9.18. On the other hand, Ramkumar and Whitcomb [9.59] have concluded that Eq. (9.58) is not a reliabile delamination growth criterion for graphite/epoxy.

In recent years much research has gone into the improvement of interlaminar fracture toughness of composites, and the results of some of this research can be seen in the G_{Ic} data of Table 9.1. For example, since the interlaminar region consists primarily of matrix material, there has been considerable interest in the use of tough matrix materials. Significant improvements in the composite G_{Ic} have been obtained by using tough matrix materials such as polysulfone [9.44], elastomer-modified epoxy [9.46], and polyetheretherketone [9.47]. It is not clear, however, that additional increases in resin matrix toughness will necessarily be

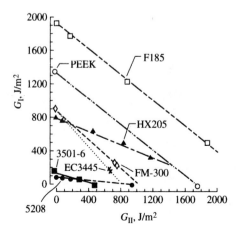

FIGURE 9.18
Comparison of predictions from Eq. (9.58) with mixed mode fracture data for several matrix resins. (*From Johnson and Mangalgiri* [9.57]. *Copyright ASTM. Reprinted with permission.*)

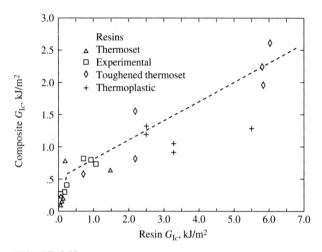

FIGURE 9.19
Mode I interlaminar strain energy release rates for steady crack growth in graphite fiber composites as a function of the neat resin strain energy release rates for several matrix resins. (*From Hunston et al.* [9.49]. *Copyright ASTM. reprinted with permission.*)

translated into correspondingly higher composite toughness [9.48, 9.49]. Figure 9.19 from Hunston et al. [9.49] shows that for resin G_{Ic} values less than about $0.4 \, kJ/m^2$ substantial gains in the corresponding graphite fiber composite, G_{Ic} are obtained by increasing the resin G_{Ic}. For resin G_{Ic} values greater than about $0.4 \, kJ/m^2$, however, the gains in the composite G_{Ic} from additional increases in resin G_{Ic} are not nearly as great. Scanning electron microscope studies of delamination fracture surfaces have shown that increased toughness of the matrix causes an increase in the delamination fracture toughness by increasing the size of the plastic zone ahead of the crack tip [9.48, 9.49]. Further increases in the size of this plastic zone are apparently prevented by the constraint of the fibers in the adjacent plies, however [9.48, 9.49].

A variety of other methods for increasing interlaminar fracture toughness of laminates have been investigated. For example, thin films, or "interleaves" made of a tough polymer resin, can be embedded between the fiber-reinforced resin laminae [9.60–9.64]. Coating the fibers with a thin, tough polymer film [9.65, 9.66], hybridization of different fiber types [9.67, 9.68], and stitching of adjacent laminae [9.69] have also been investigated. A critical review of methods for improving fracture toughness of composites through interface control has also been published [9.70]. Unfortunately, improvements in interlaminar toughness often come at the expense of degradation in other properties such as hot/wet strength and stiffness or viscoelastic creep response. Although significant progress has been made in understanding delamination, much is still to be learned. The study of delamination continues to be a very active research topic, and the reader is encouraged to consult recent journal publications and conference proceedings for the latest findings.

PROBLEMS

9.1. The thin-walled tabular shaft shown in Fig. 9.20 is made of randomly oriented, short fiber-reinforced metal matrix composite. The shaft has a longitudinal through-thickness crack of length $2a$ and is subjected to a torque $T = 1$ KN-m. If the mode II fracture toughness of the composite is $K_{IIc} = 40$ MPa-m$^{1/2}$, determine the critical crack size for self-sustaining crack growth.

9.2. (*a*) Determine the allowab'e torque, T, if the crack length for the shaft in Fig. 9.20 is $2a = 10$ mm. Use the same dimensions and fracture toughness values that were given in Problem 9.1

  (*b*) If the uniaxial yield stress for the shaft material is $Y = 1200$ MPa, and the crack is ignored, compare the answer from part (*a*) with the allowable torque based on the maximum shear stress criterion for yielding.

9.3. The tube shown in Fig. 9.20 is subjected to an internal pressure, $p = 5$ MPa, instead of a torque. Neglecting the stress along the longitudinal axis of the tube, and assuming that the mode I fracture toughness is $K_{Ic} = 10$ MPa-m$^{1/2}$, determine the critical crack size.

9.4. As in Problem 9.3, assume that the tube in Fig. 9.20 is subjected only to an internal pressure and neglect the longitudinal stress.

  (*a*) Determine the allowable internal pressure, p, if the crack length in Fig. 9.20 is $2a = 10$ mm. Use the same dimensions and fracture toughness values that were given in Problem 9.3.

  (*b*) Using the yield stress from Problem 9.2 and ignoring the crack, compare the answer from part (*a*) of this problem with the allowable internal pressure based on the maximim shear stress criterion for yielding.

9.5. Use the Whitney-Nuismer average stress criterion to estimate the allowable internal pressure for Problem 9.4 if the unnotched tensile strength of the material is $\sigma_0 = 1500$ MPa and the parameter $a_0 = 3$ mm.

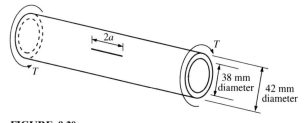

FIGURE 9.20
Thin-walled tubular composite shaft with longitudinal crack.

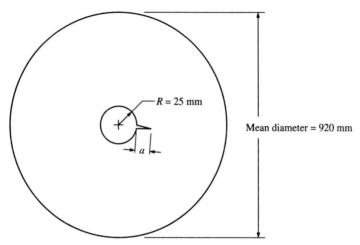

FIGURE 9.21
Spherical composite pressure vessel with single crack at the edge of entrance hole.

9.6. Repeat Problem 9.5 using the Whitney-Nuismer point stress criterion and the parameter $d_0 = 1$ mm.

9.7. The 920-mm diameter, 1.6-mm-thick spherical pressure vessel in Fig. 9.21 is a filament wound quasi-isotropic composite laminate with a single 50-mm diameter entrance hole. The vessel material has a mode I fracture toughness of $K_{1c} = 25$ MPa-m$^{1/2}$. If the vessel is to contain gas at 0.69 MPa, what is the critical length, a_c, of a single crack emanating from the edge of the hole? The Bowie equation [Eq. (9.35)] may be used for this problem, and the function $f(a/R)$ for a biaxial stress field and a single crack of length, a, at the edge of a hole of radius, R, is tabulated below for several values of a/R.

a/R	$f(a/R)$	a/R	$f(a/R)$
0.1	1.98	0.8	1.32
0.2	1.82	1.0	1.22
0.3	1.67	1.5	1.06
0.4	1.58	2.0	1.01
0.5	1.49	3.0	0.93
0.6	1.42	5.0	0.81

9.8. If the quasi-isotropic graphite/epoxy laminate in Example 7.5 has a centrally located 25-mm-diameter hole, determine the ratio of notched to unnotched uniaxial strength for the laminate using the Whitney-Nuismer average stress criterion. The parameter $a_0 = 4$ mm.

9.9. A 3-mm thick composite specimen is tested as shown in Fig. 9.4(a), and the compliance, $s = u/P$, as a function of the half-crack length, a, is shown in

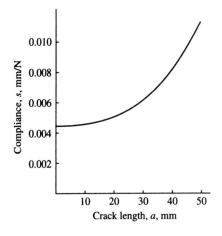

FIGURE 9.22
Variation of specimen compliance with crack length for Problem 9.9.

Fig. 9.22. In a separate test the critical load for self-sustaining crack propagation, P_c, is measured for different crack lengths, and the critical load corresponding to a crack length $2a = 50$ mm is found to be 100 N. Determine the critical mode I strain energy release rate, G_{Ic}.

REFERENCES

9.1. Gibson, R. F. (ed.), "Composites Contents," in *Journal of Composites Technology and Research,* published quarterly by the American Society for Testing and Materials, Philadelphia, PA.

9.2. Sendeckyj, G. P. (ed.), *Fracture Mechanics of Composites,* ASTM STP **593,** American Society for Testing and Materials, Philadelphia, PA (1975).

9.3. Reifsnider, K. L. (ed.), *Damage in Composite Materials,* ASTM STP **775,** American Society for Testing and Materials, Philadelphia, PA (1982).

9.4. Johnson, W. S. (ed.), *Delamination and Debonding of Materials,* ASTM STP **876,** American Society for Testing and Materials, Philadelphia, PA (1985).

9.5. Hahn, H. T. (ed.), *Composite Materials: Fatigue and Fracture,* ASTM STP **907,** American Society for Testing and Materials, Philadelphia, PA (1986).

9.6. Lagace, P. A. (ed.), *Composite Materials: Fatigue and Fracture,* ASTM STP **1012,** American Society for Testing and Materials, Philadelphia, PA (1989).

9.7. O'Brien, T. K. (ed.), *Composite Materials: Fatigue and Fracture, ASTM STP* **1110,** *American Society for Testing and Materials, Philadelphia, PA* (1991).

9.8. Friedrich, K. (ed.), *Application of Fracture Mechanics to Composite Materials,* Vol. 6 of Composite Material Series, R. B. Pipes (Series ed.), Elsevier Science Publisher, Amsterdam, The Netherlands (1989).

9.9. Nawaz, G. M. (ed.), *Delamination in Advanced Composites,* Technomic Publishing Co., Lancaster, PA (1991).

9.10. Griffith, A. A., "The Phenomena of Rupture and Flow in Solids," *Philosophical Transactions of the Royal Society,* **221A,** 163–198 (1920).

9.11. The original Griffith paper (Ref. [9.10]) was republished with corrections and commentary by J. J. Gilman in *Transactions of the American Society for Metals,* **61,** 861–906 (1968).

9.12. Sih, G. C. and Liebowitz, H., "Mathematical Theories of Brittle Fracture," in H. Liebowitz (ed.), *Fracture—An Advanced Treatise: Volume II Mathematical Fundamentals*, 67–190, Academic Press, New York (1968).

9.13. Irwin, G. R., *Fracturing of Metals*, 147–166, American Society of Metals, Cleveland, OH (1949).

9.14. Orowan, E., *Fatigue and Fracture of Metals*, MIT Press, Cambridge, MA (1950).

9.15. Westergaard, H. M., "Bearing Pressures and Cracks," *Transactions of the ASME, Series E, Journal of Applied Mechanics*, **61**, A49–A53 (1939).

9.16. Irwin, G. R., "Analysis of Stresses and Strains Near the End of a Crack Traversing a Plate," *Transactions of the ASME, Journal of Applied Mechanics*, **24**, 361–364 (1957).

9.17. Tada, H., Paris, P. C., and Irwin, G. R., *The Stress Analysis of Cracks Handbook*, Del Research Corporation, Hellertown, PA (1973).

9.18. Lekhnitskii, S. G., *Theory of Elasticity of an Anisotropic Elastic Body*, Holden-Day, Inc., San Francisco, CA (1963).

9.19. Wu, E. M., "Fracture Mechanics of Anisotropic Plates," in S. W. Tsai, J. C. Halpin, and N. J. Pagano (eds.), *Composite Materials Workshop*, 20–43, Technomic Publishing Co., Lancaser, PA (1968).

9.20. Konish, H. J., Swedlow, J. L., and Cruse, T. A., "Experimental Investigation of Fracture in an Advanced Composite," *Journal of Composite Materials*, **6**, 114–124 (1972).

9.21. Parhizgar, S., Zachary, L. W., and Sun, C. T., "Application of the Principles of Linear Fracture Mechanics to the Composite Materials," *International Journal of Fracture*, **20**, 3–15 (1982).

9.22. Alexander, R. M., Schapery, R. A., Jerina, K. L., and Sanders, B. A., "Fracture Characterization of a Random Fiber Composite Material," in B. A. Sanders (ed.), *Short Fiber Reinforced Composite Materials*, ASTM STP **772**, 208–224, American Society for Testing and Materials, Philadelphia, PA, (1982).

9.23. Sun, C. T. and Sierakowski, R. L., "Fracture Characterization of Composites with Chopped Fiberglass Reinforcement," *SAMPE Quarterly*, **11**(4), 15–21 (1980).

9.24. Gaggar, S. K. and Broutman, L. J., "The Development of a Damage Zone at the Tip of a Crack in a Glass Fiber Reinforced Polyester Resin," *International Journal of Fracture*, **10**, 606–608 (1974).

9.25. Irwin, G. R., "Fracture," in S. Flugge (ed.), *Handbuch der Physik*, Vol. **6**, 551–590, Springer, Berlin (1958).

9.26. Corten, H. T., "Fracture Mechanics of Composites," in H. Liebowitz (ed.), *Fracture—An Advanced Treatise: Volume VII Fracture of Nonmetals and Composites*, 675–769, Academic Press, New York (1972).

9.27. Cruse, T. A., "Tensile Strength of Notched Composites," *Journal of Composite Materials*, **7**, 218–229 (1973).

9.28. Paris, P. C. and Erdogan, F., "A Critical Analysis of Crack Propagation Laws," *Transactions of ASME, Journal of Basic Engineering*, **85**, 528–534 (1963).

9.29. Kunz, S. C. and Beaumont, P. W. R., "Microcrack Growth in Graphite Fiber-Epoxy Resin Systems during Compressive Fatigue," in J. R. Hancock (ed.), *Fatigue of Composite Materials*, ASTM STP **569**, 71–91, American Society for Testing and Materials, Philadelphia, PA (1975).

9.30. Spearing, M., Beaumont, P. W. R., and Ashby, M. F., "Fatigue Damage Mechanics of Notched Graphite-Epoxy Laminates," in T. K. O'Brien (ed.), *Composite Materials: Fatigue and Fracture (Third Volume)*, ASTM STP **1110**, 617–637, American Society for Testing and Materials, Philadelphia, PA (1991).

9.31. Whitney, J. M. and Nuismer, R. J., "Stress Fracture Criteria for Laminated Composites Containing Stress Concentrations," *Journal of Composite Materials*, **8**, 253–265 (1974).

9.32. Nuismer, R. J. and Whitney, J. M., "Uniaxial Failure of Composite Laminates Containing Stress Concentrations," in *Fracture Mechanics of Composites*, ASTM STP **593**, 117–142, American Society for Testing and Materials, Philadelphia, PA (1975).

9.33. Waddoups, M. E., Eisenmann, J. R., and Kaminski, B. E., "Macroscopic Fracture Mechanics of Advanced Composite Materials," *Journal of Composite Materials*, **5**(4), 446–454 (1971).

9.34. Bowie, O. L., "An Analysis of an Infinite Plate Containing Radial Cracks Originating from the Boundary of an Internal Circular Hole," *Journal of Mathematics and Physics*, **35**, 60–71 (1956).

9.35. Timoshenko, S. P. and Goodier, J. N., *Theory of Elasticity*, 2d ed., McGraw-Hill, Inc., New York (1951).

9.36. Lekhnitskii, S. G., *Anisotropic Plates*, Translated from 2d Russian ed. by S. W. Tsai and T. Cheron, Gordon and Breach Science Publishers, New York (1968).

9.37. Roderick, G. L., Everett, R. A., and Crews, J. H., "Debond Propagation in Composite-Reinforced Metals," in *Fatigue of Composite Materials*, ASTM STP **569**, 295–306, American Society for Testing and Materials, Philadelphia, PA (1975).

9.38. Kanninen, M. F., Rybicki, E. F., and Brinson, H. F., "A Critical Look at Current Applications of Fracture Mechanics to the Failure of Fiber Reinforced Composites," *Composites*, **8**, 17–22 (January 1977).

9.39. Rybicki, E. F., Schmueser, D. W., and Fox, J., "An Energy Release Rate Approach for Stable Crack Growth in the Free-Edge Delamination Problem," *Journal of Composite Materials*, **11**, 470–487 (October 1977).

9.40. Wang, S. S., "Delamination Crack Growth in Unidirectional Fiber-Reinforced Composite under Static and Cyclic Loading," in S. W. Tsai (ed.), *Composite Materials: Testing and Design*, ASTM STP **674**, 642–663, American Society for Testing and Materials, Philadelphia, PA (1979).

9.41. O'Brien, T. K., "Characterization of Delamination Onset and Growth in a Composite Laminate," in K. L. Reifsnider (ed.), *Damage in Composite Materials*, ASTM STP **775**, 140–167, American Society for Testing and Materials, Philadelphia, PA (1982).

9.42. Devitt, D. F., Schapery, R. A., and Bradley, W. L., "A Method for Determining the Mode I Delamination Fracture Toughness of Elastic and Viscoelastic Composite Materials," *Journal of Composite Materials*, **14**, 270–285 (1980).

9.43. Wilkins, D. J., Eisenmann, J. R., Camin, R. A., Margolis, W. S., and Benson, R. A., "Characterizing Delamination Growth in Graphite-Epoxy," in K. L. Reifsnider (ed.), *Damage in Composite Materials*, ASTM STP **775**, 168–183, American Society for Testing and Materials, Philadelphia, PA (1982).

9.44. Whitney, J. M., Browning, C. E., and Hoogsteden, W., "A Double Cantilever Beam Test for Characterizing Mode I Delamination of Composite Materials," *Journal of Reinforced Plastics and Composites*, **1**, 297–313 (October 1982).

9.45. Aliyu, A. A. and Daniel, I. M., "Effects of Strain Rate on Delamination Fracture Toughness of Graphite/Epoxy," in W. S. Johnson (ed.), *Delamination and Debonding of Materials*, ASTM STP **876**, 336–348, American Society for Testing and Materials, Philadelphia, PA (1985).

9.46. Daniel, I. M., Shareef, I., and Aliyu, A. A., "Rate Effects on Delamination of a Toughened Graphite/Epoxy," in N. J. Johnston (ed.), *Toughened Composites*, ASTM STP **937**, 260–274, American Society for Testing and Materials, Philadelphia, PA (1987).

9.47. Leach, D. C., Curtis, D. C., and Tamblin, D. R., "Delamination Behavior of Carbon Fiber/Poly(etheretherketone) (PEEK) Composites," in N. J. Johnson (ed.), *Toughened Composites*, ASTM STP **937**, 358–380, American Society for Testing and Materials, Philadelphia, PA (1987).

9.48. Bradley, W. L., "Relationship of Matrix Toughness to Interlaminar Fracture

Toughness," in K. Friedrich (ed.), *Application of Fracture Mechanics of Composite Materials*, Chap. 5, Vol. 6, Composite Material Series, R. B. Pipes (Series ed.), Elsevier Science Publishers, Amsterdam, The Netherlands (1989).

9.49. Hunston, D. L., Moulton, R. J., Johnston, N. J., and Bascom, W., "Matrix Resin Effects in Composite Delamination: Mode I Fracture Aspects," in N. J. Johnston (ed.), *Toughened Composites*, ASTM STP **937**, 74–94, American Society for Testing and Materials, Philadelphia, PA (1987).

9.50. Carlsson, L. A. and Gillispie, J. W., "Mode II Interlaminar Fracture of Composites," in K. Friedrich (ed.), *Application of Fracture Mechanics to Composite Materials*, Chap. 4, Vol. 6, Composite Material Series, R. B. Pipes (Series ed.), Elsevier Science Publishers, Amsterdam, The Netherlands (1989).

9.51. Russell, A. J. and Street, K. N., "Moisture and Temperature Effects on the Mixed Mode Delamination Fracture of Unidirectional Graphite/Epoxy," in W. S. Johnson (ed.), *Delamination and Debonding of Materials*, ASTM STP **876**, 349–370, American Society for Testing and Materials, Philadelphia, PA (1985).

9.52. Carlsson, L. A., Gillispie, J. W., Jr., and Pipes, R. B., "On the Analysis and Design of the End Notched Flexure (ENF) Specimen for Mode II Testing," *Journal of Composite Materials*, **20**, 594–604 (1986).

9.53. Carlsson, L. A. and Pipes, R. B., *Experimental Characterization of Advanced Composite Materials*, Prentice-Hall, Inc., Englewood Cliffs, NJ (1987).

9.54. Kageyama, K., Kikuchi, M., and Yanagisawa, N., "Stabilized End Notched Flexure Test: Characterization of Mode II Interlaminar Crack Growth," in T. K. O'Brien (ed.), *Composite Materials: Fatigue and Fracture (Third Volume)*, ASTM STP **1110**, 210–225, American Society for Testing and Materials, Philadelphia, PA (1991).

9.55. Russell, A. J., "Initiation and Growth of Mode II Delamination in Toughened Composites," in T. K. O'Brien (ed.), *Composite Materials: Fatigue and Fracture (Third Volume)*, ASTM STP **1110**, 226–242, American Society for Testing and Materials, Philadelphia, PA (1991).

9.56. O'Brien, T. K., Johnston, N. J., Raju, I. S., Morris, D. H., and Simonds, R. A., "Comparisons of Various Configurations of the Edge Delamination Test for Interlaminar Fracture Toughness," in N. J. Johnston (ed.), *Toughened Composites*, ASTM STP **937**, 199–221, American Society for Testing and Materials, Philadelphia, PA (1987).

9.57. Johnson, W. S. and Mangalgiri, P. D., "Influence of the Resin on Interlaminar Mixed-Mode Fracture," in N. J. Johnston (ed.), *Toughened Composites*, ASTM STP **937**, 295–315, American Society for Testing and Materials, Philadelphia, PA (1987).

9.58. Murri, G. B. and O'Brien, T. K., "Interlaminar G_{IIc} Evaluation of Toughened Resin Matrix Composites Using the End Notched Flexure Test," in *Proceedings of the 26th AIAA/ASME/ASCE/AHS Structures, Structural Dynamics and Materials Conference*, 197–202, American Institue for Aeronautics and Astronautics, New York (1985).

9.59. Ramkumar, R. L. and Whitcomb, J. D., "Characterization of Mode I and Mixed Mode Delamination Growth in T300/5208 Graphite/Epoxy," in W. S. Johnson (ed.), *Delamination and Debonding of Materials*, ASTM STP **876**, 315–335, American Society for Testing and Materials, Philadelphia, PA (1985).

9.60. Chan, W. S., Rogers, C., and Aker, S., "Improvement of Edge Delamination Strength of Composite Laminates Using Adhesive Layers," in J. M. Whitney (ed.), *Composite Materials: Testing and Design (Seventh Conference)*, ASTM STP **893**, 266–285, American Society for Testing and Materials, Philadelphia, PA (1986).

9.61. Evans, R. E. and Masters, J. E., "A New Generation of Epoxy Composites for Primary Structural Applications: Materials and Mechanics," in N. J. Johnston (ed.), *Toughened Composites*, ASTM STP **937**, 413–436, American Society for Testing Materials, Philadelphia, PA (1987).

9.62. Ishai, O., Rosenthal, H., Sela, N., and Drukker, E., "Effect of Selective Adhesive Interleaving on Interlaminar Fracture Toughness of Graphite/Epoxy Composite Laminates," *Composites*, **19**(1), 49–54 (1988).

9.63. Sela, N., Ishai, O., and Banks-Sills, L., "The Effect of Adhesive Thickness on Interlaminar Fracture Toughness of Interleaved CFRP Specimens," *Composites*, **20**(3), 257–264 (1989).

9.64. Lagace, P. A. and Bhat, N. V., "Efficient Use of Film Adhesive Interlayers to Suppress Delamination," in G. C. Grimes (ed.), *Composite Materials: Testing and Design (Tenth Volume)*, ASTM STP **1120**, 384–396, American Society for Testing and Materials, Philadelphia, PA (1992).

9.65. Broutman, L. J. and Agarwal, B. D., "A Theoretical Study of the Effect of an Interfacial Layer on the Properties of Composites," *Polymer Engineering and Science*, **14**(8), 581–588 (1974).

9.66. Schwartz, H. S. and Hartness, T., "Effect of Fiber Coatings on Interlaminar Fracture Toughness of Composites," in N. J. Johnston (ed.), *Toughened Composites*, ASTM STP **937**, 150–178, American Society for Testing and Materials, Philadelphia, PA (1987).

9.67. Browning, C. E. and Schwartz, H. S., "Delamination Resistant Composite Concepts," in J. M. Whitney (ed.), *Composite Materials: Testing and Design (Seventh Conference)*, ASTM STP **893**, 256–265, American Society for Testing and Materials, Philadelphia, PA (1986).

9.68. Mignery, L. A., Tan, T. M., and Sun, C. T., "The Use of Stitching to Suppress Delamination in Laminated Composites," in W. S. Johnson (ed.), *Delamination and Debonding of Materials*, ASTM STP **876**, 371–385, American Society for Testing and Materials, Philadelphia, PA (1985).

9.69. Garcia, R., Evans, R. E., and Palmer, R. J., "Structural Property Improvements through Hybridized Composites," in N. J. Johnston (ed.), *Toughened Composites*, ASTM STP **937**, 397–412, American Society for Testing and Materials, Philadelphia, PA (1987).

9.70. Kim, J. K. and Mai, Y. W., "High Stretch, High Fracture Toughness Fibre Composites with Interface Control—A Review," *Composites Science and Technology*, **41**, 333–378 (1991).

CHAPTER

10

MECHANICAL TESTING OF COMPOSITES AND THEIR CONSTITUENTS

10.1 INTRODUCTION

The purpose of this chapter is to review briefly the most widely used methods for mechanical testing of composite materials and their constituents. In previous chapters the emphasis has been on the development of analytical models for mechanical behavior of composite materials. The usefulness of such models depends heavily on the availability of measured intrinsic mechanical property data to use as input. In addition, some aspects of mechanical behavior of composites are so complex that the feasibility of proper analytical modeling is questionable, and the experimental approach may be the only acceptable solution. Much of our knowledge about the special nature of composite behavior has been derived from experimental observations. The measurement of mechanical properties is also an important element of the quality control and quality assurance processes associated with the manufacture of composite materials and structures.

Due to the special characteristics of composites, such as anisotropy, coupling effects and the variety of possible failure modes, it has been found that the mechanical test methods that are used for conventional metallic materials are usually not applicable to composites. Thus, the development and evaluation of new test methods for composites has been, and continues to be, a major challenge for the experimental mechanics community. The technology associated with composite test methods and test equipment has become just as sophisticated as that associated with the corresponding analytical methods. Many of these test methods have evolved into standards which have been adopted by the American Society for Testing and Materials (ASTM). All the ASTM standards for the testing of composites have now been conveniently compiled in one book [10.1], and several books are devoted to experimental characterization of composites [10.2–10.5].

10.2 MEASUREMENT OF CONSTITUENT MATERIAL PROPERTIES

From the earlier discussion of various micromechanical models, it should be obvious that experimentally determined constituent material properties are required as input to these models. Since the development of new composites depends so heavily on the development of new fiber and matrix materials, constituent material tests are often used for screening new materials before composites are made from them. This section deals with the test methods that are used to measure the mechanical properties of fiber and matrix materials.

10.2.1 Fiber Tests

The tensile strength and Young's modulus of reinforcing fibers under static longitudinal loading may be determined by the ASTM D 3379-75 standard test method for single filament materials [10.6]. As shown in Fig. 10.1, the fiber specimen is adhesively bonded to a thin paper, compliant

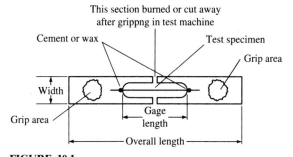

FIGURE 10.1
Single filament tensile test specimen. (*From ASTM Standard D 3379-75 [10.6]. Copyright ASTM. Reprinted with permission.*)

metal, or plastic backing strip which has a central longitudinal slot of fixed gage length. Once the specimen is clamped in the grips of the tensile testing machine, the backing strip is burned or cut away, so that the filament transmits all the applied tensile load. The specimen is pulled to failure, the load and elongation are recorded, and the tensile strength and modulus are calculated from the usual formulas. For such small specimens, however, it is important to correct the measured compliance by subtracting out the system compliance. The system compliance can be determined by testing specimens of different gage lengths, plotting the compliance vs. gage length, and extrapolating the curves to zero gage length. The compliance corresponding to zero gage length is assumed to be the system compliance [10.6]. This and other techniques for the measurement of single graphite fiber, longitudinal tensile properties have been evaluated by McMahon [10.7].

Resin-impregnated yarns, strands, rovings, and tows of carbon and graphite fibers may be tested by using ASTM D 4018-81 [10.8]. The impregnating resin is used to produce a rigid specimen (Fig. 10.2) which is easier to handle and test than a loose bundle of yarn, and which should ensure uniform loading of the fibers in the bundle. The specimen test procedure is similar to that used in D 3379-75, except that the tensile strength and modulus should be corrected to account for the portion of the tensile load carried by the resin.

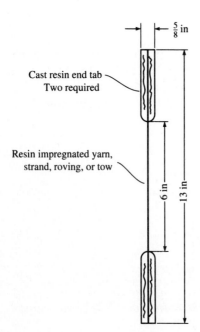

FIGURE 10.2
Resin-impregnated strand tensile test specimen. (*From ASTM Standard D 4018-81* [10.8]. *Copyright ASTM. Reprinted with permission.*)

Direct measurement of fiber properties under longitudinal compressive loading or longitudinal shear loading does not appear to be feasible. Such properties may be inferred from matrix and composite test data, however, and these tests will be discussed later in this chapter. As mentioned in Chap. 3, the transverse Young's modulus of fibers may also be inferred from matrix and composite test data, but direct measurement is possible. For example, Kawabata [10.9] has tested fibers in transverse diametral compression using the apparatus shown in Fig. 10.3. The resulting load-deflection curve is compared with the corresponding load-deflection curve from a theoretical model of the fiber under transverse compression. One of the inputs to the model is the transverse Young's modulus of the fiber, which is used as a curve-fitting parameter to match the predictions with the measurements. Kawabata's measurements on graphite and aramid fibers showed even greater anisotropy than did the inferred properties.

The methods described above are used to determine static mechanical properties of fibers. Dynamic test methods involving the use of vibration will be discussed later in this chapter.

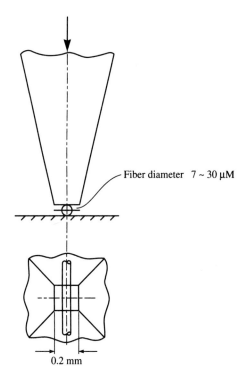

Fiber diameter 7 ~ 30 µM

0.2 mm

FIGURE 10.3
Diametral compression of fiber for measurement of fiber transverse Young's modulus. (*From Kawabata* [10.9]).

10.2.2 Neat Resin Matrix Tests

The tensile yield strength, tensile strength at break, modulus of elasticity, and elongation of neat resin matrix specimens may be determined by using the ASTM D 638-90 method for tensile properties of plastics [10.10]. Several types of "dogbone-shaped" specimens are specified under this standard, depending on the thickness of the available material and whether the material is rigid or nonrigid. Specimens may be fabricated by machining or die-cutting from sheets or plates of the material, or by molding the resin to the desired shape. For example, the type I specimen for rigid or semirigid plastics having thicknesses of 0.28 in (7 mm) or less is shown in Fig. 10.4. Since plastics may be sensitive to temperature and relative humidity, the procedure used to condition specimens should be consistent from one test to another. The so-called "Standard Laboratory Atmosphere" of 23°C (73.4°F) and 50 percent relative humidity is specified in the ASTM standard D 618-81 [10.11]. Some plastics are also strain-rate sensitive, so the speed of testing should be consistent, as specified in D 638-90. The details of the other test conditions and procedures are also given in the standard.

The ASTM D 695-90 test method [10.12] can be used to determine compressive yield strength, compressive strength, and modulus of elasticity of neat resin matrix materials. Out-of-plane buckling failures are avoided by using a very short specimen (Fig. 10.5) and a support jig on each side of the specimen (Fig. 10.6). In order to generate true axial loading on the specimen without bending, a special compression fixture with a ball-and-socket arrangement is used (Fig. 10.7).

In either the tensile test or the compressive test of the neat resin matrix material biaxial strain gages can be attached to the specimen so as to measure the longitudinal and transverse strains. The Young's modulus, E, and the Poisson's ratio, v, can then be determined from the standard definitions of those parameters. If desired, the shear modulus, G, can also be found from the isotropic relationship

$$G = \frac{E}{2(1 + v)} \qquad (10.1)$$

However, Novak and Bert [10.13] have reported that for some epoxies the values of G found from applying Eq. (10.1) to either tensile or compressive tests differ substantially from directly measured values of G. Directly measured values of G were determined from a plot of angle of twist vs. torque for solid rod torsion tests. It was found that a more accurate calculation of G could be obtained by taking into account differences between tensile and compressive values of E and v. Their approach was based on the premise that since the elastic strain energy is

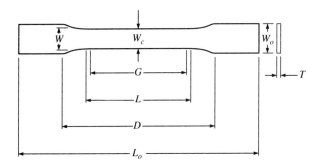

Dimensions for type I specimen

			in	mm
W	=	width of narrow section	0.5	13
L	=	length of narrow section	2.25	57
W_o	=	width overall, min.	0.75	19
L_o	=	length overall, min.	6.5	165
G	=	gage length	2.0	50
D	=	distance between grips	4.5	115
R	=	radius of fillet	3.0	76

FIGURE 10.4
Type I neat resin tensile specimen for thicknesses of 0.28 in (7 mm) or less. (*From ASTM Standard D 638-90 [10.10]. Copyright ASTM. Reprinted with permission.*)

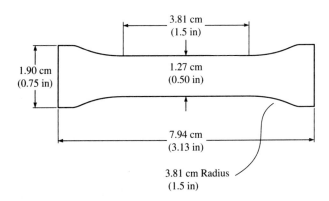

FIGURE 10.5
Neat resin compressive test specimen. (*From ASTM Standard D 695-90 [10.12]. Copyright ASTM. Reprinted with permission.*)

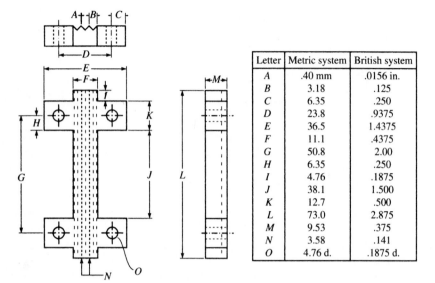

Letter	Metric system	British system
A	.40 mm	.0156 in.
B	3.18	.125
C	6.35	.250
D	23.8	.9375
E	36.5	1.4375
F	11.1	.4375
G	50.8	2.00
H	6.35	.250
I	4.76	.1875
J	38.1	1.500
K	12.7	.500
L	73.0	2.875
M	9.53	.375
N	3.58	.141
O	4.76 d.	.1875 d.

FIGURE 10.6
Support jig for D 695-90 compressive test specimen. (*From ASTM Standard D* 695-90 [10.12]. *Copyright ASTM. Reprinted with permission.*)

invariant to a rotation of coordinates, the strain energy for an isotropic material in pure shear along the x, y axes is equal to the strain energy associated with the corresponding biaxial tensile and compressive principal stresses oriented at 45° to the x, y axes. By equating these strain

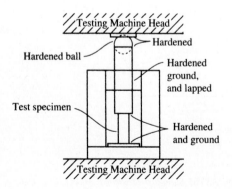

FIGURE 10.7
Compression fixture with ball-and-socket joint to minimize bending. (*From ASTM Standard D* 695-90 [10.12]. *Copyright ASTM. Reprinted with permission.*)

energy terms and using the Hooke's law for an isotropic material with different properties in tension and compression, Novak and Bert showed that the shear modulus, G, can be expressed as

$$\frac{1}{G} = \frac{(1 + v_t)}{E_t} + \frac{(1 + v_c)}{E_c} \tag{10.2}$$

where E_t = Young's modulus from tensile test
E_c = Young's modulus from compressive test
v_t = Poisson's ratio from tensile test
v_c = Poisson's ratio from compressive test

This equation, which involves both tensile and compressive properties, was found to give much better agreement with directly measured values of G than did Eq. (10.1) when Eq. (10.1) was used with either tensile compressive values of E and v. It is easily shown that when $E_t = E_c = E$ and $v_t = v_c = v$, Eq. (10.2) reduces to Eq. (10.1).

The flexural yield strength, flexural strength, and modulus of elasticity of plastics may be determined by the ASTM D 790-90 test method [10.14]. This test method includes both three-point bending (Fig. 10.8) or four-point bending (Fig. 10.9). Allowable ranges of radii for the loading noses for both methods are specified in Figs. 10.8 and 10.9, and recommended specimen dimensions are provided in tables in D 790-90 [10.14]. Test methods for measurement of mechanical properties of other constituents such as sandwich core materials and other constituent properties such as coefficient of thermal expansion, impact, creep, and fatigue response are also given in Ref. [10.1].

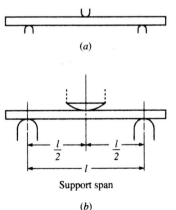

(a)

$\dfrac{l}{2}$ $\dfrac{l}{2}$

l

Support span

(b)

FIGURE 10.8
Three-point bending specimen for flexural properties of neat resin or composite. (*From ASTM Standard D* 790-90 [10.14]. *Copyright ASTM. Reprinted with permission.*)

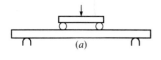

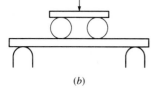

FIGURE 10.9
Four-point bending specimen for flexural properties of neat resin or composite. (*From ASTM Standard D 790-90* [10.14]. *Copyright ASTM. Reprinted with permission.*)

10.3 MEASUREMENT OF BASIC COMPOSITE PROPERTIES

This section is concerned with test methods for measurement of the basic composite mechanical properties that are needed in analysis and design. Methods for measurement of lamina properties such as tensile and compressive strength and stiffness, shear strength and stiffness, flexural strength and stiffness, fiber/matrix interfacial strength, and laminate properties such as interlaminar strength and fracture toughness are discussed. Both direct and indirect methods are reviewed. Direct methods involve the application of uniaxial, shear, or flexural loading to a lamina or laminate specimen so as to determine the basic property that governs the response to such loading. Indirect methods may involve such techniques as "backing out" of lamina properties from tests of laminates. Difficulties encountered in some of these tests are discussed, along with limitations and possible sources of error.

10.3.1 Tensile Tests

Lamina tensile strengths $s_L^{(+)}$ and $s_T^{(+)}$; Young's moduli, E_1 and E_2; and Poisson's ratios, ν_{12} and ν_{21}, may be measured by testing longitudinal (0°) and transverse (90°) unidirectional specimens according to the ASTM D 3039-76 standard test method [10.15]. The D 3039-76 specimen geometry is shown in Fig. 10.10. Laminated load transfer tabs are adhesively bonded to the ends of the specimen in order that the load may be transferred from the grips of the tensile testing machine to the specimen without damaging the specimen. Recommended dimensions for 0° and 90° specimens of various types of high modulus resin matrix composites are provided in the D 3039-76 standard, along with recommended test procedures and calculations. Typical longitudinal and transverse strain data from such a test on a $[0]_8$ graphite/epoxy composite are given in Fig. 10.11 for various stresses along with the resulting values of E_1, ν_{12}, $s_L^{(+)}$,

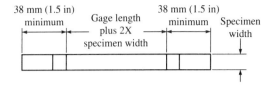

FIGURE 10.10
Composite tensile test specimen with adhesively bonded load transfer tabs. (*From ASTM Standard D 3039-76 [10.15]. Copyright ASTM. Reprinted with permission.*)

and $e_L^{(+)}$. These results show the typical fiber-dominated linearity for the longitudinal strain response and a slight nonlinearity in the transverse strain response due to the influence of the matrix.

The D 3039-76 test method works well for orthotropic specimens because a uniform state of stress is produced across the specimen as it is loaded in tension. However, nonuniformities in the stress distribution may arise when the method is used for off-axis specimens which exhibit shear coupling. Such off-axis tests would typically be used to measure such properties as the off-axis Young's modulus, E_x, and the off-axis

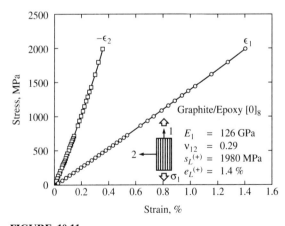

FIGURE 10.11
Longitudinal and transverse strain data at different stresses for $[0]_8$ graphite/epoxy tensile specimen. (*From Carlsson and Pipes [10.4]. Reprinted by permission of Prentice-Hall, Englewood Cliffs, New Jersey.*)

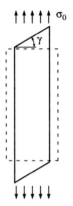

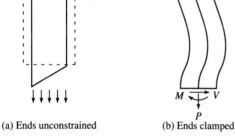

(a) Ends unconstrained　　　(b) Ends clamped

FIGURE 10.12
Effect of end conditions on deformation of an off-axis tensile specimen exhibiting shear coupling. (*From Pagano and Halpin* [10.16].)

tensile strength, $s_x^{(+)}$. Pagano and Halpin [10.16] showed that a specimen which exhibits shear coupling will deform as shown in Fig. 10.12(a) if the ends are unconstrained. But if the ends are constrained by clamping fixtures, the shear-coupling effects will produce shear forces and bending moments which distort the specimen, as shown in Fig. 10.12(b). Thus, in the latter case the specimen is no longer under a uniform state of stress, and the usual definitions of the engineering constants are not valid. Pagano and Halpin found that the distortion shown in Fig. 10.12(b) decreases with decreasing shear-coupling ratio, $\eta_{x,xy}$ [recall Eq. (2.40)], and increasing length-to-width ratio of the specimen. They also suggested that in order to minimize such effects, angle-ply laminates could be used instead of off-axis specimens, or that a test fixture which allowed free rotation of the ends of the specimen could be used with off-axis specimens.

A similar conclusion regarding end constraint effects in off-axis specimens was proposed by Jones [10.17]. Jones suggested that for long, slender off-axis specimens under a uniaxial stress, as shown in Fig. 10.13(a), the state of stress in the gage section of the specimen would be approximately

$$\sigma_y = \tau_{xy} = 0 \quad \text{and} \quad \sigma_x = E_x \epsilon_x \quad (10.3)$$

because the gage length is sufficiently far removed from the effects of the clamped ends. However, in a short, wide specimen [Fig. 10.13(b)] the proximity of the clamped ends to the gage length will cause the strains in the gage length to be approximately

$$\epsilon_y = \gamma_{xy} = 0 \quad (10.4)$$

When these strain conditions are substituted in Eqs. (2.35), the resulting

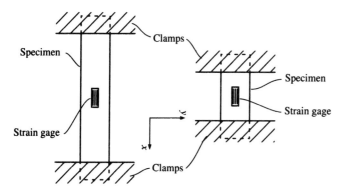

(a) Long, slender specimen   (b) Short, wide specimen

FIGURE 10.13
Tensile specimens of different length showing relationship of gage length to specimen length.

stress condition in the gage length of the short, wide specimen is found to be

$$\sigma_x = \bar{Q}_{11}\epsilon_x \qquad (10.5)$$

The conclusion here is that in the case of a long, slender specimen the off-axis Young's modulus, E_x, is measured. However, in the case of the short, wide specimen the transformed lamina stiffness, $\bar{Q}_{11}$, is measured instead of E_x. For example, Jones [10.17] points out that for a 30° off-axis test of graphite/epoxy the value of $\bar{Q}_{11}$ is more than 10 times as great as E_x. Thus, the analyses of Pagano and Halpin [10.16] and Jones [10.17] lead to the same conclusion regarding the effect of specimen length in off-axis tensile tests.

In the analysis of Jones [10.17] it is assumed that Eqs. (10.3) are valid when the specimen is "long enough" so that the end effects are not significant in the gage length. The decay of such localized effects with distance away from the source is justified by the use of Saint-Venant's principle. However, Horgan et al. [10.18–10.20] have shown that the characteristic decay length over which end effects are significant in orthotropic composites is generally several times greater than the corresponding length for isotropic materials. The decay length, λ, which is the distance from the end of the specimen over which the stress decays to $1/e$ of the value of the stress at the end, was found to be

$$\lambda \simeq \frac{b}{2\pi}\sqrt{\frac{E_1}{G_{12}}} \qquad \text{as} \quad \frac{G_{12}}{E_1} \to 0 \qquad (10.6)$$

for an anisotropic, transversely isotropic, rectangular strip [10.20], where b is the width of the strip.

In the tests described above lamina properties are measured directly by testing unidirectional specimens. A different approach involves the use of the Classical Lamination Theory (CLT) to "back out" lamina properties from laminate test data [10.21, 10.22]. For example, Rawlinson [10.21] has shown that CLT "back-out" factors for obtaining equivalent 0° tensile strengths from both angle-ply and cross-ply laminates showed good agreement with the corresponding experimentally determined factors for several graphite/epoxy composites. As shown in Fig. 10.14, Rawlinson's data for the equivalent 0° tensile strength of IM7G/8551-7 graphite/epoxy appear to be nearly the same regardless of whether unidirectional 0° specimens or various cross-ply laminate configurations are used. The one notable exception is the particular case of $[0/90]_{2s}$ cross-ply specimens without load transfer tabs. It is seen in Fig. 10.14 that the scatter in the data is generally less for the cross-ply specimens and that the data for the cross-ply specimens without load transfer tabs are generally just as good as the data for the corresponding specimens with tabs. Thus, there appears to be considerable potential for cost savings with the tests of untabbed cross-ply specimens.

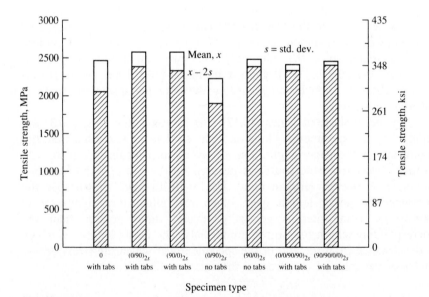

FIGURE 10.14
"Backed out" tensile strength data from seven different laminates of IM7G/8551-7 graphite/epoxy. (*From Rawlinson* [10.21]. *Reprinted by permission of the Society for the Advancement of Material and Process Engineering.*)

10.3.2 Compressive Tests

Compression testing has proved to be one of the most interesting and difficult challenges to those concerned with the testing of composites. Even though an ASTM standard for compression testing has been published, there is still much discussion regarding various alternative test methods, as shown in several review articles [10.23, 10.24]. The ASTM D 3410-87 [10.25] standard test method for compression strength and modulus actually covers three different procedures. Procedures A and B both involve the use of the specimen geometry shown in Fig. 10.15, whereas procedure C involves the use of the sandwich beam specimen shown in Fig. 10.16. In procedures A and B the specimen shown in Fig. 10.15 is placed in the fixture (such as the one for procedure B shown in Fig. 10.17), which is then placed in the testing machine and loaded to

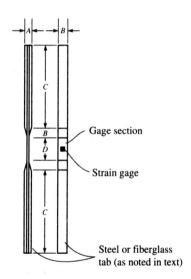

EDGE   FRONT

Dimensions	m	in
A	0.00399	0.157
B*	0.00635	0.250
C	0.05715	2.250
D	0.01270	0.500

*Procedure B allows for the use of wider specimens.

FIGURE 10.15
Compressive test specimen for D 3410-87, procedures A and B. (*From ASTM Standard D 3410-87* [10.25]. *Copyright ASTM. Reprinted with permission.*)

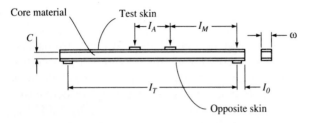

	Dimensions	
	mm	**in**
ω	25.4	1.0
l_0	25.4	1.0
l_T	508.0	20.0
l_M	203.2	8.0
l_A	101.6	4.0
C	38.1	1.5

FIGURE 10.16
D 3410-87, procedure C sandwich beam specimen for compressive test of upper composite skin. (*From ASTM Standard D 3410-87* [10.25]. *Copyright ASTM. Reprinted with permission.*)

failure. The fixtures for procedures A and B are designed to produce compression in the specimen through side-loading, as opposed to the end-loading fixture for the previously discussed D 695-90 standard [10.12] for compression testing of plastics. The fixture for procedure A, which is often referred to as the Celanese fixture, produces side-loading of the specimen through conical wedges inside the housing for the fixture. In the case of the fixture for procedure B, which was originally known as the Illinois Institute of Technology Research Institute (IITRI) fixture, the side-loading of the specimen is accomplished by pyramidal wedges inside a heavy-housing, as shown in Fig. 10.17.

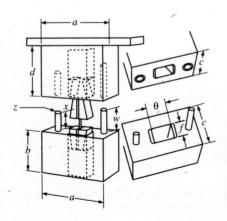

FIGURE 10.17
Assembled compressive test fixture for D 3410-87, procedure B. (*From ASTM Standard D 3410-87* [10.25]. *Copyright ASTM. Reprinted with permission.*)

The sandwich beam specimen for procedure *C* shown in Fig. 10.16 is constructed so that the upper skin consists of the [0] unidirectional composite material of interest. The sandwich beam is bonded together with a structural adhesive and the specimen is loaded in four-point bending, so that the upper skin is subjected to compressive stress. A honeycomb core material in the sandwich beam provides lateral support for the skin in order to avoid premature buckling of the skin, and thus the failure is due to compressive failure of the skin material.

In a comparison of four test methods which eventually led to the adoption of the current D 3410-87 standard Adsit [10.26] showed that what are now referred to as procedures *A, B,* and *C* gave equivalent results for compressive strength and modulus of graphite/epoxy, but that the D 695 method for plastics was inadequate for high modulus composites. Although all four methods gave acceptable results for compression modulus, the D 695 method produced premature delamination or shear failures due to end-loading of the specimens.

A number of alternative methods for measurement of compressive modulus and strength has been reported in the literature. The Wyoming-modified Celanese fixture [10.23] incorporates features of both the IITRI and Celanese fixtures. The minisandwich specimen [10.27] is smaller than that used in procedure *C* of the ASTM D 3410-90 standard and has a core consisting of the neat resin matrix material instead of a honeycomb material. Compressive properties of the [0] unidirectional lamina can

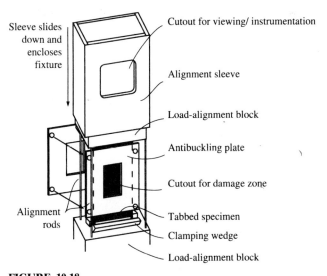

Sleeve slides down and encloses fixture

Cutout for viewing/ instrumentation

Alignment sleeve

Load-alignment block

Antibuckling plate

Cutout for damage zone

Alignment rods

Tabbed specimen

Clamping wedge

Load-alignment block

FIGURE 10.18
Compression after impact (CAI) fixture. (*From Nettles and Hodge* [10.30]. *Reprinted by permission of the Society for the Advancement of Material and Process Engineering.*)

also be "backed out" of [0/90] cross-ply laminate compression test data [10.28].

The problem of local buckling and the corresponding reduction of in-plane compressive strength after delamination due to transverse impact has been discussed in Chap. 7 (see Fig. 7.34) and in Chap. 9. Concern about this failure mode has led to the development of the compression after impact (CAI) test [10.29, 10.30]. A recent version of the CAI fixture is shown in Fig. 10.18 from Ref. [10.30]. In a typical CAI test small rectangular panel specimens are subjected to transverse impact in an impact testing machine, then the residual in-plane compressive strength is measured by loading the specimen in the CAI fixture. As mentioned in Chap. 9, the CAI strength appears to be strongly dependent on the interlaminar fracture toughness, and CAI strength has become a major issue in aerospace applications.

10.3.3 In-Plane Shear Tests

As shown earlier, the in-plane properties of a composite material are not necessarily equal to the through-thickness shear properties. Thus, test methods which will generate pure shear loading of both types are needed. This section deals only with in-plane shear test methods. According to Whitney et al. [10.3], the four most widely used test methods for measurement of in-plane shear properties of a unidirectional composite lamina are the $[\pm 45]_s$ laminate tensile test method, the off-axis tensile test method, the rail shear test method, and the torsion test method. Two of these methods have now been adopted as ASTM standards. The $[\pm 45]_s$ laminate tensile test method is described in D 3518/D3518M-91 [10.31], whereas the rail shear test method is described in D 4255-83 [10.32].

For the $[\pm 45]_s$ laminate tensile specimen (Fig. 10.19) it can be shown from laminate analysis and a transformation of stresses that the lamina shear stress, τ_{12}, along the principal material axes, is related to the uniaxial tensile stress, σ_x, acting on the laminate by

$$|\tau_{12}| = \left|\frac{\sigma_x}{2}\right| \qquad (10.7)$$

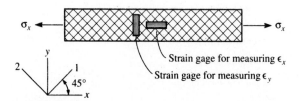

FIGURE 10.19
A $[\pm 45]_s$ laminate tensile specimen for determination of in-plane shear properties.

The laminate, strains, ϵ_x^o and ϵ_y^o, can be transformed to the lamina principal axes at $\pm 45°$, so that the magnitude of the lamina shear strain is

$$|\gamma_{12}| = |\epsilon_x^o - \epsilon_y^o| \qquad (10.8)$$

where ϵ_x^o is assumed to be a positive tensile strain and ϵ_y^o is assumed to be a negative Poisson strain. Thus, measurement of the applied stress, σ_x, and the laminate strains, ϵ_x^o and ϵ_y^o, during a tensile test of the $[\pm 45]_s$ laminate enables one to generate the shear stress-shear strain curve for the lamina material. The shear strength and the shear modulus can then be evaluated from this stress-strain curve.

Although the off-axis tensile test is not yet a standard, it is a useful method. For example, a tensile test of an off-axis specimen can be used to determine the off-axis Young's modulus, E_x, as defined by Eq. (2.38). If the values E_1, E_2, and ν_{12} are known from separate tests of longitudinal and transverse specimens, then Eq. (2.39) can be solved for the in-plane shear modulus, G_{12}. It appears that the optimum fiber orientation, θ, for best strength results may not necessarily be the same as the optimum angle for best modulus results, but the $10°$ off-axis test seems to be a good compromise [10.3].

The rail shear test standard, as described in ASTM D 4255-83 [10.32], covers two separate methods. Method A involves the use of the two-rail fixture shown in Fig. 10.20(a), whereas method B requires the use of the three-rail fixture shown in Fig. 10.20(b). In both methods a flat

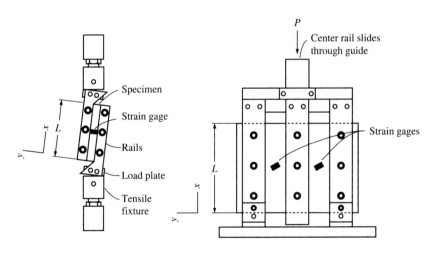

(a) Two-rail fixture for method A (b) Three-rail fixture for method B

FIGURE 10.20

Rail shear test fixtures. (*From ASTM Standard D 4255-83* [10.32]. *Copyright ASTM. Reprinted with permission.*)

rectangular plate specimen is clamped in between the rail fixtures and the fixture is subjected to uniaxial loading by a testing machine. The uniaxial loading on the fixture generates in-plane shear loading of the specimen and the resulting strains are monitored by the strain gages shown in Figs. 10.20(a) and (b). Simple equilibrium requires that the average shear stress along the specimen loading axes (x, y) for method A is

$$\tau_{xy} = \frac{P}{Lt} \tag{10.9}$$

where L = specimen length along the x direction
P = applied load along the x direction
t = specimen thickness

whereas the corresponding shear stress for method B is

$$\tau_{xy} = \frac{P}{2Lt} \tag{10.10}$$

The shear strain along the x, y directions can be determined from the measured normal strain, $\epsilon_{x'}$, along the x' axis, which is oriented at 45° from the x axis. From the strain transformation relationship for a state of pure shear along the x, y axes, we have

$$\gamma_{xy} = 2\epsilon_{x'} \tag{10.11}$$

Thus, the shear stress-shear strain data can be generated from Eqs. (10.9) to (10.11), and the corresponding modulus and strength can be found from the resulting stress-strain curve.

In-plane shear response can also be determined by testing a thin-walled composite tube in torsion and by measuring the resulting shear strain with a strain gage. If the strain gage is oriented at 45° to the tube axis, and if the applied torque creates a state of pure shear along the x, y axes, Eq. (10.11) can again be used to determine the shear strain. The shear stress can be estimated from the mechanics of materials formula for a thin-walled tube [10.33]:

$$\tau_{xy} = \frac{T}{2At} \tag{10.12}$$

where T = applied torque
t = wall thickness
A = area enclosed by median line = πR^2 for cylindrical tube
R = mean radius of tube

The Iosipescu shear test [10.34], which was first proposed for use with metals in 1967, has more recently been successfully adapted for use with composites by Adams et al. [10.35–10.38]. A drawing of the

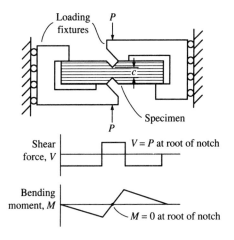

FIGURE 10.21
Iosipescu test fixture with shear and moment diagrams.

Iosipescu loading fixture and the corresponding shear and moment diagrams for the specimen are shown in Fig. 10.21. The action of the Iosipescu fixture is to produce pure shear loading with no bending at the midspan section of the specimen between the notches. The average shear stress generated in that section is simply

$$\tau = \frac{P}{ct} \qquad (10.13)$$

where P = applied load (see Fig. 10.21)
  c = distance between roots of notches
  t = specimen thickness

No subscripts were used in Eq. (10.13) because the test method can be used for either in-plane or through-thickness shear testing. Shear strains can be measured by using a strain gage to measure the normal strains at 45° to the specimen axis, then using Eq. (10.11) as before.

The in-plane shear test methods that have been discussed here, as well as all other existing shear test methods that have appeared in the literature, were evaluated by Lee and Munro [10.39]. In ranking the test methods, Lee and Munro decided that the ideal test methods were those that were easy to conduct, required small, easily fabricated specimens, and were capable of measuring both shear modulus and shear strength. Based on these criteria, the top three methods were found to be Iosipescu, [±45] tensile, and 10° off-axis tensile, respectively.

10.3.4 Interlaminar Shear Tests

In addition to the previously mentioned Iosipescu test, there is at least one other test that is used for interlaminar shear. The reader is cautioned in advance, however, that the interlaminar strength data from this test

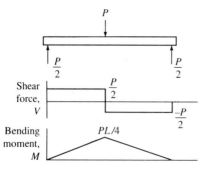

FIGURE 10.22
Short beam shear test specimen with shear and moment diagrams.

should not be used for design purposes. The ASTM D 2344-84 short beam shear test [10.40] involves the use of a short beam loaded in three-point bending, as shown in Fig. 10.22. The resulting shear stress, τ_{xy}, due to the shear force, V, and the normal stress, σ_x, due to the bending moment, M, can be estimated by using well-known mechanics of materials formulas [10.33]. While the shear stress is independent of specimen length, L, the normal stress, because of its dependence on the bending moment, is a linear function of L. Thus, failure by interlaminar shear can theoretically be induced by making the beam short enough so that under load the shear stress will reach its limiting value before the normal stress does. This is why the test is referred to as the "short beam shear test." However, Whitney [10.41, 10.42] has used the theory of elasticity analyses of short beam shear specimens to show that the actual state of stress in the vicinity of the applied load where failure initiates is much more complex than that predicted by the simple mechanics of materials formulas used in the D 2344-84 standard. Thus, the interlaminar strength derived from the D 2344-84 test is referred to only as "apparent" interlaminar shear strength and should not be used in design. According to D 2344-84, such data can be used for quality control and specification purposes, however.

10.3.5   Flexure Tests

Recall from Chap. 7 that unlike homogeneous, isotropic materials, composite laminates have flexural properties that are not necessarily the same as the corresponding tensile properties. Since many laminates are used as flexural members, there is a need to determine the flexural properties experimentally. The ASTM D 790-90 test method [10.14], which was described in Sec. 10.2.2, requires the use of beam specimens in either three-point bending (Fig. 10.8) or four-point bending (Fig. 10.9), and can be used for both unreinforced plastics and high modulus composites. Tables of recommended specimen dimensions are provided in the standard. It is important to realize, however, that in highly anisotropic composites such as unidirectional graphite/epoxy, through-thickness shear deformation can be significant unless the beam span-to-depth ratio is large enough. Thus, the specimen dimensions for these materials should be selected from the column corresponding to the highest span-to-depth ratio in the tables provided in D 790-90. It is also important to remember from Chap. 7 that the flexural modulus of highly anisotropic laminates depends on the ply stacking sequence and is not necessarily the same as the in-plane Young's modulus of the laminate.

10.3.6 Interlaminar Fracture Tests

In Sec. 9.4 the importance of interlaminar fracture was discussed and the use of the strain energy release rate to characterize the interlaminar fracture toughness was described. The most important modes of delamination seem to be modes I and II, and the corresponding fracture toughnesses are usually characterized by the strain energy release rates G_{Ic} and G_{IIc}, respectively. Although a number of test methods for measuring G_{Ic} and G_{IIc} has been reported in the literature, the most widely used methods appear to be the double cantilever beam (DCB) test for G_{Ic} [Fig. 9.16(a)] and the end-notched flexure (ENF) test for G_{IIc} [Fig. 9.16(b)].

The calculation of G_{Ic} from experimental DCB data can be carried out by using the method of Whitney et al. [10.42, 10.43], who analyzed each cracked half of the DCB specimen as though it were a cantilever beam [see Figs. 9.16(a) and 10.23]. Using the mechanics of materials beam theory, the tip deflection of the cantilever beam in Fig. 10.23 is

$$\frac{\delta}{2} = \frac{Pa^3}{3E_{fx}I} \tag{10.14}$$

where P = applied load
$\quad a$ = beam length in Fig. 10.23 is the DCB crack length in Fig. 9.16(a)
$\quad E_{fx}$ = flexural modulus of cracked half of DCB along the x direction
$\quad I$ = moment of inertia of cracked half of DCB about centroidal axis of cracked half

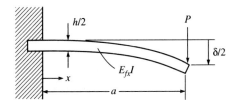

FIGURE 10.23
Cantilever beam representing half the cracked DCB specimen.

From Eq. (10.14), the compliance of the DCB specimen is found to be

$$s = \frac{\delta}{P} = \frac{64a^3}{E_{fx}th^3} \tag{10.15}$$

where $t = $ DCB specimen width, the crack width
$h = $ DCB specimen depth

The strain energy release rate, G_I, is then found by substituting Eq. (10.15) in Eq. (9.28) and differentiating with respect to the crack length, a. The result is

$$G_I = \frac{96P^2a^2}{E_{fx}t^2h^3} \tag{10.16}$$

or

$$G_I = \frac{3P\delta}{2ta} \tag{10.17}$$

where the deflection, δ, is given by Eq. (10.14). The critical strain energy release rate is then

$$G_{Ic} = \frac{96P_c^2a^2}{E_{fx}t^2h^3} \tag{10.18}$$

or

$$G_{Ic} = \frac{3P_c\delta_c}{2ta} \tag{10.19}$$

where P_c and δ_c are the critical values of the load and deflection, respectively, measured at the onset of crack growth. Whitney et al. [10.43] suggested that G_{Ic} could also be determined by rearranging Eq. (10.19) as

$$G_{Ic} = \frac{3H}{2t} \tag{10.20}$$

where $H = \dfrac{P_c\delta_c}{a}$ is a constant, and averaging H over some number of data points during continuous loading and crack extension. The average value of H is given by

$$H = \frac{1}{N}\sum_{i=1}^{N}\frac{P_{ci}\delta_{ci}}{a_i} \tag{10.21}$$

where P_{ci}, $\delta_{ci} = $ critical values of P and δ, respectively, associated with the ith crack length a_i
$N = $ total number of data points

Several other data reduction schemes for the DCB test are reviewed by Whitney [10.42].

The determination of G_{IIc} from ENF test data can be accomplished by using the method of Russell and Street [10.44, 10.45], who employed the elementary beam theory to derive the expression

$$G_{II} = \frac{9P^2a^2s}{2t(2L^3 + 3a^3)}$$ (10.22)

where the parameters P, a, t, and L are all defined in Fig. 9.16(b) and $s = \delta/P$ is the midspan compliance. The critical strain energy release rate, G_{IIc}, then corresponds to the critical load, P_c, and the associated compliance, s_c, at the onset of crack growth. The compliance, s, can be determined experimentally or calculated from the following equation, which was also derived using the elementary beam theory [10.44]:

$$s = \frac{(2L^3 + 3a^3)}{8E_{fx}th^3}$$ (10.23)

where E_{fx} is now the flexural modulus of the beam of depth $2h$, as shown in Fig. 9.16(b). Carlsson et al. [10.46] have used the Timoshenko beam theory to modify Eqs. (10.22) and (10.23), so that the effects of shear deformation are included. Several other test methods for measurement of G_{Ic} and G_{IIc} are examined in detail by Whitney [10.42] and by Carlsson and Pipes [10.4].

10.3.7 Fiber/Matrix Interface Tests

Good adhesion between the fiber and the matrix is a fundamental requirement if a composite is to be a useful structural material, and optimization of the fiber/matrix interface can only occur if reliable methods for measurement of fiber/matrix interfacial strength are available. One such method, a single-fiber fragmentation technique, has been developed and used by Drzal et al. [10.47–10.50]. The specimen, shown in Fig. 10.24, consists of a single fiber embedded in a dogbone tensile specimen of matrix resin. This specimen is loaded in tension under a microscope until the fiber breaks up into segments corresponding to the critical length, L_c, which are measured by using the microscope. If the

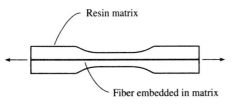

Resin matrix

Fiber embedded in matrix

FIGURE 10.24
Single-fiber fragmentation specimen developed by Drzal et al. [10.47, 10.48].

fiber tensile strength, $s_{f1}^{(+)}$, corresponding to the critical length, and the fiber diameter, d, are known, then the interfacial shear strength can be estimated from Eq. (6.9), which is repeated here as Eq. (10.24).

$$\tau_y = \frac{ds_{f1}^{(+)}}{2L_c} \qquad (10.24)$$

Since the observed lengths actually vary because of variations in fiber and matrix properties, a statistical distribution of fiber lengths must be used. If the measured values of L_c/d can be fitted by a two-parameter Weibull distribution, the mean value of interfacial shear strength can be expressed as [10.47]

$$\tau_y = \frac{s_{f1}^{(+)}}{2\beta}\Gamma\left(1 - \frac{1}{\alpha}\right) \qquad (10.25)$$

where Γ is the gamma function and α and β are the shape and scale parameters, respectively, for the two-parameter Weibull distribution.

In the so-called microbond test a single fiber is embedded in a resin droplet, and the free end of the fiber is loaded in tension until the fiber pulls out of the resin [10.50]. The interfacial strength is simply the pull-out force divided by the interfacial area. One potential difficulty with this test is that it may be difficult to reproduce the composite resin matrix cure conditions in a small droplet of resin [10.50].

While a single-fiber specimen is required for the two techniques describe above, in-situ fiber/matrix interfacial shear strength in composite specimens may be measured by using the microindentation technique [10.51, 10.52]. As shown in Fig. 10.25, this approach involves the use of a diamond microindenter to load the end of a fiber in longitudinal compression until debonding between the fiber and the matrix occurs. The experimental data for debonding load are combined with a finite element analysis in order to calculate the interfacial strength.

A comparison of the three interfacial strength measurement techniques described above has been reported by McDonough et al. [10.50], and the results for carbon fibers having different surface treatments and

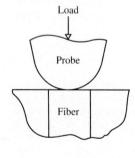

FIGURE 10.25
Microindenter test for fiber/matrix interfacial strength. (*From Mandell et al.* [10.51]. *Copyright ASTM. Reprinted with permission.*)

TABLE 10.1
Interfacial shear strength values obtained with different techniques

	Interfacial shear strength (MPa)		
Fiber type	Fragmentation	Microbond	Microindentation
AS-4	68	50	71
AU-4	37	23	55
IM6-600	47	15	43
IM6-100	40	19	37
IM6-U	22	15	27

Source: From McDonough et al. [10.50]. Reprinted by permission of the Society for the Advancement of Material and Process Engineering.

the same epoxy matrix are shown in Table 10.1. It was concluded that while the single-fiber fragmentation technique and the microindentation techniques showed good agreement, the microbond technique produced interfacial strength values that were consistently lower than those of the other two methods. The above mentioned uncertainty regarding the state of cure in the microbond droplet was given as the reason for these results.

10.4 MEASUREMENT OF VISCOELASTIC AND DYNAMIC PROPERTIES

In Chap. 8 creep, relaxation, damping, and strain rate dependence were described as being four important physical manifestations of viscoelastic behavior. All these characteristics can be determined experimentally, but in this section we will only describe test methods for measurement of creep compliance and damping. Since the complex modulus notation conveniently describes both dynamic stiffness and damping of linear viscoelastic materials, vibration test methods for measurement of the complex moduli of composites will be considered. The use of vibration tests of beams and plates to determine dynamic elastic moduli alone will also be reviewed. Wave propagation test methods will not be covered here.

10.4.1 Creep Tests

A creep test usually consists of the application of constant loading to a specimen, followed by measurement of the resulting time-dependent strains in the specimen, as shown schematically in Fig. 8.1(*a*). Although there are no standard creep test methods for composites at this time, creep test methods and creep rupture test methods have been standardized for tensile, compressive, and flexural creep of plastics [10.53]. In a creep rupture test (or stress rupture test) the time to failure is measured rather than the time-dependent strain in the specimen.

In principle, any of the previously described composite test methods can be used to characterize creep if the following provisions are made: (1) The applied loading on the specimen should be constant, (2) the resulting strains in the specimen should be measured as a function of elapsed time under load, and (3) the specimen should be kept under controlled environmental conditions for the duration of the test. Since viscoelastic behavior depends on temperature and humidity, the specimen would normally be enclosed in an environmental chamber. Stability of the measurement system electronics over long periods of time is also very important.

As shown by Halpin and Pagano [10.54], the principal creep compliances $S_{11}(t)$, $S_{22}(t)$, $S_{12}(t) = S_{21}(t)$, and $S_{66}(t)$ for a linear viscoelastic, orthotropic lamina can be determined by conducting the three tensile creep tests in Fig. 10.26. For example, in the creep test of the longitudinal specimen in Fig. 10.26(a) the constant stress, σ_1, is applied; the time-dependent longitudinal strain, $\epsilon_1(t)$, and the transverse strain, $\epsilon_2(t)$, are measured, and the longitudinal creep compliance is determined from the equation

$$S_{11}(t) = \frac{\epsilon_1(t)}{\sigma_1} \tag{10.26}$$

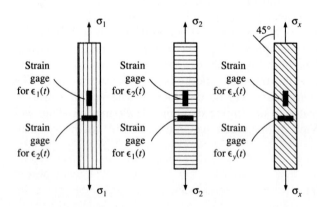

(a) Longitudinal test (b) Transverse test (c) Off-axis test
for $S_{11}(t)$ and $S_{21}(t)$ for $S_{22}(t)$ and $S_{12}(t)$ for $S_{66}(t)$

FIGURE 10.26
Tensile tests for measurement of creep compliances of an orthotropic, viscoelastic lamina.

whereas
$$S_{21}(t) = \frac{\epsilon_2(t)}{\sigma_1} \qquad (10.27)$$

For the transverse tensile creep test in Fig. 10.26(b), the constant transverse stress, σ_2, is applied, and the resulting strains are used to find the creep compliances

$$S_{22}(t) = \frac{\epsilon_2(t)}{\sigma_2} \qquad (10.28)$$

and
$$S_{12}(t) = \frac{\epsilon_1(t)}{\sigma_2} \qquad (10.29)$$

The data of Halpin and Pagano [10.54] and others from such tests have shown that $S_{12}(t) = S_{21}(t)$.

Finally, the off-axis tensile creep test in Fig. 10.26(c) can be used to determine the shear creep compliance $S_{66}(t)$. For example, if the constant uniaxial stress is σ_x and the fiber orientation $\theta = 45°$, a transformation of stresses gives the shear stress along the principal material axes as $\tau_{12} = \sigma_x/2$. The corresponding time-dependent shear strain, $\gamma_{12}(t)$, can be determined from the measured strains, $\epsilon_x(t)$ and $\epsilon_y(t)$, by modifying Eq. (10.8) as $|\gamma_{12}(t)| = |\epsilon_x(t) - \epsilon_y(t)|$. In practice, a more accurate determination of the shear strain $\gamma_{12}(t)$ is possible by using a strain gage rosette which has three strain gages oriented at 45° to each other [10.55]. The shear compliance is given by

$$S_{66}(t) = \frac{\gamma_{12}(t)}{\tau_{12}} \qquad (10.30)$$

Similar tensile creep tests of composites have been reported by Beckwith [10.56, 10.57] and Sullivan [8.25].

Uniaxial compressive creep testing presents the same difficulties that were discussed earlier in Sec. 10.3.2, and there appear to be few references dealing with such tests. For example, Irion and Adams [10.58] have used the previously discussed Wyoming-modified Celanese fixture [10.23] for compressive creep testing of unidirectional composites.

Since viscoelastic behavior is dependent on the stress-time history, preconditioning of creep specimens is recommended. Lou and Schapery [10.59] have suggested that mechanical conditioning of specimens before creep testing leads to much more repeatable test results. Specimens are mechanically conditioned by subjecting them to specified numbers of cycles of creep and recovery (loading and unloading) at a certain stress level. The actual creep tests are then conducted at stresses less than or equal to the conditioning stress. In creep tests of polymer composites where the effects of physical aging are being studied rejuvenation of the specimens at temperatures above T_g is necessary before the aging and/or creep tests begin [8.25].

In the discussion of viscoelastic behavior in Chap. 8 linear viscoelastic behavior was assumed. Experiments have shown, however, that polymer composites may exhibit nonlinear viscoelastic behavior at relatively low stress levels [10.56, 10.59]. For example, Beckwith [10.56, 10.57] has shown that the creep compliances for filament-wound S-glass/epoxy composites at various lay-ups followed a power law of the form

$$S(t) = S_0 + S_1 t^n \qquad (10.31)$$

where $S(t)$ = creep compliance

S_0 = initial elastic compliance

S_1, n = empirically determined parameters

The exponent, n, was found to be approximately equal to 0.19 for all compliances in the linear range, but at high stress levels and after multiple cycles of loading and unloading microcracking in the materials caused the exponent n to increase substantially.

Recall from Chap. 8 that for a linear viscoelastic material the time domain creep compliance is related to the frequency domain complex compliance by a Fourier transform pair, as are the relaxation modulus and the complex modulus. This relationship makes it possible to obtain time domain creep and relaxation characteristics from frequency domain test data, and vice versa. Using frequency domain complex modulus data and the Fourier transform approach, Gibson et al. [10.60, 10.61] have developed alternative techniques for determination of creep and relaxation behavior of linear viscoelastic composites in both tension and compression. A similar technique [10.62] involves using frequency domain vibration tests to determine the parameters in a spring-dashpot model (recall Sec. 8.2.2), and then substituting those same spring-dashpot parameters in the corresponding time domain creep compliance expression (recall Sec. 8.3.4). Vibration test techniques for measurement of complex moduli will be discussed in the next section.

10.4.2 Vibration Tests

The complex modulus notation, which is convenient for characterization of dynamic behavior of linear viscoelastic composites, was developed in Chap. 8. The two components of the complex modulus (stiffness and damping) of a material are generally referred to as its dynamic mechanical properties, and measurement of these properties is often referred to as dynamic mechanical analysis. Dynamic mechanical properties may be measured by using either wave propagation or vibration experiments, but only vibration test methods will be discussed here. The only relevant ASTM standards are D 4065-90a [10.63], which was developed for unreinforced plastics, and E 756 [10.64], which was developed for add-on surface damping treatments. In principle, some of the techniques described in these standards can also be used for composites. In practice, however, there are many pitfalls which must be avoided. For example, most commercially available dynamic mechanical testing machines, or dynamic mechanical analyzers, were developed for testing small specimens of unreinforced low modulus polymers, and the stiffness of the specimen mounting hardware in the machines is generally insufficient for use with high modulus composites. To reduce the composite specimen stiffness to the range required for valid data with these devices, it may be necessary to use specimen thicknesses on the order of the single ply thickness, so that testing of multi-ply laminates may not be possible. In addition, the equations used for data reduction in these machines typically do not take into account coupling effects, transverse shear effects, and other peculiarities of composite material behavior. Valid dynamic mechanical property measurements are difficult to obtain, particularly with composite materials. Only a brief overview of test methods and difficulties will be given here, as a detailed review has been published elsewhere [10.65].

The complex modulus [recall Eq. (8.70)] for a particular vibration test specimen is obtained by measuring the storage modulus and the loss factor of the specimen as it vibrates in the desired mode. Specimens usually consist of rods, beams, or plates supported in such a way as to minimize the extraneous damping due to the apparatus or the environment. Friction damping at specimen support points and transducer attachments, aerodynamic drag on the vibrating specimen, and phase lag in the instrumentation may all lead to erroneous damping data. Cross-verification of damping measurements using several different techniques is always a good way to locate potential problems.

The storage modulus is generally obtained by measuring a natural frequency of the specimen and by solving the frequency equation for the specimen. For example, Eq. (8.106) can be used to solve for the longitudinal modulus of a composite bar if the frequency, f_n, for the nth mode, the specimen length, L, and the density, ρ, are measured. Similarly, Eq. (8.117) can be used to determine the flexural modulus, E_f, of a composite beam specimen [10.66], and Eq. (8.139) can be solved simultaneously with four measured frequencies to determine the composite plate stiffnesses D_{11}, D_{22}, D_{12}, and D_{66} [10.67]. Care must be taken to make sure that the effective modulus criteria have been met and that various effects such as coupling and transverse shear have been

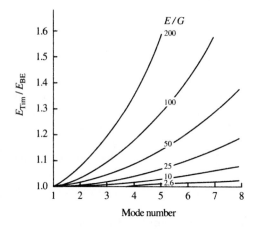

FIGURE 10.27
Correction factors required to correct modulus values from resonant frequency measurements using the Bernoulli-Euler theory to values using the Timoshenko beam theory. Factors are plotted as a function of mode number for several values of E/G and length/thickness ratio of 100. (*From Dudek* [10.68]).

accounted for where necessary. As mentioned in Sec. 8.3.2, transverse shear effects are much more significant for high modulus composites than they are for conventional materials, and Timoshenko beam theory may be required for valid results. Figure 10.27 shows correction factors which, when multiplied by modulus values from the Bernoulli-Euler beam theory, yield corrected modulus values which are consistent with Timoshenko beam theory [10.68].

Damping is conveniently characterized by using the loss factor in the complex modulus notation. For lightly damped systems the loss factor is related to the parameters that are used to characterize damping in a single-degree-of-freedom (SDOF) spring-mass system. The SDOF-damping parameters are typically estimated by curve-fitting to the measured response of specimens in either free vibration or forced vibration if a single mode can be isolated for the analyis.

In the free vibration experiment a specimen such as a rod or a beam is released from some initial displacement, or a steady-state excitation is removed, and the ensuing free vibration decay of the specimen is observed (Fig. 10.28). The logarithmic decrement, Δ, is calculated from such a decay curve by using the equation

$$\Delta = \frac{1}{n} \ln \frac{x_0}{x_n} \tag{10.32}$$

where x_0 and x_n are amplitudes measured n cycles apart, as shown in Fig. 10.28. Eq. (10.32) is based on the assumption of viscous damping, but for light damping the loss factor, η, is related to the logarithmic decrement by [10.69]

$$\eta = \frac{\Delta}{\pi} \tag{10.33}$$

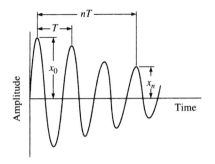

FIGURE 10.28
Free vibration decay curve for logarithmic decrement measurement.

Care must be taken to ensure that only one mode of vibration is present in the response decay curve, as the damping value should be measured for one particular mode.

One type of forced vibration test method involves fixed frequency oscillation of the specimen in a testing machine and simultaneous plotting of the resulting stress-strain hysteresis loop (Fig. 10.29). Using the dimensions a, b, and c from such hysteresis loops at a frequency, f, the components of the complex modulus can then be estimated by the equations [10.65]

$$\eta(f) = \frac{a}{b} \qquad (10.34)$$

and

$$E'(f) = \frac{b}{c} \qquad (10.35)$$

Another forced vibration technique is based on variation of the excitation frequency, simultaneous measurement of the response, and plotting of the magnitude and/or phase of the response in the frequency

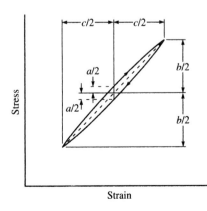

FIGURE 10.29
Hysteresis loop from fixed frequency forced oscillation test.

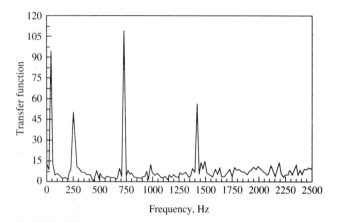

FIGURE 10.30
Typical specimen transfer function vs. frequency, or frequency response curve.

domain. The resulting frequency response curve, or frequency response spectrum (Fig. 10.30), has a number of peaks which represent natural frequencies of the specimen, and SDOF curve-fitting techniques can be applied to these peaks to extract the data needed to compute the complex modulus. The storage modulus is determined by substituting the peak frequency for a particular mode into the specimen frequency equation as described previously. The loss factor may be determined by using the half power bandwidth equation

$$\eta = \frac{\Delta f}{f_n} \tag{10.36}$$

where Δf = bandwidth at the half power points on the peak
$\quad f_n$ = peak frequency for the nth mode of vibration

Digital frequency spectrum analyzers, or Fast Fourier transform (FFT) analyzers are commonly used for this purpose. Excitation may be either variable frequency sinusoidal, random or impulsive. The impulse-frequency response method is perhaps the fastest and simplest method in this category [10.66, 10.70, 10.71]. A cantilever beam test apparatus based on the impulse-frequency response method is shown in Fig. 10.31. In this apparatus the beam specimen is impulsively excited by a hammer which has a small force transducer in its tip, while the specimen response is monitored by a noncontacting displacement sensor. Excitation and response signals are fed into the FFT analyzer, which computes and displays the frequency response function in real time. Curve-fitting to the frequency response curve and calculation of the complex modulus are

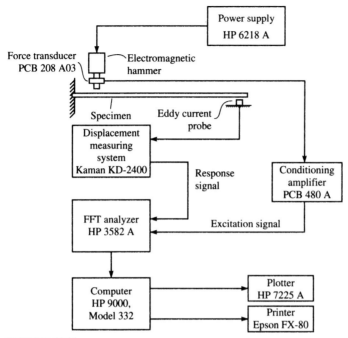

FIGURE 10.31
Cantilever beam test apparatus for impulse-frequency response method.

accomplished by a desktop computer which is interfaced with the FFT analyzer. Frequency dependence of the complex moduli can be determined by testing beams of different lengths and/or by taking data for multiple modes of vibration. The experimental data in Figs. 8.18 and 8.35 were obtained by using this method.

Impulse techniques have also been used in conjunction with laminated plate vibration models to determine the elastic constants of composite plates [10.67, 10.72, 10.73]. An impulse test apparatus based on this method has been developed for measurement of the complex extensional modulus of reinforcing fibers at elevated temperatures [10.74, 10.75]. Damping has been found to be particularly sensitive to damage and degradation in composites, and the impulse-frequency response method has been successfully used in such studies [10.76, 10.77].

In conclusion, dynamic test methods provide rapid and inexpensive alternatives to conventional static test methods in the measurement of composite stiffnesses. The dynamic test also yields information on the internal damping of the material, which is not only an important design property, but is also a useful nondestructive test parameter which can be related to the integrity of the material. It is conceivable that such test methods could be integrated into the manufacturing process itself in order to provide on-line monitoring and control of composite material properties.

PROBLEMS

10.1. Derive Eq. (10.2).

10.2. The results of longitudinal, transverse, and 45° off-axis tensile tests on samples from an orthotropic lamina are shown in Fig. 10.32. Based on these results, find numerical values for the engineering constants E_1, E_2, v_{12}, and G_{12}.

10.3. The in-plane shear modulus, G_{12}, of a graphite/epoxy lamina is to be measured by using the rail shear test shown in Fig. 10.33. The test is

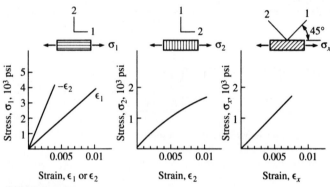

FIGURE 10.32
Stress-strain curves for longitudinal, transverse, and 45° off-axis tensile tests of an orthotropic lamina.

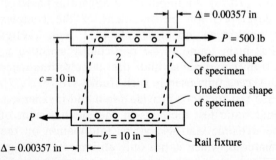

FIGURE 10.33
Shear deformation of a graphite/epoxy specimen during a rail shear test.

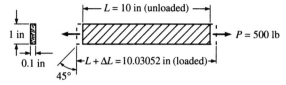

FIGURE 10.34
Extensional deformation of a graphite/epoxy specimen during a 45° off-axis tensile test.

conducted on a $10 \text{ in} \times 10 \text{ in} \times 0.1 \text{ in}$ ($254 \text{ mm} \times 254 \text{ mm} \times 2.54 \text{ mm}$) panel specimen, which deforms under the applied load as shown. Determine the value of G_{12} from these data.

10.4. The 45° off-axis test shown in Fig. 10.34 is conducted on a $10 \text{ in} \times 1 \text{ in} \times 0.1 \text{ in}$ ($254 \text{ mm} \times 25.4 \text{ mm} \times 2.54 \text{ mm}$) graphite/epoxy specimen, which deforms as shown under the applied load. It is also known from separate tensile tests that $E_1 = 32 \times 10^6 \text{ psi}$ (220 GPa), $E_2 = 1.0 \times 10^6 \text{ psi}$ (6.89 GPa), and $\nu_{12} = 0.3$. Determine the value of G_{12} from these test data.

10.5. A 45° off-axis specimen cut from a AS/3501 graphite/epoxy lamina is subjected to a tensile test. The specimen is 3 mm thick and 25 mm wide, and a tensile load of $F_x = 1 \text{ kN}$ on the specimen produces a corresponding strain $\epsilon_x = 0.0003$. It is claimed that the off-axis Young's modulus, E_x, can be determined from these results. Is this a valid claim? If so, why? If not, why not?

10.6. Describe the measurements that must be taken and the equations that must be used to determine the shear creep compliance, $S_{66}(t)$, of a unidirectional viscoelastic lamina by using a rail shear test.

10.7. Extensional vibration experiments are conducted on longitudinal, transverse, and 45° off-axis unidirectional composite specimens, and the complex moduli results for a particular vibration frequency are, respectively:

$$E_1^* = E_1'(1 + i\eta_1) = 35.6(1 + 0.004i) \text{ GPa}, \quad \text{for } \theta = 0°$$

$$E_2^* = E_2'(1 + i\eta_2) = 10.8(1 + 0.009i) \text{ GPa}, \quad \text{for } \theta = 90°$$

$$E_x^* = E_x'(1 + i\eta_x) = 11.6(1 + 0.011i) \text{ GPa}, \quad \text{for } \theta = 45°$$

Using the above data derive the equations for both parts of the complex shear modulus, $G_{12}^* = G_{12}'(1 + i\eta_{12})$, then find numerical values for both parts. Assume that all loss factors are very small ($\ll 1$), and that the major Poisson's ratio $\nu_{12} = 0.3$ is a real constant.

10.8. Using the results from Problem 10.7, derive the equations for both parts of the off-axis complex modulus, $E_x^* = E_x'(1 + i\eta_x)$, for an arbitrary angle θ; then find numerical values of both parts for an angle of $\theta = 30°$.

10.9. Describe an experiment, and give the necessary equations for measurement of the complex flexural modulus, E_{fx}^*, of a symmetric laminated beam.

10.10. Describe an experiment, and give the necessary equations for measurement of the complex extensional (or longitudinal) modulus, E_x^*, of a symmetric laminated bar.

10.11. Describe an experiment, and give the necessary equations for measurement of the complex through-thickness shear modulus, G_{23}^*, of a unidirectional, specially orthotropic, transversely isotropic beam.

10.12. Describe an experiment, and give the necessary equations for measurement of the complex Young's modulus, E_m^*, of an isotropic matrix material.

10.13. Describe an experiment, and give the necessary equations for measurement of the complex longitudinal modulus, E_{f1}^*, of a reinforcing fiber.

10.14. Describe an experiment, and give the necessary equations for measurement of the through-thickness creep compliance, $S_{32}(t)$, of a specially orthotropic, transversely isotropic lamina.

REFERENCES

10.1. *ASTM Standards and Literature References for Composite Materials,* 2d ed., American Society for Testing and Materials, Philadelphia, PA (1990).

10.2. Tarnopolski, Y. M. and Kincis, T., *Static Test Methods for Composites,* Van Nostrand Reinhold Co., New York (1981).

10.3. Whitney, J. M., Daniel, I. M., and Pipes, R. B., *Experimental Mechanics of Fiber Reinforced Composite Materials,* Society for Experimental Mechanics, Bethel, CT (1982).

10.4. Carlsson, L. A. and Pipes, R. B., *Experimental Characterization of Advanced Composite Materials,* Prentice-Hall, Inc., Englewood Cliffs, NJ (1987).

10.5. Pendleton, R. L. and Tuttle, M. E. (eds.), *Manual on Experimental Methods for Mechanical Testing of Composites,* Society for Experimental Mechanics, Bethel, CT (1989).

10.6. D 3379-75, "Standard Test Method for Tensile Strength and Young's Modulus for High-Modulus Single-Filament Materials," *ASTM Standards and Literature References for Composite Materials,* 2d ed., 34–37, American Society for Testing and Materials, Philadelphia, PA (1990).

10.7. McMahon, P. E., "Graphite Fiber Tensile Property Evaluation," in *Analysis of Test Methods for High Modulus Fibers and Composites,* ASTM STP **521**, 367–389, American Society for Testing and Materials, Philadelphia, PA (1973).

10.8. D 4018-81, "Standard Test Methods for Tensile Properties of Continuous Filament Carbon and Graphite Yarns, Strands, Rovings and Tows," *ASTM Standards and Literature References for Composite Materials,* 2d ed., 85–97, American Society for Testing and Materials, Philadelphia, PA (1990).

10.9. Kawabata, S., "Measurements of Anisotropic Mechanical Property and Thermal Conductivity of Single Fiber for Several High Performance Fibers," in J. R. Vinson (ed.), *Proceedings of the 4th Japan–U.S. Conference on Composite Materials,* 253–262, Technomic Publishing Co., Lancaster, PA (1989).

10.10. D 638-90, "Standard Test Method for Tensile Properties of Plastics," *ASTM Standards and Literature References for Composite Materials,* 2d ed., American Society for Testing and Materials, Philadelphia, PA (1990).

10.11. D 618-81, "Standard Practice for Conditioning Plastics and Electrical Insulating

Materials for Testing," *ASTM Standards and Literature References for Composite Materials,* 2d ed., American Society for Testing and Materials, Philadelphia, PA (1990).

10.12. D 695-90, "Standard Test Method for Compressive Properties of Rigid Plastics," *ASTM Standards and Literature References for Composite Materials,* 2d ed., American Society for Testing and Materials, Philadelphia, PA (1990).

10.13. Novak, R. C. and Bert, C. W., "Theoretical and Experimental Bases for More Precise Elastic Properties of Epoxy," *Journal of Composite Materials,* **2,** 506–508 (1968).

10.14. D 790-90, "Standard Test Methods for Flexural Properties of Unreinforced and Reinforced Plastics and Electrical Insulating Materials," *ASTM Standards and Literature References for Composite Materials,* 2d ed., American Society for Testing and Materials, Philadelphia, PA (1990).

10.15. D 3039-76, "Standard Test Method for Tensile Properties of Fiber-Resin Composites," *ASTM Standards and Literature References for Composite Materials,* 2d ed., American Society for Testing and Materials, Philadelphia, PA (1990).

10.16. Pagano, N. J. and Halpin, J. C., "Influence of End Constraint in the Testing of Anisotropic Bodies," *Journal of Composite Materials,* **2,** 18–31 (1968).

10.17. Jones, R. M., *Mechanics of Composite Materials,* Hemisphere Publishing Co., New York (1975).

10.18. Horgan, C. O., "Some Remarks on Saint-Venant's Principle for Transversely Isotropic Composites," *Journal of Elasticity,* **2**(4), 335–339 (1972).

10.19. Choi, I. and Horgan, C. O., "Saint-Venant's Principle and End Effects in Anisotropic Elasticity," *Journal of Applied Mechanics,* **44,** 424–430 (1977).

10.20. Horgan, C. O., "Saint-Venant End Effects in Composites," *Journal of Composite Materials,* **16,** 411–422 (1982).

10.21. Rawlinson, R. A., "The Use of Crossply and Angleply Composite Test Specimens to Generate Improved Material Property Data," in *How Concept Becomes Reality— Proceedings of the 36th International SAMPE Symposium and Exhibition,* **36,** Book 1, 1058–1068, Society for the Advancement of Material and Process Engineering, Covina, CA (1991).

10.22. Hart-Smith, L. J., "Backing Out Equivalent Unidirectional Lamina Strengths from Tests on Cross-plied Laminates," in *Materials Working for You in the 21st Century—Proceedings of the 37th International SAMPE Symposium and Exhibition,* **37,** 977–990, Society for the Advancement of Material and Process Engineering, Covina, CA (1992).

10.23. Berg, J. S. and Adams, D. F., "An Evaluation of Composite Material Compression Test Methods," *Journal of Composites Technology and Research,* **11**(2), 41–46 (Summer 1989).

10.24. Schoeppner, G. A. and Sierakowski, R. L., "A Review of Compression Test Methods for Organic Matrix Composites," *Journal of Composites Technology and Research,* **12**(1), 3–12 (Spring 1990).

10.25. D 3410-87, "Standard Test Method for Compressive Properties of Unidirectional or Crossply Fiber-Resin Composites," *ASTM Standards and Literature References for Composite Materials,* 2d ed., American Society for Testing and Materials, Philadelphia, PA (1990).

10.26. Adsit, N. R., "Compression Testing of Graphite/Epoxy," in R. Chait and R. Papirno (eds.), *Compression Testing of Homogeneous Materials,* ASTM STP **808,** 175–186, American Society for Testing and Materials, Philadelphia, PA (1983).

10.27. Crasto, A. S. and Kim, R. Y., "Compression Strength of Advanced Composites from a Novel Mini-Sandwich Beam," *SAMPE Quarterly,* **22**(3), 29–39 (April 1991).

10.28. Wilson, D. W., Altstadt, V., and Prandy, J., "On the Use of Laminate Test Methods to Characterize Lamina Compression Strength," in *Materials Working for You in the*

21st Century—Proceedings of the 37th International SAMPE Symposium and Exhibition, **37**, 606–619, Society for the Advancement of Material and Process Engineering, Covina, CA (1992).

10.29. Boeing Specification Support Standard BSS 7260, "Advanced Composite Tests," The Boeing Company, Seattle, WA, issued 1982, rev. 1988.

10.30. Nettles, A. T. and Hodge, A. J., "Compression after Impact Testing of Thin Composite Materials," in *Advanced Materials/Affordable Processes—Proceedings of the 23rd International SAMPE Technical Conference*, **23**, 177–183, Society for the Advancement of Material and Process Engineering, Covina, CA (1991).

10.31. D 3518/D 3518M-91, "Standard Practice for In-Plane Shear Stress-Strain Response of Unidirectional Polymer Matrix Composites," *ASTM Standards and Literature References for Composite Materials*, 2d ed., American Society for Testing and Materials, Philadelphia, PA (1990).

10.32. D 4255-83, "Standard Guide for Testing In-Plane Shear Properties of Composite Laminates," *ASTM Standards and Literature References for Composite Materials*, 2d ed., American Society for Testing and Materials, Philadelphia, PA (1990).

10.33. Higdon, A., Ohlsen, E. H., Stiles, W. B., Weese, J. A., and Riley, W. F., *Mechanics of Materials*, 3d ed., John Wiley & Sons, Inc., New York (1976).

10.34. Iosipescu, N., "New Accurate Procedure for Single Shear Testing of Metals," *Journal of Materials*, **2**(3), 537–566 (1967).

10.35. Walrath, D. E. and Adams, D. F., "The Iosipescu Shear Test as Applied to Composite Materials," *Experimental Mechanics*, **23**(1), 105–110 (1983).

10.36. Adams, D. F. and Walrath, D. E., "Current Status of the Iosipescu Shear Test Method," *Journal of Composite Materials*, **21**, 484–505 (1987).

10.37. Adams, D. F. and Walrath, D. E., "Further Development of the Iosipescu Shear Test Method," *Experimental Mechanics*, **27**(2), 113–119 (1987).

10.38. Adams, D. F., "The Iosipescu Shear Test Method as Used for Testing Polymers and Composite Materials," *Polymer Composites*, **11**(5), 286–290 (1990).

10.39. Lee, S. and Munro, M., "Evaluation of In-Plane Shear Test Methods for Advanced Composite Materials by the Decision Analysis Technique," *Composites*, **17**(1), 13–22 (1986).

10.40. D2344-84, "Standard Test Method for Apparent Interlaminar Shear Strength of Parallel Fiber Composites by Short Beam Method," *ASTM Standards and Literature References for Composite Materials*, 2d ed., American Society for Testing and Materials, Philadelphia, PA (1990).

10.41. Whitney, J. M., "Elasticity Analysis of Orthotropic Beams under Concentrated Loads," *Composites Science and Technology*, **22**, 167–184 (1985).

10.42. Whitney, J. M., "Experimental Characterization of Delamination Fracture," in N. J. Pagano (ed.), *Interlaminar Response of Composite Materials, Composite Materials Series, Volume 5*, R. B. Pipes (Series ed.), Elsevier Science Publishers, Amsterdam, The Netherlands (1989).

10.43. Whitney, J. M., Browning, C. E., and Hoogsteden, W., "A Double Cantilever Beam Test for Characterizing Mode I Delamination of Composite Materials, *Journal of Reinforced Plastics and Composites*, **1**, 297–313 (1982).

10.44. Russell, A. J. and Street, K. N., "Moisture and Temperature Effects on the Mixed Mode Delamination Fracture of Unidirectional Graphite/Epoxy," *DREP Technical Memo 83-22*, Defence Research Establishment Pacific, Victoria, B.C., Canada (1983).

10.45. Russell, A. J. and Street, K. N., "Moisture and Temperature Effects on the Mixed Mode Delamination Fracture of Unidirectional Graphite/Epoxy," in W. S. Johnson (ed.), *Delamination and Debonding of Materials*, ASTM STP **876,** 349–370, American Society for Testing and Materials, Philadelphia, PA (1985).

10.46. Carlsson, L. A., Gillispie, J. W., Jr., and Pipes, R. B., "On the Analysis and Design of the End Notched Flexure (ENF) Specimen for Mode II Testing," *Journal of Composite Materials*, **20**, 594–604 (1986).

10.47. Drzal, L. T., Rich, M. J., and Lloyd, P. F., "Adhesion of Graphite Fibers to Epoxy Matrices: I. The Role of Fiber Surface Treatment," *Journal of Adhesion*, **16**, 1–30 (1982).

10.48. Drzal, L. T., Rich, M. J., Koenig, M. F., and Lloyd, P. F., "Adhesion of Graphite Fibers to Epoxy Matrices: II. The Effect of Fiber Finish," *Journal of Adhesion*, **16**, 133–152 (1983).

10.49. Drzal, L. T., Rich, M. J., and Subramoney, S., "Fiber-Matrix Bonding and Its Effect on Composite Properties," in *Advanced Composites III—Expanding the Technology: Proceedings of the 3rd Annual Conference on Advanced Composites*, 305–308, ASM International, Materials Park, OH (1987).

10.50. McDonough, W. G., Herrera-Franco, P. J., Wu, W. L., Drzal, L. T., and Hunston, D. L., "Fiber-Matrix Bond Tests in Composite Materials," in *Advanced Materials/Affordable Processes: Proceedings of 23rd International SAMPE Technical Conference*, Kiamesha Lake, NY, 247–258, Society for Advancement of Material and Process Engineering, Covina, CA (1991).

10.51. Mandell, J. F., Grande, D. H., Tsiang, T. H., and McGarry, F. J., "Modified Microbonding Test for Direct In Situ Fiber/Matrix Bond Strength Determination in Fiber Composites," in J. M. Whitney (ed.), *Composite Materials: Testing and Design (Seventh Conference)*, ASTM STP **893**, 87–108, American Society for Testing and Materials, Philadelphia, PA (1986).

10.52. Caldwell, D. L., "Determination of the Interfacial Strength of Composites," in *Advanced Composites III—Expanding the Technology: Proceedings of the 3rd Annual Conference on Advanced Composites*, 299–303, ASM International, Materials Park, OH (1987).

10.53. D 2990-90, "Standard Test Methods for Tensile, Compressive, and Flexural Creep and Creep-Rupture of Plastics," *ASTM Standards and Literature References for Composite Materials*, 2d ed., American Society for Testing and Materials, Philadelphia, PA (1990).

10.54. Halpin, J. C. and Pagano, N. J., "Observations on Linear Anisotropic Viscoelasticity," *Journal of Composite Materials*, **2**(1), 68–80 (1968).

10.55. Yeow, Y. T., Morris, D. H., and Brinson, H. F., "Time-Temperature Behavior of a Unidirectional Graphite/Epoxy Composite," in S. W. Tsai (ed.), *Composite Materials: Testing and Design (Fifth Conference)*, ASTM STP **674**, 263–281, American Society for Testing and Materials, Philadelphia, PA (1979).

10.56. Beckwith, S. W., "Viscoelastic Creep Behavior of Filament-Wound Case Materials," *Journal of Spacecraft and Rockets*, **21**(6), 546–552 (1984).

10.57. Beckwith, S. W., "Creep Evaluation of a Glass/Epoxy Composite," *SAMPE Quarterly*, **11**(2), 8–15 (1980).

10.58. Irion, M. N. and Adams, D. F., "Compression Creep Testing of Unidirectional Composite Materials," *Composites*, **2**(2), 117–123 (1981).

10.59. Lou, Y. C. and Schapery, R. A., "Viscoelastic Characterization of a Nonlinear Fiber Reinforced Plastic," *Journal of Composite Materials*, **5**, 208–234 (1971).

10.60. Gibson, R. F., Hwang, S. J., and Sheppard, C. H., "Characterization of Creep in Polymer Composites by the Use of Frequency-Time Transformations," *Journal of Composite Materials*, **24**(4), 441–453 (1990).

10.61. Gibson, R. F., Hwang, S. J., Kathawate, G. R., and Sheppard, C. H., "Measurement of Compressive Creep Behavior of Glass/PPS Composites Using the Frequency-Time Transformation Method," in *Advanced Materials/Affordable Processes—*

Proceedings of 23rd International SAMPE Technical Conference, 208–218, Society for the Advancement of Material and Process Engineering, Covina, CA (1991).

10.62. Gibson, R. F. and Kathawate, G. R., "Rapid Screening of Creep Susceptibility of Structural Polymer Composites," in V. J. Stokes (ed.), *Plastics and Plastic Composites: Material Properties, Part Performance and Process Simulation*, ASME MD Vol. **29**, 161–171, American Society of Mechanical Engineers, New York (1991).

10.63. D 4065-90a, "Standard Practice for Determining and Reporting Dynamic Mechanical Properties of Plastics," *ASTM Standards and Literature References for Composite Materials*, 2d ed., American Society for Testing and Materials, Philadelphia, PA (1990).

10.64. E 756-83, "Standard Method for Measuring Vibration Damping Properties of Materials," *1990 Annual Book of ASTM Standards*, Vol. 04.06, *Thermal Insulation and Environmental Acoustics*, 763–769, American Society for Testing and Materials, Philadelphia, PA (1983).

10.65. Gibson, R. F., "Vibration Test Methods for Dynamic Mechanical Property Characterization," in R. L. Pendleton and M. E. Tuttle (eds.), *Manual on Experimental Methods for Mechanical Testing of Composites*, 151–164, Society for Experimental Mechanics, Bethel, CT (1989).

10.66. Suarez, S. A. and Gibson, R. F., "Improved Impulse-Frequency Response Techniques for Measurement of Dynamic Mechanical Properties of Composite Materials," *Journal of Testing and Evaluation*, **15**(2), 114–121 (1987).

10.67. Deobald, L. R. and Gibson, R. F., "Determination of Elastic Constants of Orthotropic Plates by a Modal Analysis/Rayleigh Ritz Technique," *Journal of Sound and Vibration*, **124**(2), 269–283 (1988).

10.68. Dudek, T. J., "Young's and Shear Moduli of Unidirectional Composites by a Resonant Beam Method," *Journal of Composite Materials*, **4**, 232–241 (1970).

10.69. Soovere, J. and Drake, M. L., *Aerospace Structures Technology Damping Design Guide: Volume I—Technology Review*, AFWAL-TR-84-3089, Vol. I, Air Force Wright Aeronautical Labs, Wright-Patterson AFB, OH (1985).

10.70. Suarez, S. A., Gibson, R. F., and Deobald, L. R., "Random and Impulse Techniques for Measurement of Damping in Composite Materials," *Experimental Techniques*, **8**(10), 19–24 (1984).

10.71. Crane, R. M. and Gillispie, J. W., Jr., "A Robust Testing Method for Determination of the Damping Loss Factor of Composites," *Journal of Composites Technology and Research*, **14**(2), 70–79 (1992).

10.72. DeWilde, W. P., Narmon, B., Sol, H., and Roovers, M., "Determination of the Material Constants of an Anisotropic Lamina by Free Vibration Analysis," in *Proceedings of the 2nd International Modal Analysis Conference*, Orlando, FL, I, 44–49 (1984).

10.73. Ayorinde, E. O. and Gibson, R. F., "Elastic Constants of Orthotropic Composite Materials Using Plate Resonance Frequencies, Classical Lamination Theory and an Optimized Three Mode Rayleigh Formulation," *Composites Engineering*, **3**(5), 395–407 (1993).

10.74. Gibson, R. F., Thirumalai, R., and Pant, R., "Development of an Apparatus to Measure Dynamic Modulus and Damping of Reinforcing Fibers at Elevated Temperature," in *Proceedings 1991 Spring Conference on Experimental Mechanics*, 860–869, Society for Experimental Mechanics, Bethel, CT (1991).

10.75. Pant, R. H. and Gibson, R. F., "Analysis and Testing of Dynamic Micromechanical Behavior of Composite Materials at Elevated Temperatures," in *Vibroacoustic Characterization of Materials and Structures*, P. K. Rajn, (ed.), NCA vol. 14, 131–146, American Society of Mechanical Engineers, New York (1992).

10.76. Mantena, R., Place, T. A., and Gibson, R. F., "Characterization of Matrix Cracking in Composite Laminates by the Use of Damping Capacity Measurements," *Role of Interfaces on Material Damping,* 79–93, ASM International, Materials Park, OH (1985).

10.77. Mantena, R., Gibson, R. F., and Place, T. A., "Damping Capacity Measurements of Degradation in Advanced Materials," *SAMPE Quarterly,* **17**(3), 20–31 (1986).

INDEX

Aboudi, J., 88, 98
Achenbach, J. D., 306, 336
Activation energy for diffusion, 141
Adams, D. F., 84–87, 90–91, 98, 144–146, 392, 401, 412–413
Adams, R. D., 71, 97, 321, 336
Adsit, N. R., 389, 411
Advanced composites, 6–7
Agarwal, B. D., 120–121, 130
Aging, 275, 300–305, 401–402
Airy stress functions, 87
Alexander, R. M., 344, 370
Aliyu, A. A., 364, 371
American Society for Testing and Materials (ASTM) standards, 375–403
Amorphous microstructure, 272–273
Angle-ply laminates, 208–212, 217–220, 229–231
Anisotropic material:
  mechanical behavior, 31, 36–56
  stress-strain relationships for, 36–56
Antisymmetric laminates, 210–212, 218–220, 229–231
Aramid fibers, 8–9, 80
Arrhenius relationship, 141
Ashby, M. F., 1–2, 33
Ashton, J. E., 61, 151
Autoclave molding, 22–24, 65
Automotive applications, 18–19
AV-8B fighter aircraft, 13–14
Average stress criterion (see Whitney-Nuismer Criterion; stress fracture criteria)
Azzi, V. D., 109, 129

"Backing out" lamina properties, 382, 386, 390
Balanced cross-ply laminate, 208
Balanced orthotropic material, 42, 47–48
Basic composites, 6–7
Beaumont, P. W. R., 348, 370
Beckwith, S. W., 301, 335, 401–402, 413
Beech Starship aircraft, 16–17
Bernoulli-Euler equation, 311–314, 404
Bert, C. W., 321, 336, 378–381, 411
Biaxial tests, 105, 111
Binder, 1, 4
Biot, M. A., 295, 335
Boeing 757/767 airliners, 13, 15–16
Boltzmann Superposition integrals, 275–276
Boltzmann Superposition Principle, 275
Boron fibers, 8–9, 80
Boundary layer region, 232
Bounds on moduli, 89–91
Brewer, J. C., 245, 268
Broutman, L. J., 120–121, 130, 155
Browning, C. E., 132–134, 137, 139, 235, 267
Buckling:
  of laminated beams, 195
  of laminated plates, 249–258

Carbon fibers, 7–9, 80
Carlsson, L. A., 397, 413
Carpet plots, 258, 260–262
Caruso, J. J., 79, 86
Celanese fixture, 388
Ceramic matrix materials, 12

Ceramics, 1–2, 12
Chamis, C. C., 63, 79–82, 91, 97, 102, 125, 130, 139, 141, 151–152, 155
Chaturvedi, S. K., 321, 337
Chen, P. E., 175, 184, 189
Chon, C. T., 169–172, 189
Christensen, R. M., 63, 90, 98, 177–181, 189, 203, 267, 277, 295–296, 306
Classical Lamination Theory, 190, 201–207, 324, 386
Coefficient of hygroscopic expansion, 145–152
  of isotropic materials, 145
  of laminates, 231
  of orthotropic materials, 146–152
Coefficient of thermal expansion, 144–152
  of isotropic materials, 144
  of laminates, 231–232
  of orthotropic materials, 146–152
Complex compliance, 292–295, 299
Complex modulus, 291–295, 298–300, 322–325, 402–408
Compliances:
  for anisotropic elastic materials, 37
  for creep of viscoelastic materials, 275–278, 282–287, 400–402
  for generally orthotropic elastic materials, 51–56
  for specially orthotropic elastic materials, 45–47
Composite density, 64–65
Composite materials:
  constituent materials for, 6–13
  definition of, 1
  fabrication of, 21–27
  history of, 1–3
  importance of, 2
  mechanical behavior of, 28–31
  reinforcement of, 1–2
  structural applications of, 13–20
  types of, 4–5
Composite shaft, 19, 21, 94
Compression-after-impact (CAI) strength, 363, 389–390
Compression-after-impact (CAI) test, 389–390
Compression testing:
  of composites, 387–390
  of fibers, 377
  of neat resin matrix, 378–381
Constrained viscoelastic layer, 329
Conti, P., 237, 268

Contracted notation, 37, 144
Convolution integral, 289
Corten, H. T., 345, 370
Coupling:
  bending-extension, 208
  bending-twisting, 212
  extension-twisting, 212
  at lamina level, 31, 44–46, 53, 207
  at laminate level, 201–270
Cox, H. L., 159–167, 170–171, 174, 177, 179, 181–182, 188
Crack deformation modes, 343, 362–363
Crack growth, 348, 357–362
Cracks in laminates:
  interlaminar, 356–367
  through-thickness, 339–356
Crasto, A. S., 122, 130
Creep:
  of anisotropic materials, 276
  compliances, 275–278
  definition of, 271–272
  of isotropic materials, 275–276
  measurement of, 399–402
  of orthotropic materials, 276–290
Critical fiber volume fraction, 117
Critical flaw size, 341
Critical length of fiber, 161–162
Critical strain energy release rate, 347, 363–367, 395–397
Critical stress, 341
Critical stress intensity factor, 342–349
Crossman, F. W., 235, 237, 268
Cross-ply laminates, 208
Cruse, T. J., 347, 370
Crystalline microstructure, 272–273

Damping:
  analysis of, 321–330
  introduction to, 271
  measurement of, 402–407
Damping ratio, 329
Daniel, I. M., 235, 267, 364, 371
Deflection:
  of laminated beams, 194
  of laminated plates, 249–255
DeIasi, R., 139–140, 142, 145, 155
Delamination, 232, 244–249, 356–367, 395–397
  (*See also* Fracture mechanics; Interlaminar fracture)
Delamination growth, 356–366

Density, 64–65, 151
DePaulis, A., 237, 268
Design of laminates, 258–262
Diffusion of moisture, 136–138, 141–142
Diffusivity, 136–138, 141–142, 151
Dirac delta function, 289
Discontinuous fiber composites:
  with aligned fibers, 157–169
  with off-axis aligned fibers, 157, 169–173
  with randomly oriented fibers, 157, 173–184
Doner, D. R., 84–87, 90–91, 98
Dong, S. B., 201, 267
Double cantilever beam (DCB) test, 362–364, 395–396
Drzal, L. T., 122, 130, 162, 188, 397, 413
Dynamic behavior, 306–330
Dynamic mechanical analysis, 403
Dynamic mechanical properties:
  definition of, 402–403
  measurement of, 403–408

Effective moduli, 35–39, 62–63
  of anisotropic viscoelastic composites, 276–277
  of continuous fiber-reinforced lamina, 35–39, 62–63
  of discontinuous fiber-reinforced lamina, 163–182
  of laminate, 220–223
E-glass fibers, 7–8, 80
Elastic coefficients (*see* Compliances; Engineering constants; Stiffnesses)
Elastic constants (*see* Compliances; Engineering constants; Stiffnesses)
Elasticity approach:
  in analysis of interlaminar stresses, 232–237
  in micromechanics, 83–90
Elastic-Viscoelastic Correspondence Principle:
  in analysis of creep and relaxation, 296–298
  in analysis of damping, 324–325
  in analysis of vibration, 298–300
  introduction to, 295–296
Elementary mechanics of materials approach:
  for coefficient of thermal expansion, 149–152
  for laminated beam, 192–201

Elementary mechanics of materials approach: (*Cont.*)
  for stiffness, 67–77
  for strength, 114–126
End-notched flexure (ENF) test, 363–364, 395–397
Engineering constants:
  of generally orthotropic materials, 51–53
  of laminates, 220–223
  measured values of, 48
  of specially orthotropic materials, 43–46, 48
Epoxy, 12
Erdogan, F., 348, 370

Fabrication processes, 21–27
Failure surfaces, 103–104, 127
Fast Fourier transform, 406–407
Fatigue crack growth, 348, 357–362
Fiber coatings, 366
Fiber/matrix interface, 4, 11, 72, 85, 122, 158–165, 397–399
Fiber/matrix interfacial strength, 122, 161–162, 397–399
Fiber/matrix interphase, 94–95, 326–327
Fiber microbuckling, 118–121
Fiber packing geometries, 65–67
Fiber testing:
  dynamic, 407
  static, 375–377
Fiber volume fraction, 64–93
Fibers:
  aramid, 8–9
  boron, 8–9
  chopped, 5
  glass, 7–8
  graphite (carbon), 7–8
  packing of, 65–67
  properties of, 8, 80
  silicon carbide, 8, 10
  tensile strength of, 3, 8, 9
  types of, 7–11
  woven, 5
Fick's law, 136–138, 141
Filament winding, 25–28, 265
Filament wound pressure vessel, 32, 56–58, 290–291
Filler materials, 12–13
Finite difference method:
  in analysis of interlaminar stresses, 234–235
  in micromechanics, 84–85, 124

Finite element method:
  in analysis of damping, 326–329
  in analysis of fiber/matrix interface, 165, 182
  in analysis of interlaminar stresses, 235–237
  in fracture mechanics, 357
  in micromechanics, 85–91
First ply failure, 122, 238–244
Flexural modulus:
  of laminated beams, 194, 395
  of laminated plates, 221–222
Flexural vibration (*see* Vibrations)
Flexure tests:
  of composite, 394–395
  of matrix, 381
Fourier coefficients, 254
Fourier heat conduction equation, 136
Fourier transform, 292–294, 402
Fracture mechanics:
  in analysis of through-thickness cracks, 339–356
  in interlaminar fracture, 356–367
  strain energy release rate approach, 345–349
  stress fracture approach, 349–356
  stress intensity factor approach, 341–345
  (*See also* Delamination; Interlaminar fracture)
Fracture toughness, 341–367
Frequency response spectrum, 406
Fung, Y. C., 291, 335

Galerkin method, 255, 320
Generally orthotropic material, 41–42, 48–56
Gibson, R. F., 97, 133, 135, 154, 164, 166–168, 188, 235–237, 268, 321, 326–327, 336, 402, 413–414
Glass fibers, 7–8, 80
Glass transition temperature, 132, 139–140, 271, 302
Graphite/epoxy, 7, 13, 48, 66–67
Graphite fibers, 7–9, 80
Greszczuk, L. B., 121, 130
Griffith, A. A., 2–3, 33, 339–341, 369
Griffith Criterion, 339–341
  (*See also* Griffith, A. A.)

Hahn, H. T., 56, 61, 91–92, 97, 111, 121, 129, 240, 268

Half power bandwidth, 406
Halpin, J. C., 52, 56, 61, 63, 90–91, 98, 151, 167–168, 176–177, 184, 189, 239, 250, 267, 278, 297, 335, 384–385, 400–401, 411, 413
Halpin-Tsai equations:
  for analysis of continuous fiber composites, 90–91, 95
  for analysis of discontinuous fiber composites, 167–169, 177, 182
  for analysis of transport properties, 151
  for analysis of viscoelastic creep, 297–298
Hashin, Z., 63, 90, 98, 102, 111–112, 128, 150, 155, 179, 189, 298, 335
Hatcher, D., 257–258, 268
Hearmon, R. F. S., 306, 336
Heaviside function, 288–289
Herakovich, C. T., 235, 268
Hereditary law, 275–276
Higher-order lamination theory, 203
Hill, R., 108–111, 129, 179, 189
Hoff, N. J., 201, 267
Holes in laminates, 349–356
Hopkins, D. A., 77, 97, 151
Horgan, C. O., 385, 411
"Hot-wet" conditions, 133, 143
Hull, D., 116, 121, 129
Hunston, D. L., 366, 372
Hwang, S. J., 165–168, 188, 235–237, 268
Hybrid composites, 5, 11, 93–94, 366
Hygroscopic forces and moments, 228–230
Hygrothermal degradation:
  of lamina hygrothermal properties, 152
  of lamina mechanical properties, 132–144
  of laminate properties, 227
Hygrothermal effects, 131–152, 226–232
Hygrothermal expansion, 132, 144–149, 231–232
Hygrothermal stresses:
  in a lamina, 147
  in a laminate, 227–231
Hysteresis loop, 271–272, 405

IITRI fixture, 388
Improved mechanics of materials models, 77–83
Impulse-frequency response method, 406–408
Ineffective length of fiber, 160–161
In-plane shear modulus:
  definition of, 45

In-plane shear modulus: (*Cont.*)
  of fiber, 377
  of laminate, 220–221
  measurement of, 390–393
  of neat resin matrix, 378–381
  prediction of, 75–93
Interfacial shear strength, 122, 161–162,
    397–399
Interfacial shear stress, 158–165, 169–172
Interlaminar damping, 327–329
Interlaminar fracture:
  fracture mechanics approach, 356–367
  measurement of, 395–397
  mechanics of materials approach,
    244–249
  (*See also* Delamination; Fracture
    mechanics)
Interlaminar fracture toughness, 356–367,
    395–397
Interlaminar shear tests, 393–395
Interlaminar stresses, 232–237, 244–249
Interleaves, 366
Interphase, 326–327
Invariants, 53–56, 176, 213–215
Inverse rule of mixtures:
  for in-plane shear modulus, 76, 92
  for transverse modulus, 74, 92
Iosipescu, N., 392–393, 412
Iosipescu shear test, 392–393
Irion, M. N., 401, 413
Irwin, G. R., 341, 345, 347, 370

Janas, V. F., 303, 336
Johnson, W. S., 364–365, 372
Jones, R. M., 91, 97, 384–385, 411

Kanninen, M. F., 357, 371
Kardos, J. L., 184, 189
Kawabata, S., 79, 98, 377, 410
Kelly, A., 9–10, 115, 129, 159–165, 170
Kelvin-Voigt model, 283–284, 286
Kerwin, E. M., Jr., 325–328, 337
Kevlar® fibers, 8–9, 80, 102
Kies, J. A., 123, 130
Kim, R. Y., 122, 130, 235, 244, 268
Kim-Soni Criterion, 244–246
Kinra, V. K., 321, 337
Kirchhoff deformation hypothesis, 203
Konish, H. J., 344, 370
Kowalski, I. M., 79, 98

Kriz, R. D., 79, 97
Kunz, S. C., 348, 370

Lagace, P. A., 245, 268
Lamina:
  hygrothermal behavior of, 131–152
  stiffnesses of, 46–58, 62–93, 156–183
  strength of, 99–126, 183–184
  stress-strain relationships for, 46–58
  types of, 4–5
Laminated beams:
  buckling of, 195
  deflections of, 194
  flexural modulus of, 194–199
  interlaminar stresses in, 196–198
  strength of, 195–196
  stresses in, 195–196
  theory of, 193–201
Laminated plates (*see* Laminates)
Laminates:
  antisymmetric, 210–212
  buckling of, 255–258
  compliances of, 215–226
  creep of, 297–298, 399–402
  deflection of, 249–255
  delamination of, 232–238, 244–249
  design of, 258–262
  engineering constants of, 220–223
  hygrothermal degradation of, 227
  hygrothermal effects in, 226–232
  hygrothermal expansion of, 231–232
  interlaminar stresses in, 232–238,
    244–249
  orientation code, 191–192
  quasi-isotropic, 212–215
  software for analysis of, 259
  stiffnesses of, 194–195, 206–215
  strength of, 237–249
  stress distributions in, 217–220, 227–231
  symmetric, 208–210
  types of, 207–215
  vibration of, 299–300, 306, 402–408
Laplace transform, 279–281, 288
Leach, D. C., 364, 371
Leaf springs, 18–19
Lee, E. H., 295, 335
Lee, S., 393, 412
Lees, J. K., 183–184, 189
Leknitskii, S. G., 201, 249, 267, 352–353,
    371
Liessa, A. W., 249, 268

Load transfer length, 160–161
Logarithmic decrement, 329, 404
Longitudinal modulus:
  definition of, 44
  measurement of, 382–383
  prediction of, 69–73
Long-term creep, 302–305
Loos, A. C., 141–142
Loss factor, 293, 299, 323–329, 404–406
Loss modulus, 293, 299, 322
Lou, Y. C., 401, 413
Low velocity impact, 248

McCrum, N. G., 4, 17, 33
McCullough, R. L., 303, 336
McDonough, W. G., 398–399, 413
McMahon, P. E., 326, 410
Macromechanics, 28–31, 63
Madhukar, M. S., 122, 130
Manera, M., 177, 189
Mangalgiri, P. D., 364–365, 372
Mantena, P. R., 329, 337
Master curve, 302–305
Material property symmetry, 35, 40–43
Matrix materials:
  properties of, 81
  types of, 11
Maximum Distortional Energy Criterion, 108
Maximum Strain Criterion (*see* Multiaxial strength criteria)
Maximum Stress Criterion (*see* Multiaxial strength critiera)
Maxwell model, 281–283
Method of subregions, 78–79, 94
Microbond test, 398–399
Microindentation test, 398–399
Micromechanics:
  of continuous fiber lamina stiffness, 28–31, 63–93
  of continuous fiber lamina strength, 114–126
  of discontinuous fiber lamina stiffness, 156–186
  of discontinuous fiber lamina strength, 161, 183–184
  of hygrothermal properties, 149–151
  of transport properties, 151
Minimechanics, 63
Minisandwich specimen, 389
Mixed mode delamination, 363–365

Modal strain energy approach, 326
Mode shapes:
  of beams, 313
  of laminated plates, 319–320
  of rods, 309–310
Modified stiffness matrices, 239–243
Mohr's circles:
  for stiffness transformations, 55–56
  for stress transformations, 55
Moiré technique, 235
Moisture absorption:
  effect of material type, 133–141
  effect of stress, 141
  effect of temperature, 140–142
  effect of time, 136–138
Moisture concentration, 134–138, 145–146
Moisture effects, 132–144
  (*See also* Moisture absorption)
Moisture profiles, 136–137
Molecular composites, 1
Momentary creep, 303–305
Monoclinic material, 40
Moon, F. C., 306, 336
Multiaxial strength criteria, 101–114
  Maximum Strain Criterion, 106–108, 184
  Maximum Stress Criterion, 103–106, 183
  Quadratic interaction criteria, 108–112
  Tsai-Hill Criterion, 109, 188
  Tsai-Wu Criterion, 110
Munro, M., 393, 412
Murri, G. B., 364, 372
Mutual influence coefficients, 53

National Aerospace Plane, 18
Natural frequencies:
  of beams, 313
  of laminated plates, 319
  of rods, 309
Neat resin matrix tests, 378–382
Newaz, G. M., 248, 268
Nielsen, L. E., 175, 189
Non-Fickian diffusion, 138
Notches in laminates, 349–356
Novak, R. C., 378–381, 411
Nuismer, R. J., 244, 268, 349–356, 370–371

O'Brien, T. K., 246–248, 268, 360–365, 371
Off-axis coordinate system, 41, 49–56

Off-axis properties:
coefficients of hygrothermal expansion, 147–148
creep, 401
stiffness, 48–56, 170–173
strength, 105–114, 183
Off-axis test, 105–106, 383–385, 390–391, 401
Ogale, A. A., 303, 336
Open mold process, 22–23
Orowan, E., 341, 370

Pagano, N. J., 53, 61, 176–177, 189, 192, 196, 232–237, 267, 278, 335, 384–385, 400–401, 411, 413
PAN fibers, 8, 80
Parallel axis theorem for stiffness, 265
Parhizgar, S., 344, 370
Paris, P. C., 348, 370
Particle-reinforced composites, 1, 4, 89, 95
Paul, B., 89–91, 98
Pipes, R. B., 232–237, 267, 397, 413
Pister, K. S., 201, 267
Pitch fibers, 8–9
Planar isotropic lamina, 58
Plasticization of polymers, 132, 139
Point stress criterion (see Whitney-Nuismer Criteria)
Poisson's ratio:
analogy with shear-coupling ratio, 53
definition of, 44
of laminate, 221
major Poisson's ratio, 44
measurement of, 382–383
micromechanical prediction of, 75
minor Poisson's ratio, 44
(See also Engineering constants)
Polyacrylonitrile (PAN) fibers, 8
Polyetheretherketone (PEEK), 12, 365
Polyimide (PI), 12
Polymer matrix materials, 1–3, 9, 12, 22
Polyphenylene sulfide (PPS), 12
Polysulfone (PS), 12, 365
Power law creep model, 402
Preform, 27
Prepreg tape, 22, 24, 26
Principal material coordinates, 41
Pultrusion, 20, 26–27, 29

Quadratic Delamination Criterion, 245–246

Quadratic interaction criteria (see Multiaxial strength criteria)
Quasi-elastic analysis, 288–291
Quasi-isotropic laminates, 184, 212–215

Radius of curvature:
of laminated beam, 193
of laminated plate, 222
Rail shear test, 391–392
Random fiber packing, 66, 90
Randomly oriented discontinuous fiber composites, 173–184
Rawlinson, R. A., 386, 411
Rayleigh-Ritz method, 255, 320
Reinforced reaction injection molding (RRIM), 23, 26
Reissner, E., 201, 267
Rejuvenation before aging, 401–402
Relaxation (see Creep)
Relaxation modulus, 226, 283–287, 296
(See also Compliances; Creep)
Relaxation peaks, 323
Relaxation spectrum, 286
Relaxation time, 283–287, 301–303, 322–323
(See also Retardation time)
Representative volume element (RVE), 62, 68–69, 73–74, 77–79, 83–87, 92, 157–158, 162–165, 169, 186
Resin transfer molding (RTM), 23, 27, 30, 174
Retardation time, 284–287, 322
(See also Relaxation time)
Riley, M. B., 87, 97
Rosen, B. W., 90, 98, 118–119, 130, 162, 188
Ross, C. A., 306, 336
Rotary inertia effects, 314
Rule of mixtures:
for complex modulus, 298
for density, 64
for longitudinal modulus, 70–73, 163–164
for longitudinal stress, 70, 163
for major Poisson's ratio, 75
for relaxation modulus, 296
Russell, A. J., 397, 412
Rybicki, E. F., 235, 268, 357, 361, 371

Saint-Venant end effects, 385
Saint-Venant's theory (see Multiaxial strength criteria)

Sandwich structures, 5–6, 314, 388–389
Schapery, R. A., 150–151, 274, 278,
    287–288, 295–296, 335, 401, 413
Schroeder, R., 86, 89, 91, 98
Scotchply® composites, 102
Self-consistent model, 87
Self-similar crack growth, 356
Semiempirical micromechanics models,
    90–93
Sendeckyj, G. P., 63, 97, 102
Shear coupling, 31, 44–46, 53
Shear-coupling coefficients, 53
Shear-coupling ratio, 53, 221, 384
Shear crippling, 119
Shear lag model, 162, 169
Sheet-molding compound (SMC), 24–26,
    133–135, 174
Shen, C. H., 134, 137–138, 151, 155
Shift factor, 301–303
Short beam shear test, 393–394
Short fiber composites (see Discontinuous
    fiber composites)
Sierakowski, R. L., 249, 268, 306, 320, 336,
    344, 370
Silicon carbide fibers, 8, 10
Simplified micromechanics equations,
    79–82, 86, 91
Sims, D. F., 297, 335
Single fiber fragmentation test, 397–399
Single filament tensile test, 375
Smith, C. B., 201, 267
Soni, S. R., 244, 268
Spearing, M., 348, 370
Specially orthotropic material, 41–46
Specific damping capacity, 329
Specific heat, 136, 151
Specimen length effect, 383–385
Spencer, A., 82–83, 91, 98
Spring-dashpot models, 279–288, 402
Springer, G. S., 134, 137–138, 142, 151, 155
Standard linear solid (see Zener model)
Stavsky, Y,, 201, 267
Stiffnesses:
    of anisotropic materials, 36, 40
    bending, 206
    coupling, 206
    extensional, 206
    of generally orthotropic materials, 40–41,
        51
    of isotropic materials, 43
    of laminates, 206
    of planar isotropic materials, 58

Stiffnesses: (Cont.)
    of specially orthotropic materials, 41–42,
        47
Stinchcomb, W. W., 79, 97
Storage modulus, 293, 299, 322, 403–404
Strain concentration factor, 122–124
Strain energy, 39, 70–75, 77
Strain energy method, 324–329
Strain energy release rate:
    application to composites, 347–349,
        360–366
    definition of, 345–347
    measurement of, 395–397
Strain rate dependence, 271–272, 281
Strains:
    creep, 275–276
    engineering, 36
    hygrothermal, 144–148, 228–231
    laminate, 193, 202–204
    at a point, 36–38
    sinusoidally varying, 291
    tensor, 36
    transformation of, 50
    volume-averaged, 38–39
Street, K. N., 397, 412
Strength:
    of fibers, 8, 80, 114
    in-plane shear, 125–126
    interlaminar, 244–249
    of laminate, 237–249
    longitudinal, 115–122
    of matrix, 81
    micromechanics models for, 114–126
    multiaxial criteria for, 101–114
    of orthotropic lamina, 99–129
    of randomly oriented fiber composites,
        183–185
    transverse, 122–125
Stress fracture criteria, 349–356
    (See also Whitney-Nuismer Criteria)
Stress intensity factor:
    application to composites, 343–345,
        357–360
    definition of, 342–343
Stress partitioning parameter, 92–93
Stress rupture test, 400
Stress-strain relationships:
    general, 36, 40
    for generally orthotropic materials,
        50–58
    hygrothermal, 144–148
    at a point, 36–38

Stress-strain relationships: (*Cont.*)
  for specially orthotropic materials, 41–42, 45
  viscoelastic, 275–300
  volume-averaged, 38–39

Structural reaction injection molding (SRIM), 23, 27
Struik, L. C. E., 302, 336
Suarez, S. A., 165–167, 170, 172–173, 188, 299–300
Sullivan, J. L., 303–305, 336, 401, 413
Sun, C. T., 165, 169–172, 181–183, 188–189, 344, 370
Surface energy, 340–341
Symmetric laminates, 208–210, 217–223

Temperature distribution, 136–137
Tennyson, R. C., 112, 129
Tensile testing:
  of composites, 382–386
  of fibers, 375–376
  of neat resin matrix, 378–381
Thermal conductivity, 136, 151
Thermal forces and moments, 228–230
Thermoplastic molding, 27, 29
Thermorheologically simple material, 287
Through-thickness cracks (*see* Fracture mechanics)
Time-temperature superposition (TTS), 301–304
Timoshenko beam equation, 314, 404
Torquato, S., 90, 98
Torsion tube test, 392
Tough matrix materials, 365–366
Transverse modulus:
  definition of, 45
  measurement of, 44–45
  prediction of, 73–75, 77–83, 85–93
Transverse shear effects, 196–198, 203, 313–314, 404
Transversely isotropic materials, 41, 46
Tsai, S. W., 53, 55, 56, 61, 90–91, 98, 109–110, 151, 176–177, 189, 240–241, 268
Tsai-Hahn equations, 91–92
Tsai-Hill Criterion, 109, 188
Tuttle, M., 257–258, 268
Tyson, 159–165, 170

Ungar, 325–328, 337
Ungar-Kerwin equation, 325–328

Unit step function, 288–289

Vibration decay, 404–405
  (*See also* Damping)
Vibrations:
  of composite beams, 311–315
  of laminated plates, 315–321
  of specially orthotropic bars, 307–311
  test methods, 402–408
Vinson, J. R., 249, 268, 306, 320, 336
Viscoelastic behavior, 270–306
Void fraction, 65
Von Mises Criterion, 108

Waals, F. M., 177–181, 189
Waddoups, M. E., 106, 129, 349–350, 371
Wang, A. S. D., 235, 237, 268
Wang, S. S., 357–359, 361, 371
Wavelength, 277, 308
Wave propagation:
  analysis of, 397–311
  introduction to, 273
Weibull distribution, 398
Weng, G. J., 181–182, 189
Westergaard, H. M., 342, 370
Whiskers, 3, 174
Whiteside, J. B., 139–140, 142, 145, 155
Whitney, J. M., 87–88, 97, 118, 202, 235, 237, 244, 249, 255, 267, 306, 318–320, 349–356, 364, 370–371, 390, 394–395, 410, 412
Whitney-Nuismer Criteria, 349–356
Wilkins, D. J., 364, 371
Williams-Landel-Ferry (WLF) equation, 302–306
Wolfenden, A., 321, 337
Wu, E. M., 102–103, 110–111, 128, 344, 370
Wyoming-modified Celanese fixture, 389, 401

X-29 aircraft, 13–14

Young's modulus (*See* Engineering constants; Stiffnesses)

Zener model, 284–287, 322–323

Given Properties E_1 E_2 ν_{12} G_{12} α_1 α_2 X_T X_C Y_T Y_C.
which layer fails 1ST? Max laminate Force?
Put everything in terms of t_{TOTAL}

1) Find $\bar{Q}$'s for ea ply P 51
2) Find α_x α_y α_{xy} p 147
3) Find A matrix p 206
4) Find A^{-1}
5) N^T in terms of total thickness of laminate

$$N^T = \int [\bar{Q}]_k [\bar{\alpha}]_k \Delta T\, dz \quad \therefore \quad \frac{1}{N} \sum_{k=1}^{N} [\bar{Q}]_k [\bar{\alpha}]_k \Delta T\, t$$

$\underset{\text{\#plies}}{\underbrace{t}}$ ← t

↑ total laminate thickness.

6) Find strains

$$\left\{ \begin{bmatrix} N_x \\ {}_0N_y \\ {}_0N_{xy} \end{bmatrix} + \begin{bmatrix} N_x^T \\ N_y^T \\ N_{xy}^T \end{bmatrix} \right\} = [A] \begin{bmatrix} \varepsilon_{xo} \\ \varepsilon_{yo} \\ \gamma_{xyo} \end{bmatrix}$$

mean

p 206 / p 215 / p 217

↑
only A if
symmetrical
& No Bending

Aside:
$[\varepsilon] = \varepsilon_{x_o} + K_x Z$
p 204

7) Solve the above for $[\varepsilon^o] = [\varepsilon]$ in terms of N_x
8) Find stresses per ply (p 204)

$$\begin{bmatrix} \sigma_x \\ \sigma_y \\ \tau_{xy} \end{bmatrix}_k = [\bar{Q}]_k \begin{bmatrix} \varepsilon_x^o + K_x Z \\ \varepsilon_y^o \\ \gamma_{xy} \end{bmatrix}$$

9) Transform into σ_1 σ_2 τ_{12} p 50
10)

$$T = \begin{bmatrix} .5 & .5 & 1 \\ .5 & .5 & -1 \\ -.5 & .5 & 0 \end{bmatrix} \qquad T^{-1} = \begin{bmatrix} .5 & .5 & -1 \\ .5 & .5 & 1 \\ .5 & -.5 & 0 \end{bmatrix}$$